SUPERALLOYS

Edited by **Mahmood Aliofkhazraei**

Superalloys
http://dx.doi.org/10.5772/59358
Edited by Mahmood Aliofkhazraei

Contributors

Zenon Aleksander Opiekun, Murthy G.V.S., Vidhi Acharya, Ramesh Surla, Shawqi Al Dallal, Antonio Dos Santos, Erween Abd Rahim, Norazlan Md Warap, Zazuli Mohid, Lavinia Ardelean, Lucien Reclaru, Cristina Maria Bortun, Laura Cristina Rusu, Víctor Manuel López-Hirata, Igor Shishkovsky, Xishan Xie, Reza Mehrabi, Mathias Galetz, Lembit Kommel, Kazimierz Ducki, Nageswara Rao Muktinutalapati, Manikandan Manoharan, Arivazhagan Natarajan

Published by InTech

Janeza Trdine 9, 51000 Rijeka, Croatia

Publishing Process Manager Sandra Bakic

Technical Editor InTech DTP team

Cover InTech Design team

First published November, 2015

Printed in Croatia

Additional hard copies can be obtained from orders@intechopen.com

Superalloys, Edited by Mahmood Aliofkhazraei

p. cm.

ISBN 978-953-51-2212-8

For EU product safety concerns:
IN TECH d.o.o., Prolaz Marije Krucifikse Kozulić 3, 51000 Rijeka, Croatia, info@intechopen.com or visit our website at intechopen.com.

Meet the editor

Dr. Mahmood Aliofkhazraei works in the Corrosion and Surface Engineering Group at the Tarbiat Modares University, Iran. He is the head of Aliofkhazraei research group (www.aliofkhazraei.com). Dr. Aliofkhazraei has received several honors, including the Khwarizmi award and the best young nanotechnologist award of Iran. He is a member of the National Association of Surface Sciences, Iranian Corrosion Association, and National Elite Foundation of Iran. His research focuses on materials science, nanotechnology and its use in surface and corrosion science.

Contents

Preface

Superalloy, or high-performance alloy, is an alloy that exhibits several key characteristics: excellent mechanical strength, resistance to thermal creep deformation, good surface stability, and resistance to corrosion or oxidation. The crystal structure is typically face-centered cubic austenitic. Superalloy development has relied heavily on both chemical and process innovations. Superalloys develop high temperature strength through solid solution strengthening. An important strengthening mechanism is precipitation strengthening which forms secondary phase precipitates such as gamma prime and carbides. Oxidation or corrosion resistance is provided by elements such as aluminium and chromium. Since these alloys are intended to be used for high temperature applications, in addition to these materials being able to withstand loading at temperatures near their melting point, their creep and oxidation resistance are of primary importance. Ni based superalloys have emerged as the material of choice for these applications. The properties of these Ni based superalloys can be tailored to a certain extent through the addition of many other elements, both common and exotic, including not only metals, but also metalloids and nonmetals; chromium, iron, cobalt, molybdenum, tungsten, tantalum, aluminium, titanium, zirconium, niobium, rhenium, yttrium, vanadium, carbon, boron or hafnium are some examples of the alloying additions used. Each of these additions has been chosen to serve a particular purpose in optimizing the properties for high temperature application.

This book collects new developments about superalloys. I like to express my gratitude to all of the contributors for their high quality manuscripts. I hope open access format of this book will help all researchers and that they will benefit from this collection.

Dr. Mahmood Aliofkhazraei
Tarbiat Modares University
Iran
www.aliofkhazraei.com

Section 1

Superalloys

Chapter 1

Thermal-Assisted Machining of Nickel-based Alloy

Erween Rahim, Norazlan Warap and Zazuli Mohid

Additional information is available at the end of the chapter

http://dx.doi.org/10.5772/61101

Abstract

Nickel-based alloy can be found in different industrial applications especially in aircraft engines and hot end components of various types of gas turbines with its high strength, strong corrosion resistance and excellent thermal fatigue properties and thermal stability compared to conventional materials. However, nickel-based alloy is one of the extremely difficult-to-cut materials. During the machining process, the interaction between the tool and the workpiece causes severe plastic deformation and intense friction at the tool-workpiece interface. Because of the increasing demands in industries, any improvement of conventional machining processes or any other deployment of additional technique is directly related to higher productivity. Thermal-assisted machining (TAM) has become an effective alternative to the conventional machining of these difficult-to-cut materials. Various types of heating methods and the beneficial effects on machining of nickel-based alloys are discussed in this chapter. Finally, TAM was proven as an efficient technique to increase the machinability of nickel-based alloys in terms of tool life, surface roughness and cutting force.

Keywords: thermal assisted machining, Hot machining, Nickel-based alloy

1. Introduction

The machinability index of work materials in machining are assessed in term of various aspects and criteria. Currently, the evaluation index includes tool life, cutting force, power requirements and surface finish. However, in this decade, sustainable machining is the most important criteria that need to be considered with the development of new techniques and methods for machining. The machining technique can be assumed to be sustainable when it proves beneficial to the economy, environment and society. Machinability index of nickel-based alloy compared to carbon steel is 35 to 100 [1]. This means that machining of nickel-based alloy is more difficult than free-machining of carbon steel. The machining difficulties of nickel-based alloys originate from its excellent material properties with higher ductility and material strength at elevated temperatures. These will cause significant increments on the cutting temperature and decrements on tool life during the cutting process. Table 1 shows a summary of previous researches for various preheating methods.

Process	Preheating Methods	Workpiece/ Cutting Tool	Findings
Turning [2]	Laser	Inconel 718/carbide	With increased preheating temperatures to 620°C came a 25% decrease in specific cutting energy, an increase of 200%–300% tool life and improved surface roughness
Turning [3]	Plasma	Inconel 718/carbide	Improved surface roughness by 250%, decreased cutting force of around 30%–50% and extended tool life of up to 170% over conventional machining
Turning [4]	Laser	Inconel 718/carbide	In comparison with conventional cutting, the resultant force in LAM condition decreased by 24%–46%. However, the chip thickness increased 40% under LAM conditions by applying preheating temperatures of 800°C
Turning [5]	Oxyacetylene gas flame	Stainless steel/carbide	With preheating temperatures of 400°C, the material removal rate was maximized. Otherwise, cutting speed (31 m/min) was the most significant effect to minimize the surface roughness
Turning [6]	Electricity resistance	Ti-15333 alloy/carbide	It shows the significant reduction of cutting force in the range of 80%–85% when the preheating temperature applied was approximately 300°C
Turning [7]	Laser	42CrMo4 steel/carbide	The material strength was reduced when the preheating temperature was increased to 700°C. It was proved that the cutting force could decrease by as much as 40%
Milling [8]	Coil	AISI D2/cubic boron nitrate PCBN	By applying preheating temperatures of approximately 50°C–150°C, the surface roughness and charter result decreased compared to machining at room temperature
Turning [9]	Laser	Inconel 718/coated carbide and SiAlON ceramic	The surface finish improved by more than 25%, and the material removal rate increased by approximately 800%

Table 1. Summary of Preheating Methods

http://dx.doi.org/10.5772/61101

Generally, either conventional or advanced machining techniques have their own advantages and disadvantages. It must be chosen correctly based on the purpose and type of processes to be prepared. Selecting the right process can be maximizing production capacity and quality. Hybrid machining is another technique to enhance the machining performance and capability to cut material while minimizing the negative effects on the products and tools. Well-selected processing parameters and the combination between conventional and advanced machining techniques can efficiently increase the machining performance. A machining technique with the assistance of heat induction on nickel-based alloy was developed to counter this issue. Various heat sources such as laser, plasma, induction coil and flame heating were successfully applied in the machining process. The influence of heat on the materials' characteristics is one of the issues addressed in determining the heating and machining parameters. Categorized as nonferrous metals, nickel-based alloys show similar machining characteristics as other materials such as titanium and stainless steel. The equipment and methods used for heating gives significant influence to machinability.

The purpose of thermal-assisted machining processes is to reduce the cutting forces, prolong tool life, reduce chatter and improve surface finish. The reduction of total cutting length is associated with low shear yield strength of the work material at the primary and secondary shear planes. During the thermally assisted machining process, the plasticity of the work material increases, leading to the increment of chip-tool contact length, which plays an important role in reducing the normal stress acting on the tool [10]. Heating also reduces the intensity of chip serration, facilitating lower fluctuations of the cutting forces, which consequently reduce the dynamic stresses applied to the tool, thereby facilitating lower tool wear in heat-assisted machining.

2. Nickel-based alloy properties and its constraints

Nickel-based alloys are an important material in various industries, especially those involving high-temperature applications. They are widely used in gas turbine engines, aircrafts, nuclear reactors, submarines, steam power plants, petrochemical equipment and other higher-temperature applications. The excellent properties of nickel-based alloys at elevated temperature contribute to poor machinability and can be summarized as [11] (i) a major part of their strength is maintained during machining with higher ductility and yield strength; (ii) during the cutting process, work hardening rapidly occurs and contributes to tool wear and reduction of tool life; (iii) cutting tools suffer from high abrasive wear owing to the presence of hard abrasive carbides in the super alloy; (iv) chemical reactions occur at high cutting temperatures and leads to high diffusion wear rates; (v) adhesion of microchips onto the cutting tool frequently occurs during the cutting process and causes severe notching as well as spalling on the rack face and consequent pull-out of the coating materials; (vi) difficulties in controlling the formation of tough and continuous chips during the cutting process contributing to the degradation of the cutting tool performance; (vii) low thermal conductivity and poor thermal diffusivity of nickel-based alloys generate high temperatures on the tool tip and consequently increases the cutting temperature.

The alloying elements contained in the materials possess high strength and toughness over a wide temperature range and excellent fatigue strength, oxidation and corrosion resistance. However, the alloy was designed originally as a solid solution alloy and it has been shown that the precipitation of intermetallic phases occurs during the preheating process. For nickel-based alloys, strength is influenced by body-centred tetragonal (BCT) γ″ - Ni_3Nb and face-cantered cubic (FCC) γ′- $Ni_3(Al,Ti)$ precipitates. This metastable phase precipitates on preheating temperature above 873 K [12]. The equilibrium intermetallic phase forms in alloy element when the formation of orthorhombic δ or $Ni_3(Nb, Mo)$ phase occurs entirely and the stability of this phase will be sustained at elevated temperatures.

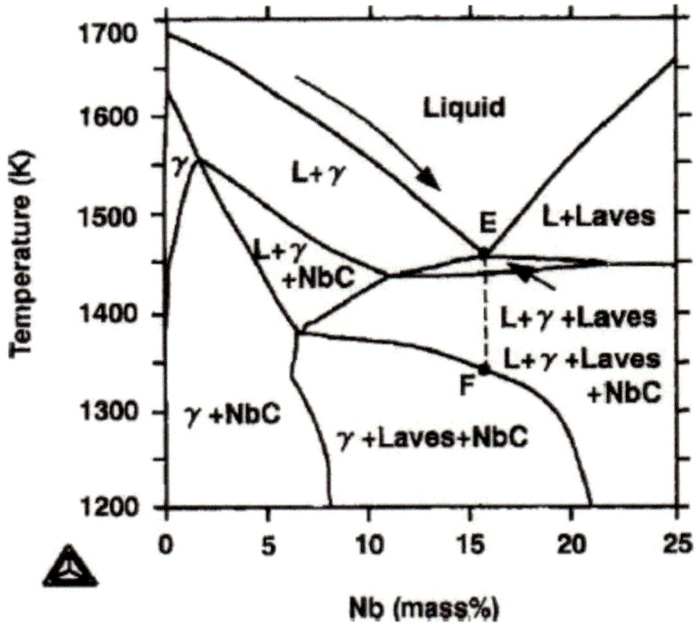

Figure 1. Phase diagram of Inconel 718 alloy calculated on the basis of Thermo-Calc software [12].

Moreover, Luo et al. [13] calculated a phase diagram for a nickel-based alloy database with Thermo-Calc, a multielement thermodynamic calculation software package and the results obtained are shown in Figure 1. According to this phase diagram, with a niobium concentration (5%), liquid phase is formed by a reaction between g and NbC in the temperature range of approximately 1420–1510 K and above, the NbC entirely disappears and leaves coexistent γ phase and liquid phase. The temperature at the liquid phase formed at 1440 K and the temperature at massive niobium carbides at 1530 K dissolved and disappeared. Both phases at different temperatures can be predicted from the phase diagram in Figure 1. Furthermore, the γ phase eutectic reaction occurs at a niobium concentration in the range of 11%–22%. At this phase, it means that the specimen has no change in the nature and initial properties and the phase diagram seems to suggest that the niobium in the liquated microstructure is far more concentrated than in the solid γ phase.

In thermal-assisted machining, it can be assumed that it is effective when the preheating temperature is controlled under deformation temperature or in metastable phase. Each category of nickel-based alloys has different deformation temperatures where it is based on the alloy element contained in the materials' composition. Deformation temperature, strain rate and the deformation heat treatment are the main factors in determining which structural mechanism would control the flow stress value. Otherwise, utilization of proper temperature will affect the microstructure deformation in determining whether it is impressed with the effect of softening or hardening. However, deformation temperature and strain rate are the main variables in controlling the structural softening processes during hot deformation of the material.

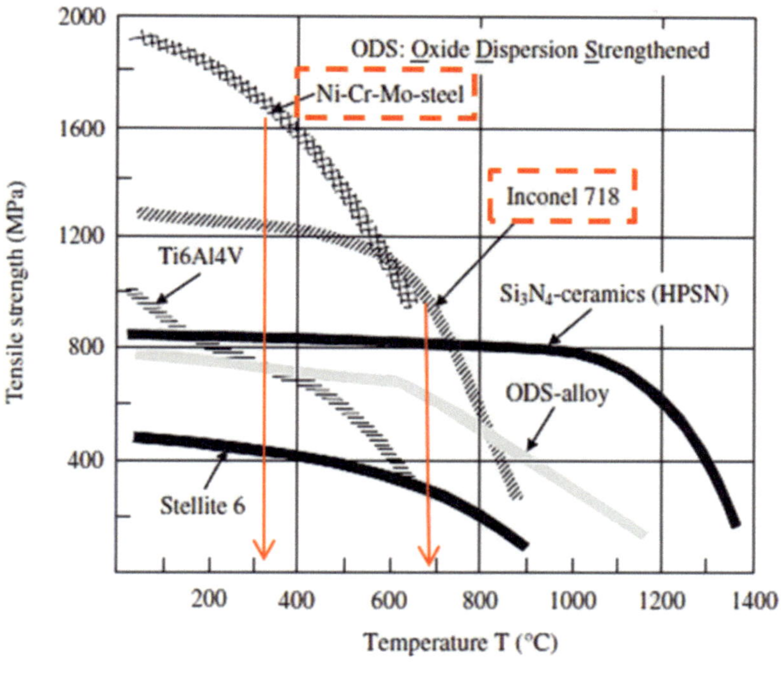

Figure 2. Effect of preheating temperature on the ultimate tensile strength for various materials [14].

Figure 2 shows the effect of preheating temperature on the ultimate tensile strength of various types of materials. In the category of nickel-based alloys, Ni-Cr-Mo steel exceeded the maximum ultimate yield strength at lower preheating temperatures of around 250°C–350°C. However, Inconel 718 with alloying elements of more than 50% have durable tensile strengths at elevated temperatures of around 600°C–700°C. In addition, the high temperature oxidation resistance has become more important, due to the fact that the demand for the development of higher temperature and more reliable Inconel 718 components in modern industry is increasing. Currently, it is well recognized that poor oxidation resistance of any thermoresistance component can pose a potential risk to its service reliability leading to severe degradation of service life.

3. Temperature prediction

Prediction of heat generation and distribution initiated by laser irradiation using finite element analysis (FEA) software has been widely reported by many researchers. Saodari and Majumdar [15] used FEA to analyse the heating rate, heat-affected zone and the shape and size of the molten pool using a Gaussian laser beam. In addition, they also analysed the effect of mesh size to obtain accurate prediction results. Furthermore, Ren et al. [16] used FEA to analyse the effect of heat generated on residual stress during laser processing.

Further investigations have been done to find more accurate data about temperature distribution. Mohid et al. [17] reported the effect of absorptivity, A, and Gaussian distribution constant, K values, on the accuracy of the numerical analysis results. The investigation shows that the A and K values have a significant effect on the characterization of melting pool and heat-affected zone (HAZ) pattern. The calculation of the moving heat flux, the position and the magnitude of the heat flux were confirmed every time for pulsed laser heating. The distribution of heat flux is dependent on the pulsed length time. The Gaussian distribution theory on laser energy was used as a heat source on the upper surface. It can be expressed as shown in Equation 1.

$$q(x,y) = \frac{AKP}{S}\exp\left[\frac{-K(x^2+y^2)}{b^2}\right] \tag{1}$$

where x and y is the location from the laser beam centre, and b is the efficient beam radius. The laser power used is assumed as P. The absorptivity A is determined based on the type of material and surface roughness. However, the value of constant K depends on the beam's intensity. Several assumptions need to be considered to facilitate the simulation work. The assumption consists of (i) the material is homogenous, (ii) simulation is work on transient mode and (iii) the keyhole formation can be neglected.

Figure 3(a) and (b) show the thermal conductivity and specific heat, respectively, of Inconel 718 measured by various investigators. The spread in the data is relatively small, indicating that all of the data is precise and most likely accurate. Figure 3(c) shows the experimental value of spectral absorptivity for Inconel 718 [18]. The laser beam energy absorptivity of Inconel 718 is very low for CO_2 lasers, while it is higher for shorter wavelength lasers, such as Nd:YAG and diode lasers. The error of HAZ depth increases when the K value is 3. This is due to the laser intensity on Gaussian distribution in simulation, which was determined by the total energy density. The greater value of K produced higher laser beam density and this does not coincide with the actual situation during irradiation.

In FEA, the model can be assumed precise when the depth and width of the HAZ in the numerical simulation are comparable with the actual experiment. In this case, an error of less than 10% in the HAZ geometry compared to substantial geometry is acceptable to validate the model. The recorded HAZ temperature in the simulation reached 850 K. This represents the borderline of the HAZ area. The actual specimens that were exposed to the laser irradiation were cut perpendicular to the scanning direction at a distance of 15 to 20 mm from the starting point. Basically, the location is selected based on the stability of the heat generated and by heat absorption into the materials during the irradiation process.

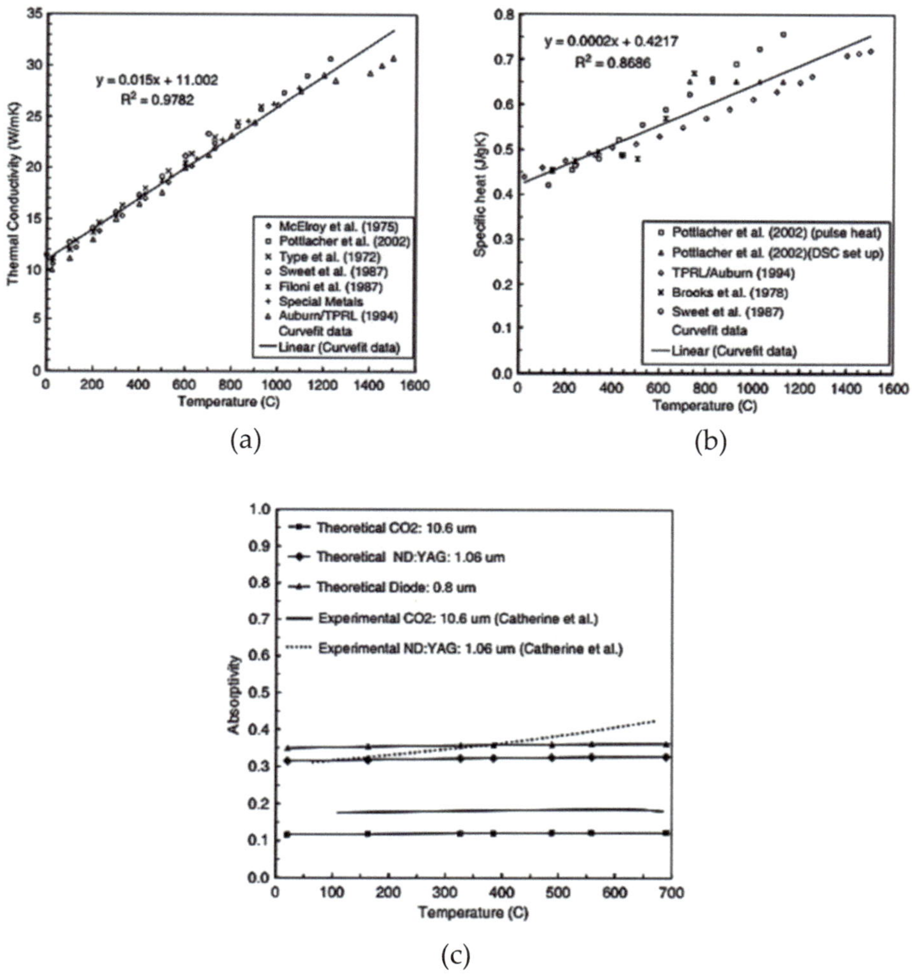

Figure 3. Comparison of temperature dependence state by various researchers on (a) thermal conductivity (b) specific heat and (c) theoretical and experimental normal spectral absorptivity at different laser wavelength of Inconel 718 [2].

Figure 4(a) and (b) show the results of experimental and simulation data for HAZ depth and width under a constant laser average power, P_{avg} 5.65 W and different absorptivity, A and K

values. It was noted that the error of melting width becomes smaller when a *K* value of 2.5 is applied. However, absorptivity also gives a significant effect on the formation of HAZ. Higher absorptivity is produced by the black surface. In the machining process, the workpiece is normally well prepared with good surface condition and dimension. Shining workpiece surface condition will reduce the absorptivity. Figure 4(b) shows the effect of absorptivity in simulation results. By increasing the *A* value from 30% to 40%, the error is higher than 5%. However, when the value of *A* is fixed at 35%, the error decreases to less than 5%. Furthermore, laser power also plays a significant role in the formation of the HAZ and melting region. As the laser power increases, the HAZ and melting region were formed a few microns underneath the surface. Finally, when the values of *A* and *K* were fixed to 35% and 2.5%, respectively, the result of HAZ width and depth are comparable with a percentage error of less than 10% as shown in Figure 4(c). A comparison image between simulation and actual specimenis shown in Figure 5.

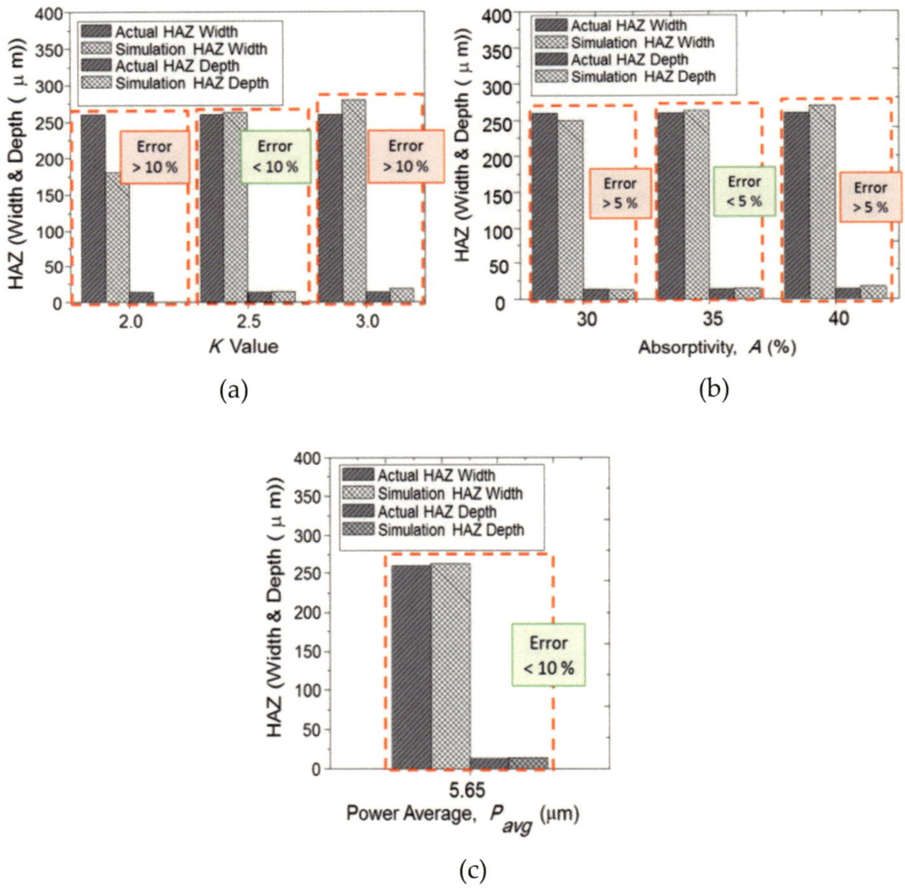

Figure 4. Effect of (a) *K* value (b) absorptivity, *A* in the formation of HAZ pattern and (c) model validation when the result of HAZ shape between actual scanning is comparable with simulation.

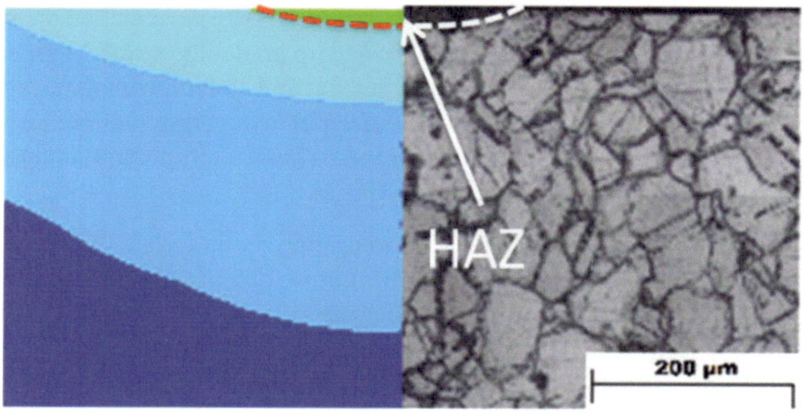

Figure 5. Comparison of HAZ shape size (P_{avg} = 5.65 W, A = 35% and K = 2.5) [19].

4. Heating methods in thermal-assisted machining

Thermal-assisted machining was introduced to overcome the machining issues regarding tool life, surface integrity and workpiece mechanical property changes. Theoretically, metal's strength and hardness decreases with increasing temperature. Heating metal below the deformation temperature will soften and reduce hardness and strength without major changes in its properties after it is chilled to room temperature. The same theory can be applied to all hot-machining processes such as hot forging, hot bending and hot stamping.

The methods of inducing heat energy into the workpiece gives different characteristics on the temperature distribution. Heat induction using a heating coil was reportedly applicable in machining [8, 20]. Oxyacetylene flame is one of the methods of applying heat to a workpiece. In micromachining, a concentrated and small heat source such as plasma and laser are preferable. In recent years, plasma and lasers have been reported to be the most promising heating techniques in thermal-assisted machining. Using laser or plasma beam is better in terms of heat distribution control. The focused and restricted heating area and easy-to-control scanning parameters will minimize thermal effects on workpieces. In thermal-assisted lathe machining, the type of heat source does not give a significant effect on machining performance compared to thermal-assisted milling. In the case of lathe machining, the heating area rotates at high speed and is repetitively heated as the specific point rotation through the laser or plasma beam focused area. The temperature gradually increases and fewer variations of gradual temperature can be seen along the cutting path. On the other hand, in thermally assisted milling processes, the heat source moves along with the tool and the heating efficiency is highly influenced by the scanning parameters, laser beam spot-to-cutting tool distance and spot size. The cutting area will be heated once and heat conduction and convection gives a significant effect on the temperature distribution characteristics.

Using a laser beam as the heat source exhibits significant effect on the temperature distribution characteristics. In general, continuous wave beam with a Gaussian distribution is preferable for the heating process. The workpiece can be heated up gradually with fewer thermal shock effects. When pulsed wave mode laser is used, heating and chilling will occur repeatedly on the workpiece. These phenomena will cause the material to undergo a hardening process and bring adverse effects on the machining performance. Thus, it is important to understand the method of preheating and its influences.

Various heat sources have been investigated as preheating media such as CO_2, Nd:YAG, diode laser and excimer laser. CO_2 laser has a wavelength of 10.6 μm and is ideal for optimum absorption, especially on ceramics, which is widely used. However, it has limitations where it requires a beam transfer method using a mirror as well as lower flexibility compared to a solid laser using fiber optic cables such as Nd:YAG laser. Table 2 summarizes the various types of preheat and heat sources with its different advantages and disadvantages. However, to have a user-friendly machine where the heat concentration is a priority, Nd:YAG laser is recommended compared to other laser sources.

Heat source	Advantages	Disadvantages
Laser	• High degree of heat concentration • Easy control of heat source	• Costly equipment • Absorption rate on different materials
Induction coil	• Easy to use • High-capacity preheating	• Impossible on high-concentration preheating • Limited tool mobility
Gas flame	• Low initial investment cost	• Impossible on high-concentration preheating
Plasma	• High degree of heat concentration	• Impossible to precisely control
Electricity	• Simple equipment • Even heat distribution	• Impossible to precisely control

Table 2. Heat Source/Heating Method Used

5. Plasma Enhancement Machining (PEM)

The ability to control the workpiece material at a constant degree of localized heating in the allocated cutting zone is the main critical issue contributing to the success of PEM. Direct current (d_c) is used to provide arcs resulting from reaction between electrode sparking and plasma gas to generate thermal or equilibrium plasma. Thoriated tungsten cathode and cooled nozzles are the main components producing plasma arc as shown in Figure 6. The nozzle can be assumed as a positive polarity (anode) when the workpiece material is nonconducting. Whereas, when the workpiece material is conductive, the nozzle can be assumed to have a negative polarity (cathode). Compared with a welding setup, there is a similarity in terms of concept and operating mechanism. In case of machining of super alloy materials, high localized energy at low gas flow rate is suited to transfer arcs with typical peak temperatures of approximately 16,000 K.

Leshock et al. [21] performed an experiment to evaluate PEM system performance on turning operation. The main components of the experimental setup consisted of 7 HP lathes, a plasma heating system and control unit. A special enclosure was designed and attached to a turning chuck to minimize and prevent turbulent airflow generated by its rotation. Resulting from the previous study, the turbulent airflow could affect the concentration of plasma arc irradiation when cutting is near the chuck. Inconsistent results are produced from the turbulent airflow interference. To counter this issue, a copper nozzle was fitted with a 3.18 mm diameter orifice. Thoriated tungsten cathodes with a 20° included angle were used throughout the experiment. Various measurement methods and temperature values were used in the actual process and offline to evaluate the performance of hot turning. Finally, PEM significantly reduced the cutting forces and improved surface roughness over a wide range of cutting conditions.

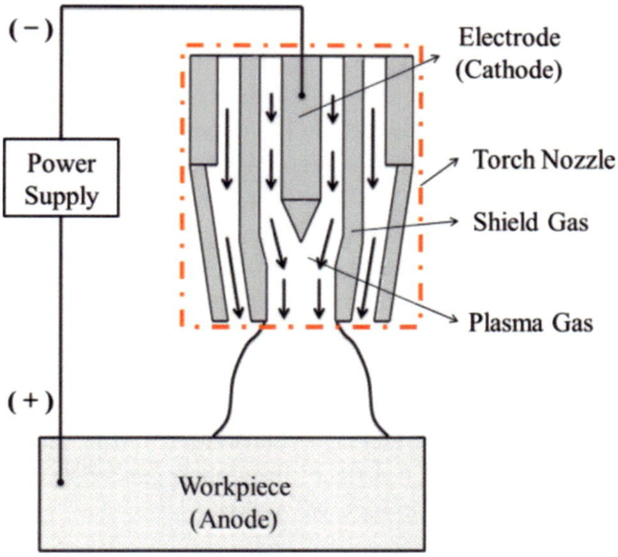

Figure 6. Plasma arc (transferred arc) generator.

Madhavulu and Ahmed [21] observed in their work that the major advantages of the plasma-assisted turning process are the increased metal removal rates, the lower spindle power requirement and the possibility of machining hard and tough metals even when fully hardened and heat-treated. The authors revealed that PEM leads to a 1.8 times gain in metal removal rate and 1.67 times prolonged tool life. However, the energy consumption used in the PEM system is far larger than the energy required for a machining process in conventional systems.

In addition, Leshock et al. [21] conducted a numerical and experimental study on PEM of Inconel 718. The surface temperature generated by the plasma system was measured by using an infrared radiation thermometer and the results between experimental and numerical analyses were compared to produce a numerical model. Results comparisons showed good agreement where PEM demonstrated a 30% reduction in the resultant cutting force, improved surface roughness and 40% increased tool life compared with conventional turning. On the other hand, Wang et al. [3] combined plasma and cryogenic cooling of the workpiece within the conventional turning process of Inconel 718. They found that the surface roughness was reduced by 250%, the cutting forces were decreased by approximately 30%–50% and tool life increased by up to 170% over conventional cutting. However, PEM are widely used only in turning process. It is difficult to associate plasma heating in end-milling processes because the feed is relatively low and the workpiece will be melted due to the heat generated by the plasma arc.

6. Induction heating

Induction heaters provide alternating electric current to an electric coil (the induction coil). The induction coil becomes the electrical (heat) source that induces a high-frequency alternating electrical current into the workpiece to be heated. The heat is restricted to localized areas or surface zones immediately adjacent to the coil. This happens because the alternating current (a_c) in the induction coil has an invisible force field (or magnetic flux) around it. The induction coil actually functions as a primary transformer, with the workpiece to be heated becoming the secondary transformer. The force field surrounding the induction coil induces an equal and opposing alternating electric current in the workpiece [8]. The workpiece will be heated up due to the resistance to the flow of this induced high-frequency alternating electric current. The rate of workpiece heating is dependent on the frequency and intensity of the induced current, the specific heat of the material, the magnetic permeability of the material, and the resistance of the material to the flow of current. The induced currents are sometimes referred to as eddy currents, with the highest intensity current being produced within the area of the intense magnetic fields. The heating system consists of three major components high-frequency transformer (invertors), matching box (transformer and condenser) and cooling unit. The overall experimental setup is shown in Figure 7.

In contrast, Luo et al. [23] conducted experiments on conduction heating on end-milling cutting tools. The experiment is different compared to custom setups where the preheating temperature is applied on the cutting tool. This technique is more focused on machining nonconductive material such as elastomers, rubbers and plastics. Generally known nonconductive materials have lower melting points and it is impossible to apply preheating temperatures on those materials. Results from the experiment show that preheating the cutting tool will make the material softer and the machining accuracy becomes much better than conventional cutting. On the other hand, Kizaki et al. [24] proved that thermal-assisted machining is not applicable in the drilling process. As an alternatif technique, the tool was heated to reduce the vibration during drilling on zirconia material. The results showed that by preheating the cutting tool up to 500°C, the workpiece temperature could increase to 150°C–400°C. Finally, the cutting experiment demonstrated an improvement in machinability.

http://dx.doi.org/10.5772/61101

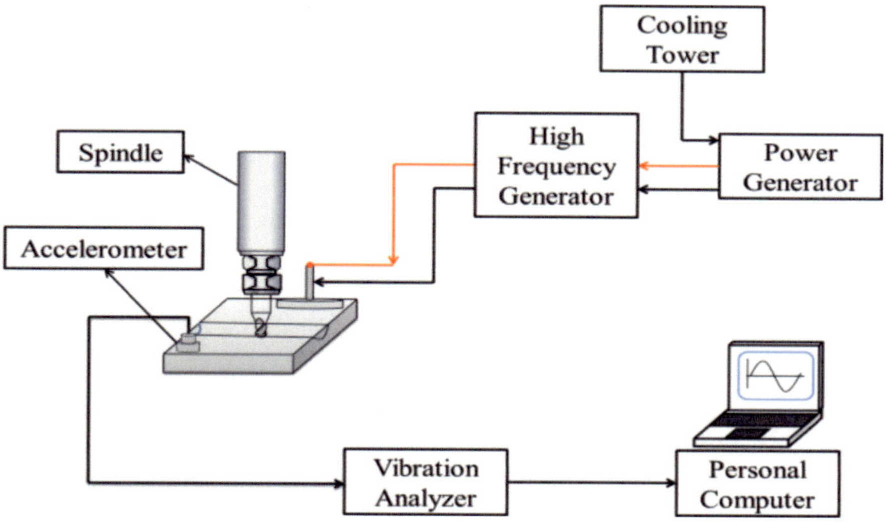

Figure 7. Overall experimental setup for induction heating machining.

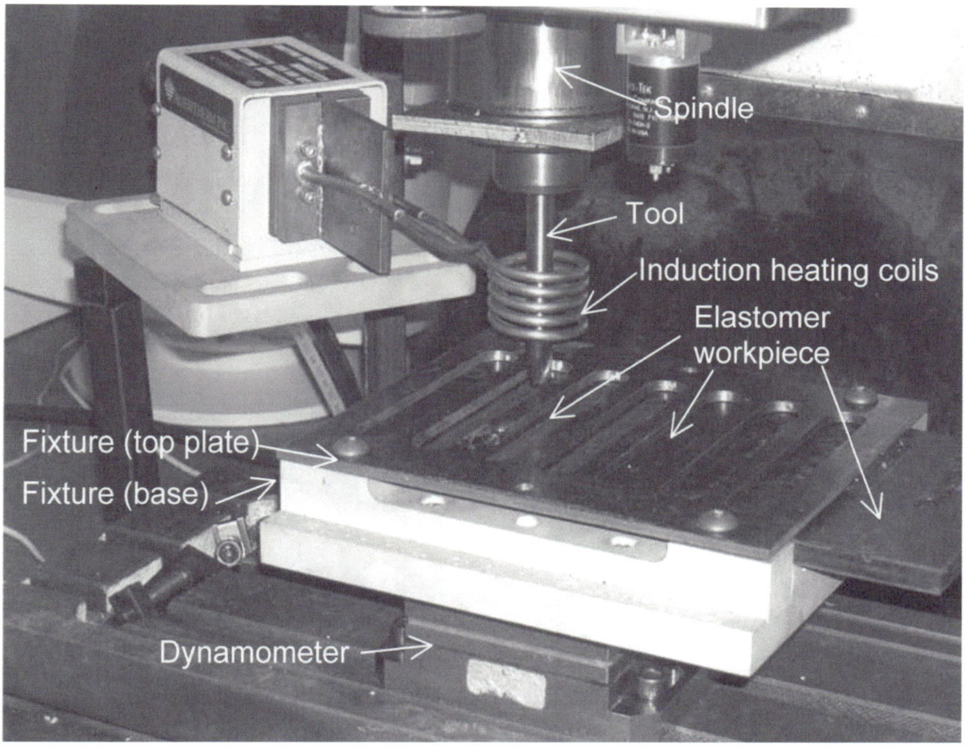

Figure 8. Experimental setup of elastomer end-milling with tool induction heating [23].

7. Laser-assisted milling

Laser-assisted milling is a hybrid machining process that combines preheating and mechanical removal processes. Many studies have been conducted using several types of lasers and milling machines. Most studies reported positive results with large improvements on tool life and lower cutting forces. However, in cases where material property preservation is crucially important, tool selection and heating temperature need to be perfectly mastered.

In the laser-assisted micromilling (LAMM) process, appropriate parameters need to be determined to ensure that the machining process can be performed at a higher level to produce microsize and highly accurate machined parts. The determination of machining parameters can be referred from the basic theory of LAMM as shown in Figure 9. The heating location generated by laser beam irradiation, T_{bc} and cutting tool, T_c must be well-determined to confirm that the heat generated is in the recommended temperature range for better softening effect (700°C). Mohid et al. [25] predicted the temperature distribution on pulsed laser mode by using ANSYS APDL software. From their results, it was concluded that by using a 140-W power laser, a beam-to-cutting tool distance of between 0.8 and 1.9 mm and a depth of 0.005 to 0.117 mm could be obtained. Yang et al. [26] have developed a 3D transient finite element method to predict the depth and width of HAZ on ductile material. It was found that the laser parameters, especially laser power, have strong influence on the depth and width of HAZ. In addition, Kim and Lee [27] used FEA to predict the preheating temperature on Inconel 718 and AISI 1045 materials to obtain the depth of cut.

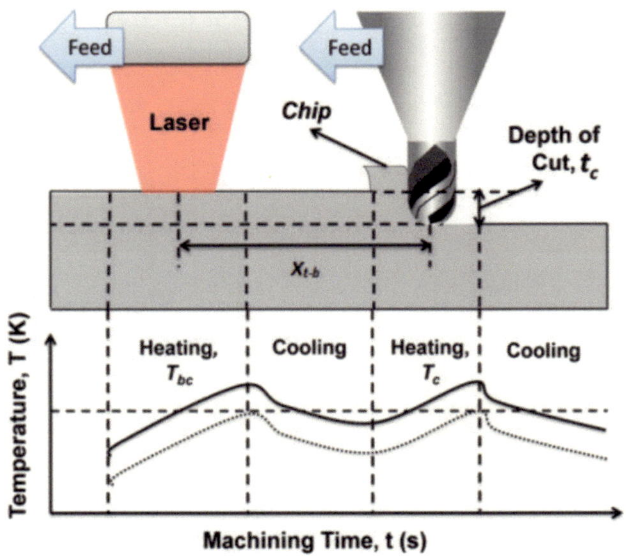

Figure 9. Theoretical analysis of LAMM.

Finite element analysis (FEA) was used to predict the temperature distribution on the top and bottom surfaces. The FEA model was developed by using ANSYS APDL software to predict heat distribution during the laser irradiation process. At the same time, it creates the HAZ pattern underneath the workpiece surface. Based on workpiece temperature distribution, the range of laser spot-to-cutting tool distance $X_{t\text{-}b}$ and the depth of cut t_c can be determined prior to the actual machining process. It is essential to determine the initial tool engagement temperature as shown in Figure 10(a) and (b). It is intended to ensure that the preheating temperature will not impair the tool performance. This argument is supported by Kim and Lee [27], who reveal that by applying temperatures between 650°C and 900°C, the material strength will be significantly reduced. However, Rahim et al. [28] mentioned that the most prominent effect on cutting force, surface texture and tool wear, was defined by $X_{t\text{-}b}$. From their findings, they concluded that it is necessary to control the irradiation temperature. However, in this study, the $X_{t\text{-}b}$ was fixed at 600 μm in order to avoid the laser beam irradiation into the cutting tool. The workpiece temperature obtained from their study was approximately 400 K.

Higher average laser powers generated higher temperatures at the workpiece's surface. A melting region will be produced when the temperature generated reaches the melting point while a HAZ region forms when the temperature exceeds the deformation temperature. Rapid cooling at the upper surface will occur and make the material harder due to hardening effects. Meanwhile, the size of the thermally affected zone is small. The temperature gradient reaches high values at the upper surface and the heat distributed to the material underneath depends on the conduction rate value. However, the temperature value is expected to be considerably higher when the temperature is distributed into the solid bulk phase. This causes the temperature at the surface to decay rapidly once the laser beam passes over this region. This process is affected by the thermal conductivity of the substrate material [29].

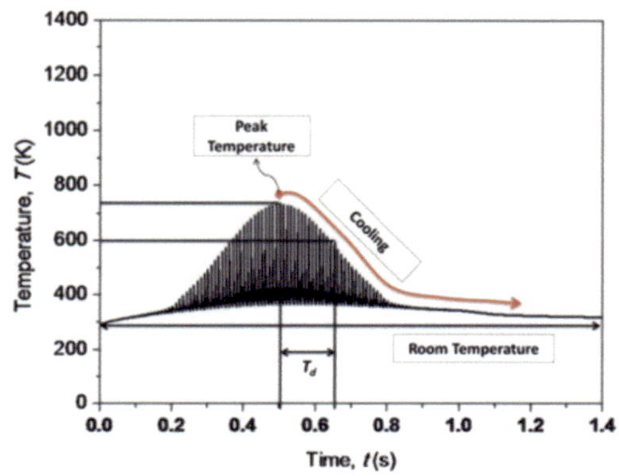

(a)

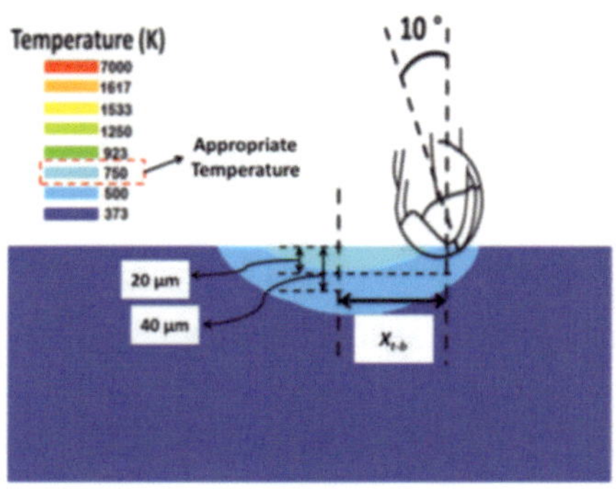

(b)

Figure 10. Results of temperature distribution when P_{avg} = 4.16 W, t_p = 1 ms and f_r = 70 mm/min, (a) recorded temperature at the centre irradiation line, (b) prediction of tool location and the depth of cut when using P_{avg} = 4.16 W, t_p = 1 ms, f_r = 70 mm/min, A = 32% [19].

The actual experimental setup of the LAM process is shown in Figure 11. An air-bearing spindle with a maximum rotation per minute of 60,000 is installed into the micromilling machine. All the laser-assisted micromilling tests were carried out on hardened Inconel 718

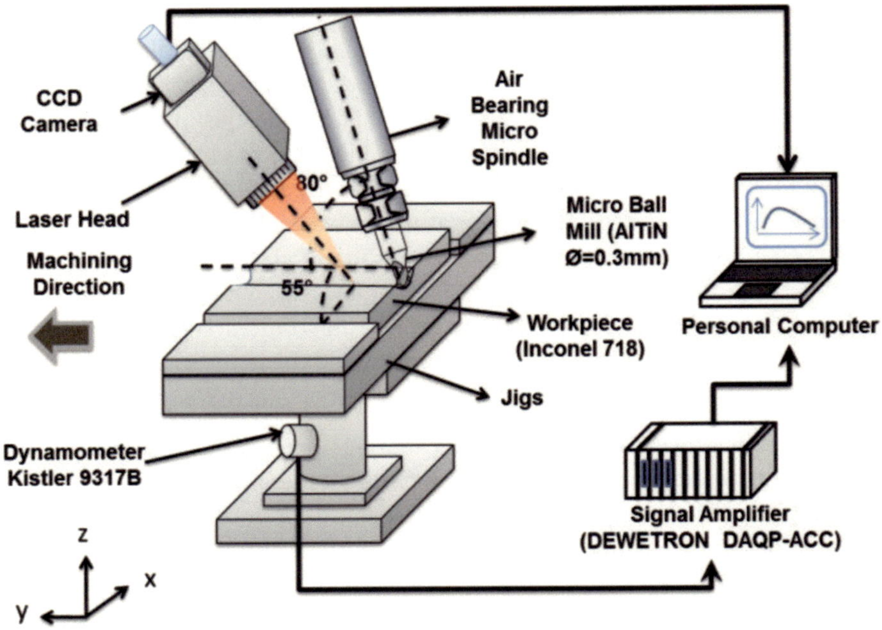

Figure 11. Experimental setup for laser-assisted milling [30].

plates (21–23 HRC) with 15 mm length, 40 mm width and 5 mm thickness. Commercially available AlTiN-coated carbide ball end mills (two flutes) with a diameter of 300 μm were used in cutting tests. The workpiece surface was preheated using Nd:YAG-pulsed laser with a 1064 nm wavelength. The laser head was inclined to 55° to avoid the deflection of laser irradiation on the cutting tool. The irradiated heat induced into the cutting tool will alter its properties. The dynamometer Kistler 9317B integrated with a DAQP-ADD card was used to measure the cutting force.

8. Laser-assisted turning

Laser-assisted turning process has attracted researchers for decades due to the demand for the machining of hard-to-cut materials. This technique was proven to reduce the cutting forces, obtain smoother machining surface and eliminate the production of continuous chips when the materials were produced under high cutting speeds.

Figure 12 shows the relative position of the laser beam, cutting tool and workpiece in the LAT process [31]. L_1 represent the distance between the focusing lens and the workpiece. The distance depends on the laser spot size being enough to cover the chamfer surface. While L_2 represent the distance between the tools' cutting edge and the laser spot point and it was determined by considering the heat effect on the tool. The actual initial cutting temperature is important to measure or predict. It will give a significant effect on cutting force results. In actual turning cutting processes, the force component consists of feed force, radial force and cutting force. The actual reduction of cutting forces is dependent on the force component and cutting tool condition. Nevertheless, the force component was reduced by up to 30%. The determination of cutting force is required for [10] (i) estimation of cutting power consumption; (ii) machine development, fixture and tool system; (iii) evaluation of role of the various machining parameters (speed (v_c), feed (f), depth of cut (DOC) and tool); (iv) understanding the behaviour and machinability characteristics of the work materials; (v) condition monitoring of the cutting tools and machine tools.

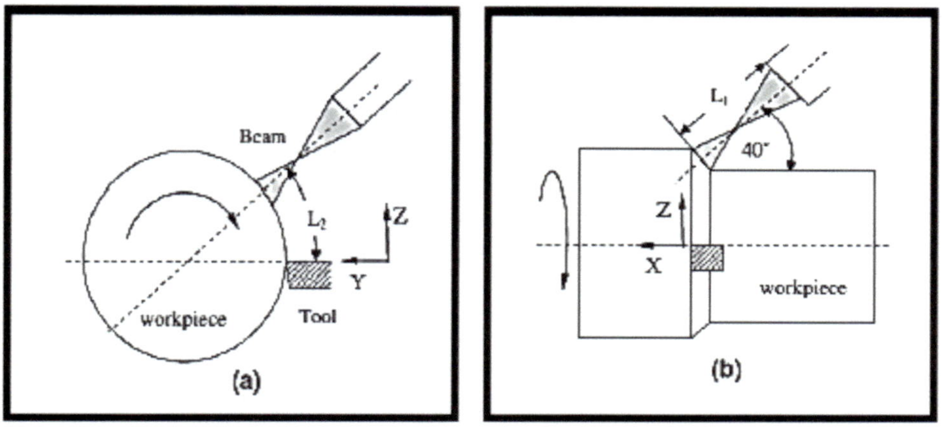

Figure 12. Relative position of laser beam, workpiece and cutting tool in LAT: (a) end view and (b) side view [32].

9. Influence of heat on nickel-based alloy

A huge demand on super alloys, especially Inconel 718, was significantly embarked by both heavy and microscale industries. Lasers and plasma are commonly used in thermally assisted machining of nickel-based alloys. It is a big challenge to determine the best method and optimum parameters to improve the machinability of super alloy materials in the machining process. Inconel has excellent material properties for high fatigue endurance limit, high yield strength, high corrosion resistance and high working temperature. The material can withstand machining temperatures of up to 700°C. This material is categorized as hard-to-machine because of the existence of carbide particles in its microstructure. Heating the material to higher than specific deformation temperatures will consequently harden the material with denser microstructure.

Since this material is weak in thermal conductivity, most of the heat generated in the cutting process will remain in the chip. The rest of the heat will be transferred into the base material and the cutting tool. This will lead to undesirable temperature increments during the machining process and further soften the material. However, in thermal-assisted machining, temperatures generated by friction between the tool and the workpiece are not favourable since it fluctuates with cutting speed and cutting depth changes. The accumulated heat energy could initiate microstructure changes onto the cutting surface. A well-controlled cutting temperature is needed to avoid the material from melting or being too soft, which promotes plastic deformation rather than ductile deformation during the cutting process. Applying cooling medium on the cutting tool during the plasma-assisted turning manages to reduce notching wear by 100%. The surface roughness can be improved by 150% to 250% for the tool life [3].

9.1. Cutting force

Figure 13 shows the effect of the depth of cut and laser-assisted micromilling (LAMM) of Inconel 718. In this experiment, the laser beam-to-cutting tool distance, $X_{t\text{-}b}$ was set at 600 μm. The results show that the thrust force significantly increases by increasing the t_c from 20 to 40 μm. When the t_c is increased from 20 to 40 μm, the centre part of the tool was rubbed on the workpiece surface and it consequently increased the cutting force. This is due to the absence of a cutting edge at the centre of the ball mill cutting tool. This phenomenon is prominent to produce rubbing and plunging processes. Otherwise, material properties also give significant effects due to the yield strength of the material, and it contributes to the increment of thrust force.

Irradiating Inconel 718 using an average laser power, P_{avg} of 4.16 W and $X_{t\text{-}b}$ of 600 μm, no melted region was observed but a HAZ effect is seen underneath the machined surface. Based on a review of the results shown in Figure 10(a) and (b), the peak temperature can be increased by up to approximately 800 K. This temperature is sufficient to reduce the material's strength. This argument is supported by Kim and Lee [13], and reveals that applying temperatures between 650°C and 900°C will reduce the material's strength significantly. Resulting from this irradiation, the thrust force was significantly reduced in all cases compared between LAMM and conventional cutting. However, in the case of LAMM using a micro ball mill tool at t_c of 40 μm, the thrust force drastically declined due to the increment of cutting tool effective-diameter and the changes in workpiece properties.

In addition, the changing trend of cutting force was also influenced by the spindle speed. The increment of spindle speed, N from 12,500 to 17,500 rpm, contributes to the higher cutting force. However, the trend shows contradiction when the spindle speeds decreased from 12,500 to 7500 rpm. Because higher spindle speeds produce higher contact ratios between the cutting edge and workpiece. However, at lower rpm, the cutting process is more effective in terms of material removal rate but with higher friction. Therefore, higher cutting temperatures were generated. This trend corresponds with the results presented by Arndt [33]. The author mentioned that the high cutting temperatures were due to the reactions of high shear resistance at high rpm. Meanwhile, the temperature still increases at low rpm, thus producing a strain-hardening effect.

Figures 13 and 14 show a comparison of cutting forces resulting between conventional machining and LAMM, respectively. It can be observed that the LAMM recorded the lowest value of force components at all tested conditions. Kong et al. [34] reported that laser-assisted machining successfully demonstrated the reduction of cutting force by around 30%–70% over the conventional machining process. Because the heat generated by cutting and the lasers approached the melting point, and thus reduced the shear resistances. In addition, the laser beam's preheating process on the workpiece material is in the deformation temperature range and managed to initiate a softening effect [9].

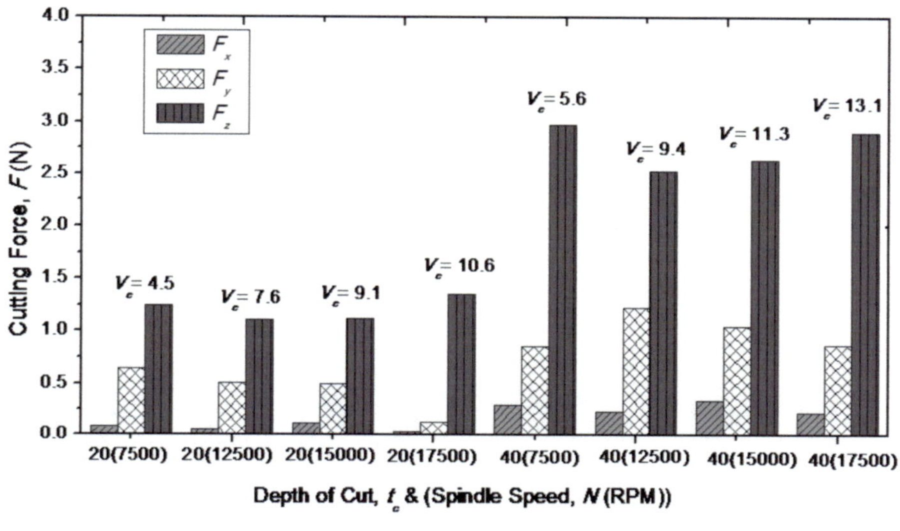

Figure 13. Results of cutting force at different cutting speeds, v_c, depth of cut, t_c and spindle speed, N in conventional micromilling process [30].

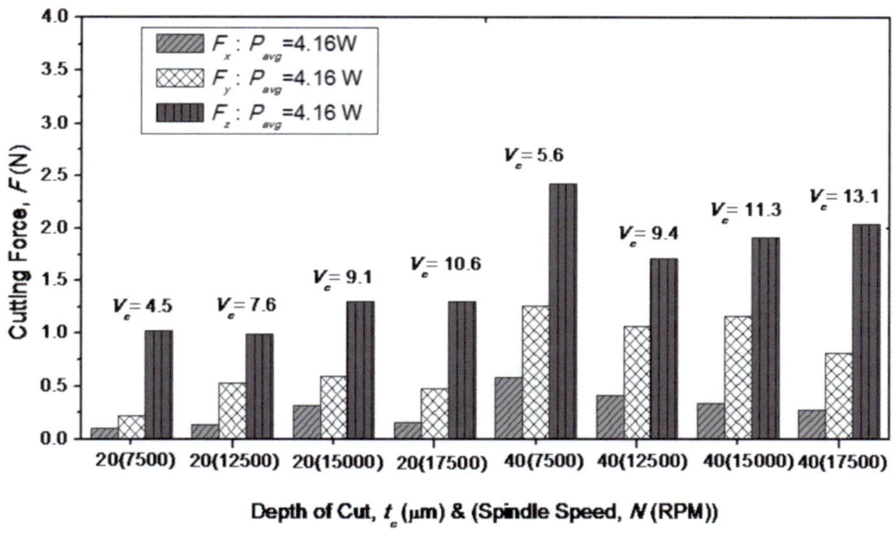

Figure 14. Results of cutting force at different cutting speeds, v_c, depth of cut, t_c and spindle speed, N in LAMM process [30].

9.2. Tool wear

Figure 15 shows a variation of wear patterns at all tested conditions. The image shows evidence that adhesion occurred on the flank face. Laser parameter and depth of cut give prominent effects on tool wear. The adhesion and coating delamination was critically increased when the depth of cut, t_c was increased from 20 to 40 μm under conventional machining processes. In addition, machining of ductile materials such as Inconel 718 using small ball mill diameters are inherently difficult processes. Once the adhesion is peeled off, it will cause a delamination of the coating and the tool loses its sharpness. This phenomenon is due to the material properties of Inconel 718 having a higher ductility and yield strength. According to Kuram and Ozcelik [35], most of the dominant wear mechanisms during micromilling of ductile materials were from abrasion and adhesion.

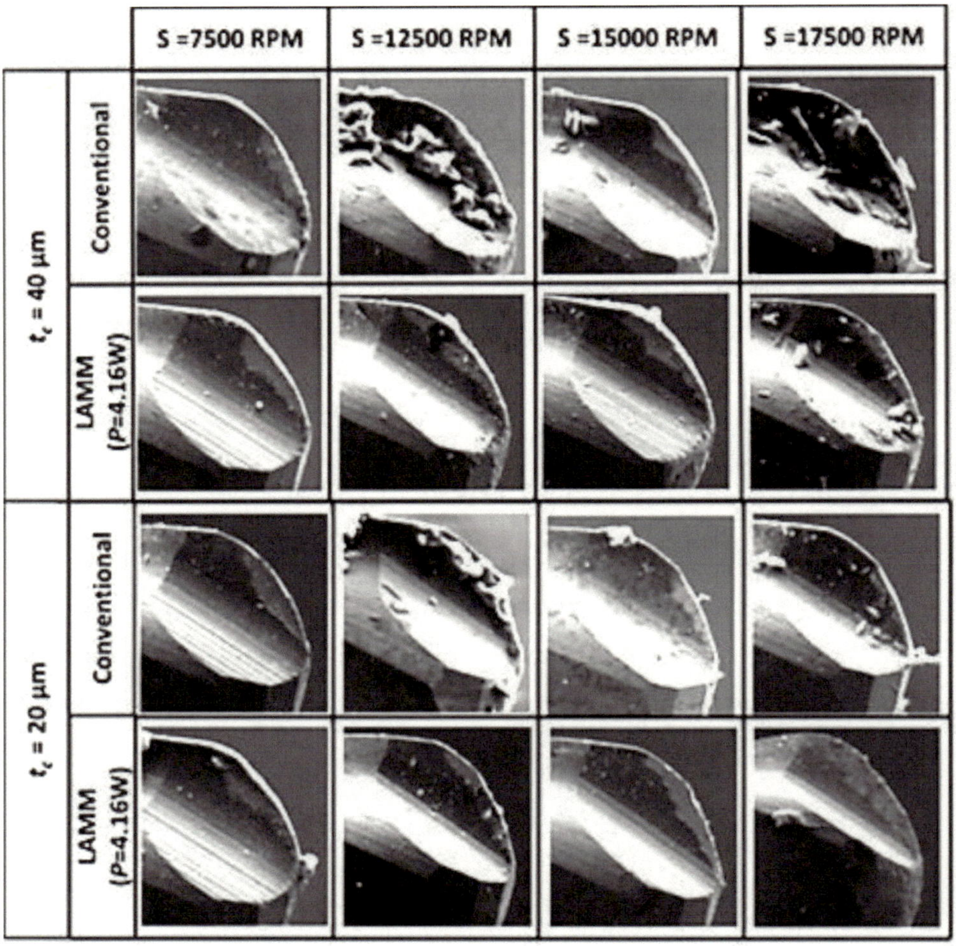

Figure 15. Tool wear on rack face of micro ball mill [30].

From these results, LAMM shows an improvement in terms of coating delamination and adhesion. The laser preheating method has shown its effectiveness in reducing the material strength and yield strength. During the irradiation process, surface material was preheated up to 900 K over the phase transformation of material. In this phase, the microstructure changes from γ' to γ'' as it dissolves the bonds between the particles [36]. These changes assist in avoiding material adherence at the tool rack face. Temporarily, adhered material on the cutting edge promotes the blunt cutting edge. However, in terms of ball end mill cutting tools, the effective of the cutting process at the bottom surface is poor and compels the rubbing process. Furthermore, the ineffectiveness of chip evacuation causes chips to re-enter the cutting region.

However, Kong et al. [34] investigated the effectiveness of coating material on laser-assisted milling of K24 nickel-based alloy. The amounts of average flank wear increases for both conventional machining and LAM with increasing cutting times are shown in Figure 16. However, the evolution of the flank wear is different between the two methods. The wear increases fast with the cutting time before 2.7 min, and is relatively stable from 2.7 to 16.8 min, after 16.8 min, rapid wear is visible again in LAM. The approximate machined times are 17.3 and 31.4 min for conventional machining and LAM tests, respectively, when the maximum wear criterion is set at $VB_{ave} = 0.3$ mm.

The results of laser-assisted milling experiments indicate that abrasive and adhesive wears were the most dominant wear mechanisms as shown in Figures 17 and 18. The TiAlN-coated tools exhibited the highest wear resistance at normal cutting speeds of 30 m/min. The triple-layer CVD-coated tool failure mode was concentrated on nonuniform flank wear due to the adhesion and depth-of-cut notching. TiCN performs poorly due to their inferior adhesion characteristics with the base material.

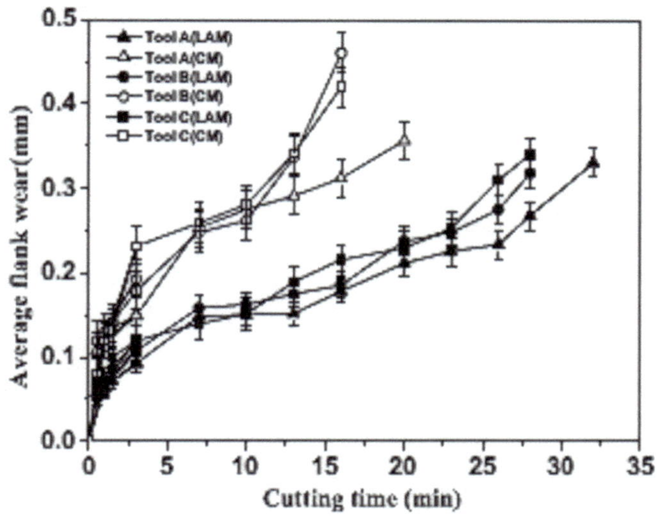

Figure 16. Results of the flank wear, in conventional machining and LAM, as a function of the cutting time [34].

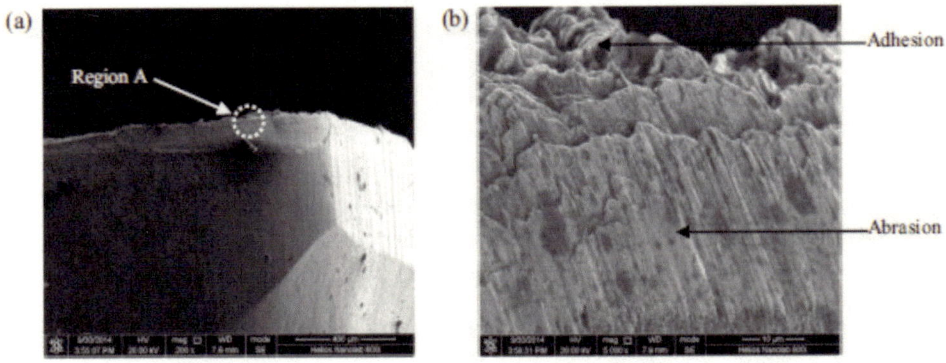

Figure 17. SEM of a TiAlN-coated tool at cutting speed $V = 30$ m/min and feed rate $f = 0.10$ mm/tooth with laser assist: (a) flank face at 10 s and (b) magnified SEM [34].

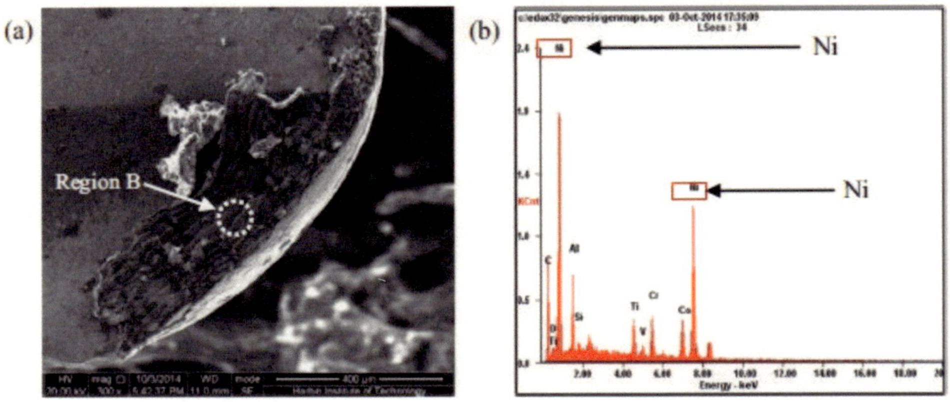

Figure 18. SEM of a TiAlN-coated tool at cutting speed $V = 30$ m/min and feed rate $f = 0.10$ mm/tooth with laser assist: (a) rack face at 10 s and (b) EDS of TiAlN-coated tool [34].

9.3. Surface roughness

Surface roughness in micromachining is recommended to be measured using noncontact measuring tools for accurate measurement. Figure 19 shows a comparison of the results of surface roughness between LAMM and conventional machining. The surface roughness values were measured using atomic force morphology (AFM) in three different points in order to obtain the roughness average, R_a.

It can be seen that the LAMM produces better surface roughness compared to the conventional machining process. In conventional machining processes, Inconel 718 retains its material properties with higher ductility and yield strength. Therefore, machinability is limited and the resulting rougher cutter mark at the bottom groove. Furthermore, conventional machining also produced a burr on the upper side of the groove due to the build-up edge and ineffective chip removal. Higher ductility of material exacerbates the phenomenon.

In contrast, LAMM manages to produce a finer surface finish compared to conventional machining. The effective laser beam-to-cutting tool distance and lower power average creates an appropriate preheating temperature to reduce the yield strength. The variation of yield strength has consequently produced finer cutter marks at the bottom surface of the groove. According to Kiswanto et al. [37], feed rate and machining time gave significant effects on the surface roughness and burr formation. Longer machining times contributed to the tool wear or delamination. Consequently, poor surface finish was produced at the centre of the groove path.

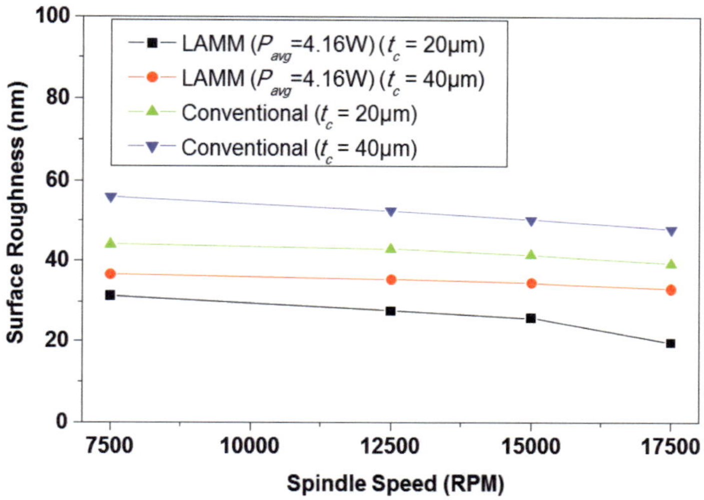

Figure 19. Comparison of surface roughness between LAMM and conventional machining processes [19].

10. Conclusion

The application of advanced materials in various parts fabrication has consequently increased the demand for new and high-performance machining techniques. Continuous study in material removal processing technique has shown that heating the workpiece is effective for increasing the machinability. The heat energy can be applied using several methods, either on the material or on the cutting tool, depending on the workpiece material behavior. The thermal-assisted machining technique is proven applicable on a wide range of material types including polymers and metals. The machinability of nickel-based alloys is proven to be improved by applying thermal machining technique. However, many studies were concentrated on turning process. A lot of issues can be explored and studied on milling process especially in microscale machining.

Author details

Erween Rahim*, Norazlan Warap and Zazuli Mohid

*Address all correspondence to: erween@uthm.edu.my

Advanced Machining Research Group, Universiti Tun Hussein Onn Malaysia, Parit Raja, Batu Pahat, Johor, Malaysia

References

[1] N. Waterman and M. Ashby, The Materials Selector, London: Chapman & Hall, 1997.

[2] M. Anderson, R. Patwa and Y. C. Shin. Laser-assisted machining of Inconel 718 with an economic analysis. *International Journal of Machine Tools & Manufacture*, vol. 46, pp. 1879–1891, 2006.

[3] Z. Y. Wang, K. P. Rajurkar, J. Fan, S. Lei, Y. C. Shin and G. Petrescu. Hybrid machining of Inconel 718. *International Journal of Machine Tools & Manufacture*, vol. 43, pp. 1391–1396, 2003.

[4] B. Shi, H. Attia, R. Vargas and S. Tavakoli. Numerical and experimental investigation of laser-assisted machining of Inconel 718. *Machining Science and Technology*, vol. 12, pp. 498–513, 2008.

[5] V. Ganta and D. Chakradhar. Multi objective optimization of hot machining of 15-5PH stainless steel using grey relation analysis. *Procedia Materials Science*, vol. 5, pp. 1810–1818, 2014.

[6] R. Muhammad, A. Maurotto, M. Demiral, A. Roy and V. V. Silberschmidt. Themelly enhanced ultrasonically assisted machining of Ti alloy. *CIRP Journal of Manufacturing Science and Technology*, vol. 7, pp. 159–167, 2014.

[7] G. Germain, P. D. Santo and J. L. Lebrun. Comprehension of chip formation in laser assisted machining. *International Journal of Machine Tools & Manufacture*, vol. 51, pp. 230–238, 2011.

[8] A. K. M. N. Amin, S. B. Dolah, M. B. Mahmud and M. Lajis. Effects of workpiece preheating on surface roughness, chatter and tool performance during end milling of hardened steel D2. *Journal of Materials Processing Technology*, vol. 201, pp. 466–470, 2008.

[9] H. Attia, S. Tavakoli, R. Vargas and V. Thomson. Laser-assisted high-speed finish turning of superalloy Inconel 718 under dry condition. *CIRP-Annal-Manufacturing Technology*, vol. 59, pp. 83–88, 2010.

[10] P. Y. Kok, A. K. M. N. Amin and A. I. Faris. Performance of preheating, cryogenic cooling and combined approachess in improving machinability of stainless steel in turning. In: *Proceedings of International Conference on Mechanical Engineering-ICME,* Dhaka, 2005.

[11] E. O. Ezugwu, Z. M. Wang and A. R. Machado. The machinability of nickel-based alloys: a review. *Journal of Materials Processing Technology,* vol. 86, pp. 1–16, 1999.

[12] M. Sundararaman, L. Kumar, G. E. Prasad, P. Mukhopadhyay and S. Banerjee. Precipitation of an intermetalic phase with Pt2Mo-type structure in alloy 625. *Metallurgical and Materials Transactions A,* vol. 30A, pp. 41–52, 1999.

[13] X. Luo., K. Shinozaki, S. Yoshihara, H. Kuroki and M. Shirai. Theoretical analysis of grain boundry liquation at HAZ of Inconel 718 alloy. *Quarterly Journal of the Japan Welding Society,* vol. 18-1, pp. 102–111, 2000.

[14] W. Konig and A. K. Zaboklicki. Laser-assisted hot machining of ceramics and composite materials. *International Conference on Machining of Advanced Materials,* vol. 847, pp. 455–463, 1993.

[15] S. Dharani and M. Pradip. Finite element analysis of laser irradiated metal heating and melting processes. *Optics & Laser Technology,* vol. 42, pp. 855–865, 2010.

[16] X. D. Ren, Q. B. Zhan, S. Q. Yuan, J. Z. Zhou, Y. Wang, N. F. Ren, G. F. Sun, L. M. Zheng, F. Z. Zhai, H. M. Yang and W. J. Dai. A finite element analysis of thermal relaxation of residual stress in laser shock processing Ni-based alloy GH4169. *Materials and Design,* vol. 54, pp. 708–711, 2014.

[17] Z. Mohid, N. M. Warap, M. I. S. Ismail, R. Ibrahim and E. A. Rahim. Determination of heat flux intensity distribution and laser absorption rate of AISI D2 tool steel. *Applied Mechanic and Materials,* vols. 465–466, pp. 730–734, 2014.

[18] Y. C. Shin and J. N. Kim. Plasma enhanced machining of Inconel 718. In: *Proceedings of the Manufacturing Science and Engineering, Symposium on Advanced Machining and Finishing Processes of Ceramics, Composites and High Temperature Alloys in ASME IMECE,* Atlanta, GA, 1996.

[19] E. A. Rahim, N. M. Warap, Z. Mohid, R. Ibrahim and N. Rafai. Numerical analysis of laser preheating for laser assisted micro milling of Inconel 718. *Applied Mechanics and Materials, vol 773-774, pp. 332-336, 2015.*

[20] A. J. Shih, M. A. Lewis and J. S. Strenkowski. End milling of elestomer-fixture design and tool effectiveness for material removal. *Journal of Manufacturing Science and Engineering,* vol. 126, no. 1, pp. 115–123, 2004.

[21] C. E. Leshock, J. N. Kim and Y. C. ShinPlasma enhanced machining of Inconel 718: modeling of workpiece temperature with plasma and experimental results. *International Journal of Machine Tools & Manufacture,* vol. 41, pp. 877–897, 2000.

[22] M. Madhavulu and B. Ahmed. Hot machining process for improved metal removal rates in turning operation. *Journal of Materials Processing Technology,* vol. 44, pp. 199–206, 1994.

[23] J. Luo, H. Ding and A. J. Shih, Induction-heated Tool Machining of Elastomers-Part 2: Chip Morphology, Cutting Forces and Machined Surface, *Machining Science and Technology: An International Journal,* vol. 9, pp. 567-588, 2005.

[24] T. Kizaki, K. Harada and M. Mitsuishi. Efficient and precise cutting of zirconia ceramics using heated cutting tool. *CIRP Annals - Manufacturing Technology,* vol. 63, pp. 105–108, 2014.

[25] Z. Mohid, N. M. Warap, S. Hassan, M. I. S. Ismail, R. Ibrahim and E. A. Rahim, Numerical Analysis of Laser Heating for Laser Assisted Micro Milling Application, *Applied Mechanics and Materials,* Vols. 465-466, pp. 720-724, 2014.

[26] J. Yang, S. Sun, B. Milan and W. Yan, Experimental investigation and 3D finite element prediction of the heat affected zone during laser assisted machining of Ti6Al4V alloy, *Journal of Materials Processing Technology,* vol. 210, pp. 2215-2222, 2010.

[27] D. H. Kim and C. M. Lee, A study of cutting force and preheating-temperature prediction for laser-assisted milling of Inconel 718 and AISI 1045 steel, *International Journal of Heat and Mass Transfer,* vol. 71, pp. 254-274, 2014.

[28] E. A. Rahim, Z. Mohid, N. M. Warap, M. R. Ibrahim and a. M. I. S. Ibrahim, A Prediction of Laser Sport-to-Cutting Tool Distance in Laser Assisted Micro Milling, in *International Conference of Micro Manufacturing,* Singapore, 2014.

[29] M. W. Norazlan, Z. Mohid and E. A. Rahim. Laser assisted machining of titanium alloys. *Materials Science Forum,* vol. 763, pp. 91–106, 2013.

[30] E. A. Rahim, N. M. Warap, M. Z. and R. Ibrahim. Investigation on laser assisted micro ball milling of Inconel 718. *Applied Machanics and Materials,* vol. 660, pp. 79–83, 2014.

[31] S. Sun, M. Brandt and M. S. Dargusch. The effect of laser beam on chip formation during machining of Ti6Al4V alloy. *The Minerals, Metals & Society and ASM International,* vol. 41, p. 1681, 2010.

[32] S. Sun, J. Harris and M. Brandt. Parametric investigation of laser-assisted machining of commercially pure titanium. *Advanced Engineering Materials,* vol. 10, pp. 565–572, 2008.

[33] G. Arndt. Ultra-high-speed machining: a review and analysis of cutting forces. *Proceedings of the Institution of Mechanical Engineers,* vol. 187, pp. 625–634, 1973.

[34] X. Kong, L. Yang, H. Zhang, K. Zhou and Y. Wang. Cutting performance and coated tool wear mechanisms in laser-assisted milling K24 nickel-based superalloy. *International Journal Advanced Manufacturing Technology,* vol. 9, p. 6606, 2014.

[35] E. Kuram and B. Ozcelik. Multi-objective optimization using Taguchi based grey relational analysis for micro milling of Al 7075 material with ball nose end mill. *Measurement,* vol. 46, pp. 1849–1864, 2013.

[36] W. Hongjie, I. Kenji, T. Makoto and I. Akio. Microstructure of Inconel 718 alloy subjected to rapid thermal and stress cycle - Joint performance controlling factor in friction welding of Inconel 718 alloy. *Welding International,* vol. 26, no. 2, pp. 152–158, 2009.

[37] G. Kiswantoa, D. L. Zariatina and T. Ko., "The effect of spindle speed, feed-rate and machining time to the surface roughness and burr formation of aluminum alloy 1100 in micro-milling operation," *Jou nal of Manufacturing Processes,* vol. 16, no. 4, pp. 435-450, 2014.

Chapter 2

Spectroscopic and Optoelectronic Properties of Hydrogenated Amorphous Silicon-Chalcogen Alloys

Shawqi Al Dallal

Additional information is available at the end of the chapter

http://dx.doi.org/10.5772/61103

Abstract

Hydrogenated amorphous silicon-chalcogen alloy thin films have been the subject of growing interest during the past two decades. Thin films of these alloys are usually prepared by the decomposition of SiH_4 and H_2S or H_2Se gas mixtures in a radiofrequency plasma glow discharge at a substrate temperature of 250°C. The alloy composition is varied by changing the gas volume ratio R_V = [chalcogen/silane]. Infrared spectroscopy is used to explore the bonding structure of the alloy. The material exhibits hydrogen-induced bands, normally observed in a-Si:H spectra and other chalcogen-induced bands resulting from bonding chalcogen atoms with hydrogen and silicon. Analysis of the vibrational spectra of this material reveals the presence of significant levels of Si-chalcogen-SiH_n configurations. Optical and electrical measurements show that increasing the chalcogen content results in an increase of the optical (Tauc) gap and a decrease in dark conductivity and photoconductivity. Subgap absorption measurements are employed to probe the Urbach energy and defect density. Upon increasing the chalcogen content, a broadening of band tails and an increase in defect density is observed. These results are shown to be consistent with photoluminescence measurements carried out on these materials.

Keywords: Amorphous silicon alloys, Sulphur, Selenium, Infrared spectroscopy, Photoconductivity, Photoluminescence

1. Introduction

There has been an intensive research work during the past two decades to search for stable and low defect density material for use in tandem with a-Si:H-based photovoltaic devices [1–5]. a-Si:H-based alloys proved to be attractive materials with a wide range of applications including solar cells, photoreceptors, displays, and imaging devices [6–9]. Hydrogenated carbon alloys (a-Si, C:H) have received the most interest [10–12]. However, substantial work has been conducted to study alternatives to this material with comparable properties. Nano-structured hydrogenated silicon (ns-Si:H) prepared by plasma-enhanced chemical vapour deposition at 27.12 MHz was found to exhibit a wide range of applications including solar cells, integrated arrays of printed circuit devices [13,14], and thin film transistors [15].

Hydrogenated polymorphous silicon also proved to be a viable alternative material to produce low-cost solar cells [16]. This material is made up of small clusters containing nanocrystals imbedded in the silicon amorphous matrix. This structure is different from traditional hydrogenated amorphous and microcrystalline silicon thin films. It exhibits better electronic properties and is a more optically stable material. No attempt has been made so far to incorporate chalcogen into this material, and this remains an area of potential interest.

a-Si,S:H [17–20] and a-Si,Se:H [21,22] alloys proved to be of particular interest because they exhibit a juxtaposition of chalcogenide and silicon-based semiconductors with a wide range of optical and electronic properties.

In this paper, we provide a review of the preparation, spectroscopic, and optoelectronic characterization of amorphous silicon-chalcogen alloys thin films. Sulphur and selenium are revealed to be among the most appropriate elements to form alloys with silicon of potential applications in optoelectronic devices [23,24]. Addition of sulphur or selenium to the amorphous silicon matrix to form a-Si_x S_{1-x} H or a-Si_x Se_{1-x} H alloys and the possibility of changing the composition gives rise to a wider range of properties of this material. Of particular relevance are the nature and properties of defect states of various compositions and their influence on the optical parameters and charge transport phenomenon. The study of the infrared (i.r.) vibrational spectra and its correlation with the electronic and optical properties of the material reveals interesting information regarding the structure of the alloy and the nature of defect states in the bandgap. The compositional dependence of hydrogenated silicon-chalcogen alloy will be shown to have an important impact on its optical bandgap and transport properties. The glow discharge experimental technique can also be employed to prepare superlattices of a controlled number of layers as revealed by small angle x-ray diffraction technique [25].

2. Experimental

Hydrogenated amorphous silicon-sulphur and silicon-selenium alloy thin films have been prepared by capacitively coupled radiofrequency (rf) glow-discharge decomposition of (5%

SiH_4 + 95% He) and (2% H_2S or H_2Se + 98% He) gas mixtures. Helium dilution of the reactant gases is employed for safety purposes. Handling of H_2S or H_2Se gases requires extreme precautions because of its toxicity and bad smell. Thin films are normally deposited on 7059 Corning glass or on high-resistivity silicon substrates held at 250°C. The plasma power was controlled at 0.4 Torr, whereas the power density was varied between 5 and 50 W. The power density of 20 W plasma power translates to about 110 mW/m^2. For a total flow rate of reactant gases of typically 30 cm^3/min, a growth rate of about 2 Å/s for equal flow rates of silane and hydrogen-chalcogen gas was obtained. The decomposition rate was found to be a function of the gas volume ratio R_V ≡ [chalcogen/silane]. Increasing R_v above a threshold value leads to a rapid decrease in the growth rate.

3. Results and discussion

Infrared transmission measurements on a-$Si_{1-x}S_x$:H and $Si_{1-x}Se_x$:H alloys with different compositions are depicted in Figs. (1) to (6). The spectra of these two alloys will be discussed separately and then compared to elucidate their nature, origin, and bonding environment.

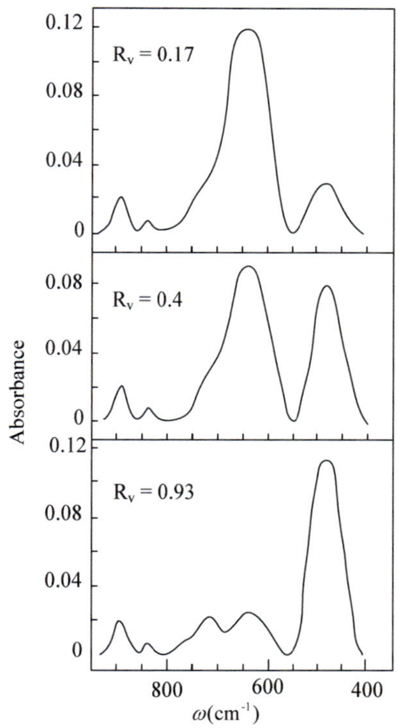

Figure 1. Compositional dependence of the infrared absorption spectra of a-Si,S:H alloy in the range 400–1000 cm^{-1}. Deposition parameters were: T_s = 250°C, P = 0.5 Torr, plasma power = 20 W (Ref. 15).

3.1. Infrared spectra of a-Si_{1-x} S_x:H alloys

Figs. (1) and (2) show the infrared (i.r.) spectra of a-$Si_{1-x}S_x$:H film in the 400–4000 cm^{-1} range. We can distinguish two types of bands in these spectra. The first is the hydrogen-induced band, and covers the spectral range from 500 to 900 cm^{-1} and from 1900 to 2300 cm^{-1}. The second is the sulphur-induced band appearing in the same spectral range. In the following section, we will discuss separately the characteristics and origin of these bands.

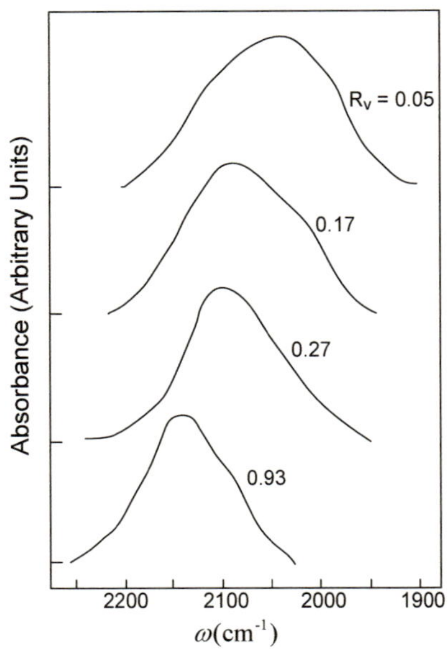

Figure 2. Evolution of the infrared transmission spectra of a-Si,S:H alloys in the 1900–2300 cm^{-1} range with composition. Spectra are normalized to the same maximum intensity. Deposition parameters were: T_s = 250°C, P = 0.5 Torr, plasma power = 20 W (Ref. 15).

3.1.1. *Si-H-induced bands in the 500–900 cm^{-1} region*

Referring to Fig. (1), the two bands at 840 and 885 cm^{-1} were ascribed by Lucovsky et al. (1979) to the bending modes of SiH_2 bonds. These authors have also demonstrated that isolated dihydrides (SiH_2) appear as a single band at 885 cm^{-1}, whereas polysilane chains give rise to two bands at 890 and 850 cm^{-1}. Because of the relative weakness of the band appearing at 840 cm^{-1} in a-$Si_{1-x}S_x$:H alloy, the band at 885 cm^{-1} was attributed primarily to bending modes of isolated dihydrides [26]. In a-Si:H, the SiH_3 radicals causes a degenerate deformation and symmetric deformation modes at 907 cm^{-1} and 862 cm^{-1} respectively. The absence of these modes in a-Si,S:H alloys' spectra suggests negligible SiH_3 bonding in these alloys [19].

The prominent band at 640 cm^{-1} is dominant in a-Si:H as well as in a-$Si_{1-x}S_x$:H alloys. It was attributed to the wagging mode of Si-H bonds and to the rocking and wagging modes of SiH_2 bonds [27–29].

3.1.2. *Stretching mode spectra in the 1900–2300 cm^{-1} region*

Figure (2) shows the absorption spectra in the Si-H bond stretching region for different compositions (R_v). This region is normally characterized by the presence of two broad bands at 2000 and 2100 cm^{-1}. These bands are attributed to the Si-H stretching vibrational modes [29]. The stretching mode of Si-H monohydrides appears at 2000 cm^{-1} and that of SiH_2 isolated dihydrides and polysilane chains $(SiH_2)_n$ appears in the region 2090–2100 cm^{-1}. These modes appear in Fig. (2) with a slightly shifted peak position. The mode at 2121–2140 cm^{-1} was assigned to SiH_3 stretching vibrations [26,27].

3.1.3. *Sulphur-induced bands in the stretching mode spectral region*

Investigation of the stretching region, Fig. (3), shows a systematic shift of the peak position in this region to higher wave numbers as the sulphur content increases. Since the electronegativity of sulphur is larger than that of silicon, this shift was attributed to induction effects induced by sulphur atoms on the Si-H stretching mode frequency in the Si-S-H group, where sulphur and hydrogen are back bonds to the same silicon atom [19]. The frequency of the perturbed Si-H stretching mode can be estimated from the work of Lucovesky [30],

$$v\left(S_iH\right) = a + b\sum_{i=1}^{3} SR\left(X_i\right) \tag{1}$$

and

$$v\left(S_iH_2\right) = c + d\sum_{i=1}^{2} SR\left(X_i\right) \tag{2}$$

where a, b, c, and d are constants, and $SR(X_i)$ is the electronegativity of the substituted element as defined by Sanderson [31]. For silicon, SR(Si) = 2.62, and for sulphur SR(S) = 4.1 [31]. Using these values and the values of the constants given by Lucovesky, it was found that the substitution of one silicon atom by one sulphur atom in $(Si)_3SiH$ species shifts the Si-H stretching frequency from 2000 cm^{-1} (without sulphur) to 2064 cm^{-1} (with sulphur). For SiH_2 species, the substitution of silicon with sulphur shifts the 2100 cm^{-1} peak to 2127 cm^{-1}, whereas substitution of two silicon atoms with sulphur atoms in the above species shifts the 2000 and the 2100 cm^{-1} bands to 2116 and 2165 cm^{-1}, respectively. The calculated frequencies of these bands were found close to the position of the peak of the stretching band and other pronounced features in a-Si,S:H spectra in this region. It was suggested that the bands in this region are produced by mixing of vibrational modes of SiH, SiH_2, $(Si)_2$ SSiH, $SiSSSiH_2$ and S_2SiH_2 species [19].

3.1.4. *Sulphur-induced bands in the 400–500 cm^{-1} region*

This region is characterized by a dominant band appearing between 480 and 490 cm^{-1}, as shown in Fig. (1). This band is degenerate with the Si TO-like vibrational mode in amorphous silicon,

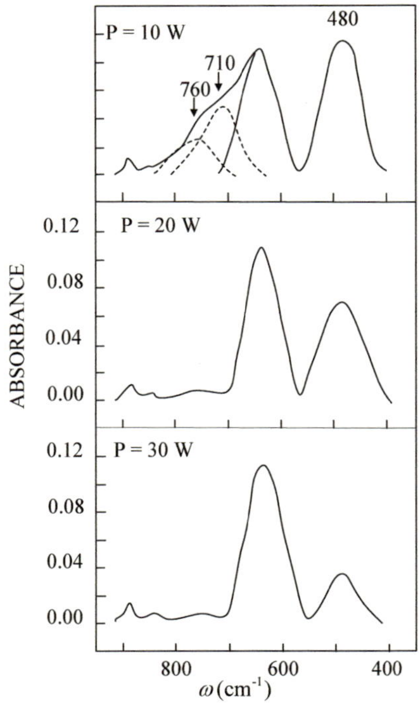

Figure 3. Variation of the infrared absorption spectra of a-SiS:H alloy with r.f. plasma power. For these samples, 10 cm^3 min^{-1} hydrogen flow rate was added to the gaseous mixture. Other deposition parameters were: T_s = 250, P =2 Torr, R_v = 0.6 (Ref. 15).

and is normally i.r. inactive but becomes activated by alloying the Si matrix with an element, such as sulphur, whose elecronegativity is sufficiently different from Si so as to be able to induce charge transfer through the back bonds. In fact, many authors have reported the association of an i.r band with lattice vibrations at 510 cm^{-1} in a-Si,F:H alloys [32,33], and at 500 cm^{-1} in a-Si,Cl:H alloys [34]. Both F and Cl are more electronegative than Si (2.65). This difference in electronegativity is responsible for the shift in the phonon vibrational mode peak to higher wave numbers. Bonding of sulphur atoms to the Si-matrix is also expected to activate the Si-TO-like mode in a-Si,S:H alloys [19]. However, since the electronegativity of S (4.1) is less than that of F or Cl, only a small shift of the phonon band is expected to occur, and the strength of the resulting i.r. band becomes relatively weak.

It was shown that S-S stretching vibrations normally occur as a very weak band in the 400–500 cm^{-1} spectral region [35], and therefore are unlikely to be responsible for the 480–490 cm^{-1} band. Fig. (1) shows that the intensity of this band is appreciably enhanced by increasing the sulphur content in the gas phase. It is therefore believed that bonded sulphur in the form Si-S is responsible for this enhancement [19]. Furthermore, it was found that this band is i.r. and Raman active [36]. Therefore, the 480–490 cm^{-1} band was assigned to the stretching frequencies of the Si-S bonds. In fact, a band near 480 cm^{-1} was first observed by Ebsworth et al. [37] in disilyl sulphide and was attributed to the stretching frequencies of the Si-S bonds.

Of particular interest is the possibility of the presence of SiS_2 species in a-Si,S:H alloys. Studies reported for glassy SiS_2 have shown the existence of three IR active modes at 597, 481, and 395 cm^{-1} [39]. Among the above vibrations, only the 481 cm^{-1} frequency is detected in a-Si,S:H, and therefore, the presence of SiS_2 species in this alloy is unlikely [19].

The effect of plasma power on the i.r. absorption spectra of a-Si,S:H is shown in Fig. (3). An important feature displayed by these spectra is the decrease in the intensity of the 480 cm^{-1} band as the plasma power increases. At low plasma power, a dominating broad band in the 600–800 cm^{-1} region appears. In fact, this region consists of a superposition of several bands, making the analysis more difficult to achieve. However, the broad band can be deconvoluted into three different band at 640, 710, and 760 cm^{-1}, as shown in Fig. (3). The band at 710 cm^{-1} was more pronounced in samples prepared with additional hydrogen flow [19]. Its spectral position is close to the calculated (701 cm^{-1}) and observed (725.8 cm^{-1}) values of the vibrational frequency of the S_2 molecule [39]. Thus, the band at 710 cm^{-1} was assigned tentatively to isolated S_2 molecules trapped in the silicon-sulphur matrix. In fact, the 710 cm^{-1} band is drastically suppressed by increasing the plasma power where the molecule is either dissociated or transferred to the gaseous phase. The band at 760 cm^{-1} is close to the calculated (726 cm^{-1}) and observed (749 cm^{-1}) vibrational frequencies of the SiS molecule [39–41]. However, we think that more experimental data are required to probe the nature of this band. An important aspect of these spectra is the decrease of the 480 cm^{-1} band intensity as the plasma power increases. Figure (4) depicts the variation of the 480 cm^{-1} band intensity with plasma power. It decreases steadily and then levels off at high plasma power. This trend suggests a decrease in the amount of incorporated sulphur atoms with increasing plasma power. This observation is supported by the optical gap measurements which show similar behavior with increasing plasma power.

3.2. Infrared spectra of a-Si,Se:H alloys

Infrared spectroscopy was employed to study the structure of a-Si,Se:H alloys in the 200–4000 cm^{-1} range. Figure (5) depicts the i.r. spectra of a-$Si_{1-x}Se_x$:H alloys in the 200–900 cm^{-1} region. As in the case of a-Si,S:H, the bands at 840 and 885 cm^{-1} are ascribed to the bending modes of SiH_2 bonds [26], whereas the band at 640 cm^{-1} is ascribed to the wagging mode of Si-H bonds and to the rocking and wagging modes of the SiH_2 bonds [27–29]. The band appearing in the region between 480 and 500 cm^{-1} is the Si TO-like phonon mode [27]. The intensity of the band at 390 cm^{-1} was observed to grow with increasing the selenium content in the gas phase. It was first observed by Ebsworth et al. [37] and Tenhover et al. [38] in disilyl selenide and was ascribed to the vibrational mode of the Si-Se bonds. The correlation between R_v in a-Si, Se:H alloys and the intensity of this band is consistent with the above assignment [26].

Figure (6) shows the stretching modes region of a-Si,Se:H alloys. It is dominated by two bands at 2000 and 2100 cm^{-1}, attributed in a-Si:H to the Si-H and SiH_2 configurations, respectively. Incorporation of Se atoms gives rise to the appearance of pronounced features at 2035, 2140, and 2165 cm^{-1}. Referring to the work of Lucovesky discussed in Section 3.1.3, the Si-H stretching mode is affected by the presence of substituted elements of higher electronegativity. Thus, since the electronegativity of Se atoms (4.1) is higher than that of Si (2.65), a shift of the vibrational frequency of the Si-H stretching modes to higher wave number is expected.

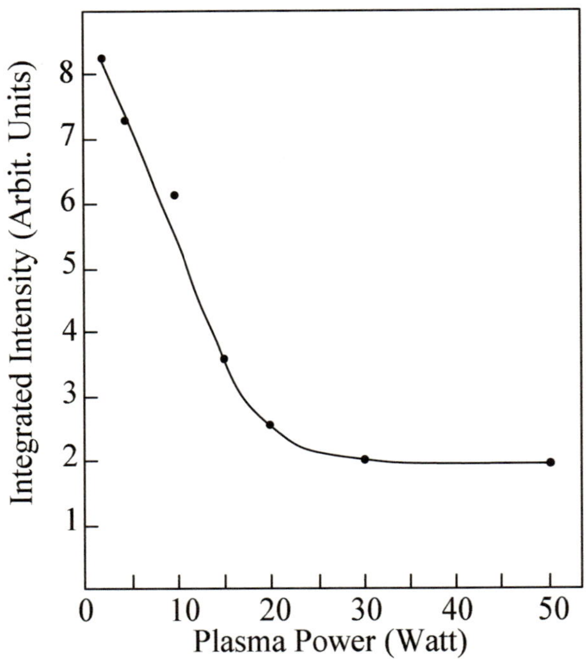

Figure 4. Variation of the integrated intensity of the 480 cm^{-1} band with plasma power (Ref. 15).

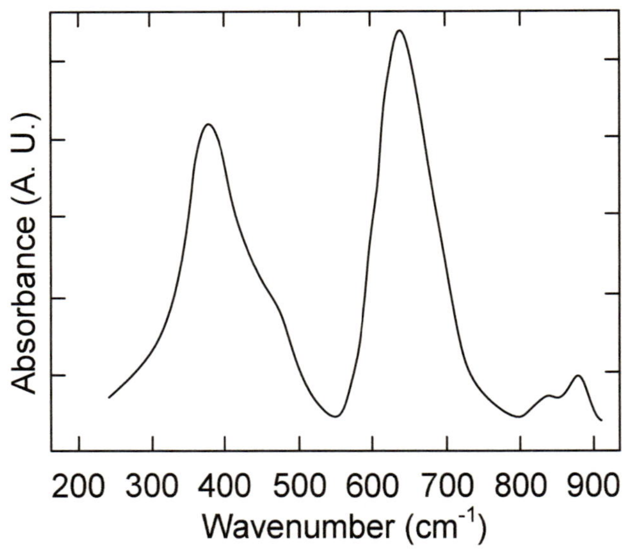

Figure 5. Infrared absorption spectrum of a typical a-Si,Se:H alloy film in the range 250–900 cm^{-1} (Ref. 18).

Analysis of this shift reveals that the bands in the 2000–2100 cm^{-1} region are due to a mixture of the stretching vibrational modes of $(Si)_xSe_{3-x}SiH$ and $(Si)_ySe_{2-y}SiH_2$ species [21].

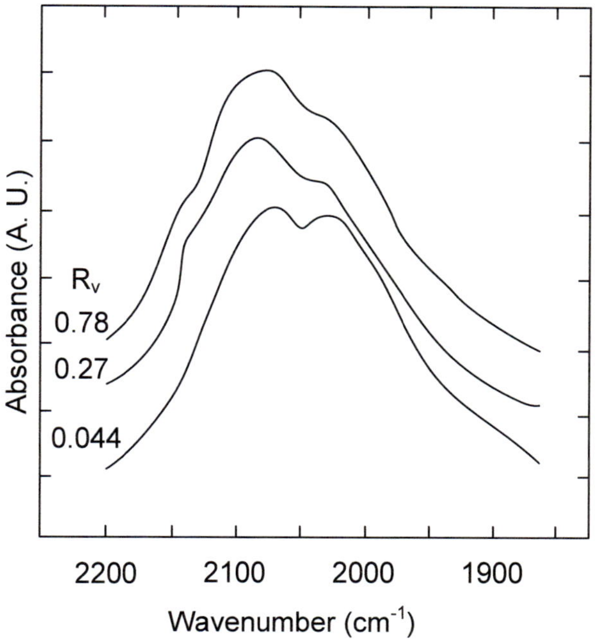

Figure 6. Compositional dependence of the infrared spectra of a-Si,Se:H alloys in the stretching mode region (1800–2200 cm^{-1}) for R_v = 0.44, 0.27, and 0.78 (Ref. 17).

3.3. Optical absorption

The optical (Tauc) gap E_{opt} is determined from optical transmission measurements. Assuming a parabolic density of states for the band tail and an energy independent matrix element for optical transitions between the initial and final states, the absorption coefficient $\alpha(E)$ is given by the relation

$$\alpha h\nu = A\left(h\nu - E_{opt}\right)^2 \tag{3}$$

where A is a constant, and E_{opt} is the optical (Tauc) gap. E_{opt} can be determined by the extrapolation of $(\alpha\, h\, \nu)^{1/2}$ towards the base line. This technique proves to be fruitful for measuring E_{opt} for any given sample and was used to determine the compositional dependence of E_{opt} as given in the next section.

3.3.1. Compositional dependence of energy gap in amorphous silicon-chalcogen alloys

The optical (Tauc) gap was measured for samples prepared at different gas volume ratio for both a-Si,S:H and a-Si,Se: alloys, as depicted in Figs. (7) and (8). These figures reveal that the variation of E_{opt} with R_v is not linear. It increases rapidly with the initial introduction of H_2S or H_2Se in the gaseous phase and then levels off at intermediate values of R_v. It increases again for R_v above 0.6 in a-Si,S:H or R_v above 0.8 in a-Si,Se:H. The E_{opt} dependence on R_v has an inverted s-shaped behavior. The same type of variation was also reported by Saito et al. [42] in a-Si,C:H, indicating that the compositional dependence of E_{opt} is universal in these types of alloys. A model was developed to explore the nature of this variation [43]. The overall compositional dependence of E_{opt} in silicon-sulphur alloys can be derived from Fig. (6) and can be written as

$$y_s = 1.958 - 0.135e^{-x/a_s} + 0.135e^{(x-x_{fs})/b_s} \tag{4}$$

where for a-Si,S:H alloys a_s=0.15, b_s=0.157, and x_{fs}=0.942.

Similarly, the compositional dependence of amorphous silicon-selenium alloy (Fig. 7) can be written as

$$y_{se} = 2.026 - 0.2e^{-x/a_{se}} - e^{(x-x_{fse})/b_{se}} \tag{5}$$

where a_{se}=0.37, b_{se}=0.38, and x_{fse}=2.1.

The significance of the (a) parameter can be evaluated by considering the ratio

$$\frac{a_{se}}{a_s} = 2.467 \tag{6}$$

The above ratio compares closely well with the ratio of the atomic weights of sulphur and selenium

$$\frac{\text{atomic weight of selenium}}{\text{atomic weight of sulphur}} = \frac{78.96}{32.066} = 2.462 \tag{7}$$

To verify the possible bias of the data, the authors [43] applied the same model to a-Si,C:H reported by Saito et al. [42]. The best fit of their data on a-Si,C:H can be approximated by the relation

$$y_C = 1.92 - 0.57\left(e^{-x/0.057} - e^{(x-0.68)/0.08}\right) \tag{8}$$

Thus, a_c=0.057, and

$$\frac{a_c}{a_s} = 0.38 \tag{9}$$

The above relation may be compared with the ratio of the atomic weights of carbon and sulphur

$$\frac{\text{atomic weight of carbon}}{\text{atomic weight of sulphur}} = \frac{12.010}{32.066} = 0.3745 \tag{10}$$

The above ratio is again very close to the predicted value, which provides an extra degree of confidence in the model. This behavior must have a deeper origin at the atomic scale. The atomic weight ratio can be understood by considering processes occurring immediately after the chalcogen atoms are deposited on the film surface. Careful analysis of these processes yield

$$\frac{a_{se}}{a_s} = \frac{\mu_s}{\mu_{se}} = \frac{m_{se}}{m_s} \tag{11}$$

where $\mu = e\tau_{col} / m$ is the mobility of atoms during the diffusion process to occupy their final site, and τ_{col} is the time between successive collisions with the background silicon matrix. Eq. (11) provides a strong hint as to the nature of the processes leading to the observed compositional dependence of the energy gap.

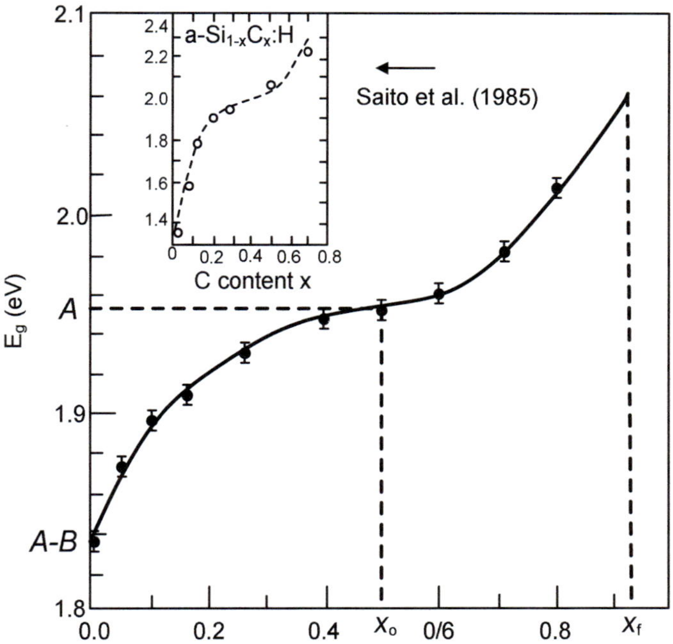

Figure 7. Compositional dependence of the energy (Tauc) gap of a-Si,S:H thin films. A, B, x_o, and x_f are model parameters. The inset shows the compositional dependence of a-Si,C:H reported by Saito et al. [42] (Ref. 17).

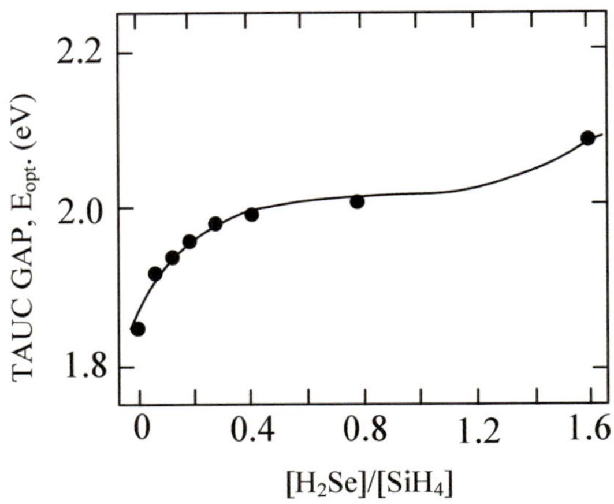

Figure 8. Compositional dependence of the optical (Tauc) gap of a-Si,Se:H (Ref. 17).

3.3.2. Optical subgap absorption, Urbach energy, and defect density

The technique used by employing Eq. (3) proves to be fruitful for determining E_{opt} for given samples, but was found to be limited when probing the structure or deep states in the band tail. For this purpose, other techniques are usually employed, as explained below.

Constant photocurrent method (CPM) is revealed as an accurate approach to measure the optical absorption spectra in the band tail [29]. Below E_{opt}, the spectra exhibits an exponential tail given by

$$\alpha = \alpha_o \exp\left(hv l E_u\right) \tag{12}$$

where E_u is the Urbach energy. A plot of the spectra measured for two films of a-Si,Se:H alloys is shown in Fig. (9). The distribution of localized states in the gap was investigated in detail [21,22]. In the following, we shall discuss separately the subgap absorption spectra in amorphous silicon-sulphur and amorphous silicon-selenium alloys respectively.

i. a-Si,S:H alloys: The subgap absorption spectra were measured via constant photocurrent method (CPC) and photothermal deflection spectroscopy (PDS) [22]. The compositional dependence of the Urbach energy E_u and of the defect density N_d are shown in Fig. (10) [22]. E_u rises from 60 to 110 meV as R_v increases from 0 (no sulphur) to 1.1. This variation suggests a substantial increase in the defect density and consequently in the width of the band tails. The increase in N_d is corroborated by the results of ESR measurements of the Si dangling bond spin density N_s. For a-Si,S:H alloys, it was found that N_s increases from 5×10^{16} cm^{-3} in nonalloyed material to about 1×10^{18} cm^{-3} in a 2.2 energy gap material [22]. The alloy also exhibits a substantial LESR signal throughout the entire compositional range, indicating that a substantial

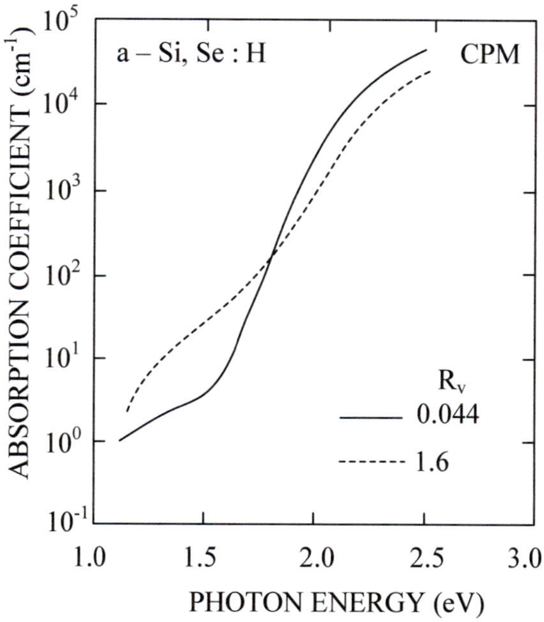

Figure 9. Absorption coefficient as a function of photon energy for two different compositions (R_v = 0.044 and R_v =1.6) (Ref. 18).

portion of the dangling bond defect density in these alloys may have negative correlation energy [22].

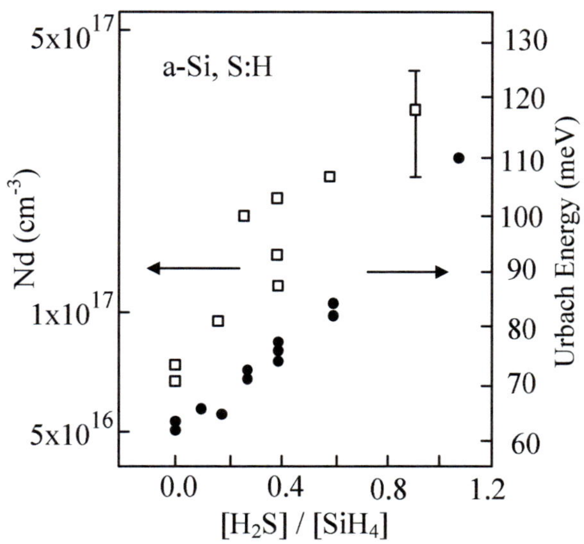

Figure 10. CPM-derived defect density and Urbach energy in a-Si,S:H (Ref. 18).

ii. a-Si,Se:H alloys: The compositional dependence of the Urbach energy E_u and the defect density N_d are shown in Fig. (11). E_u increases rapidly from 60 meV in nonalloyed material (a-Si:H) to 112 meV in selenium-rich samples (R_v = 1.6). This trend suggests a substantial increase in the width of the band tail, as in the case of a-Si,S:H alloys. This observation is consistent with the general trend of increased structural disorder. N_d was obtained from the magnitude of the subgap defect absorption shoulder using the same factor as that used for a-Si:H [44]. Figure (10) shows that N_d increases linearly with R_v from 6×10^{16} cm^{-3} in unalloyed a-Si:H samples to about 2×10^{17} cm^{-3} for R_v = 1.6. The increase in N_d is corroborated preliminary by the ESR measurements of the spin density. It is difficult to compare the spin densities with defect densities obtained from CPM measurements since it is not clear whether the dangling bonds have predominantly positive or negative correlation energies [22].

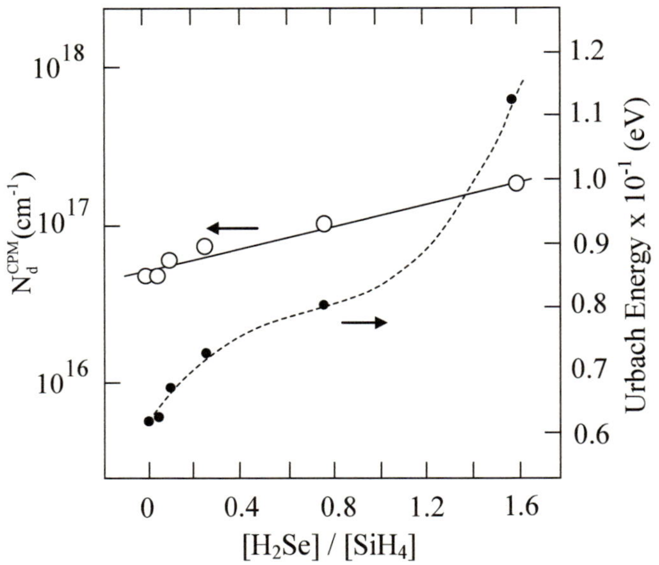

Figure 11. Compositional dependence of the Urbach energy and defect density in a-Si,Se:H alloys (Ref. 17).

3.4. Dark conductivity and photoconductivity

Photoconductivity is a valid probe of the electronic quality of materials. Measurements of the electrical conductivity (σ_d) and photoconductivity (σ_{ph}) for both a-Si,S:H and a-Si,Se:H alloys as a function of the gas volume ratio (R_v) provide a good indication about the effect of defect density in these materials. Figure (12) shows the variation of σ_d and σ_{ph} with composition (R_v) in a-Si,S:H alloys [19]. σ_d decreases monotonically with increasing sulphur content. Room temperature activation energies on the order of 0.95 eV or less are typical for samples containing a high concentration of sulphur (>10 at.%) [19]. For the same range of R_v, the photoconductivity drops by about five orders of magnitude.

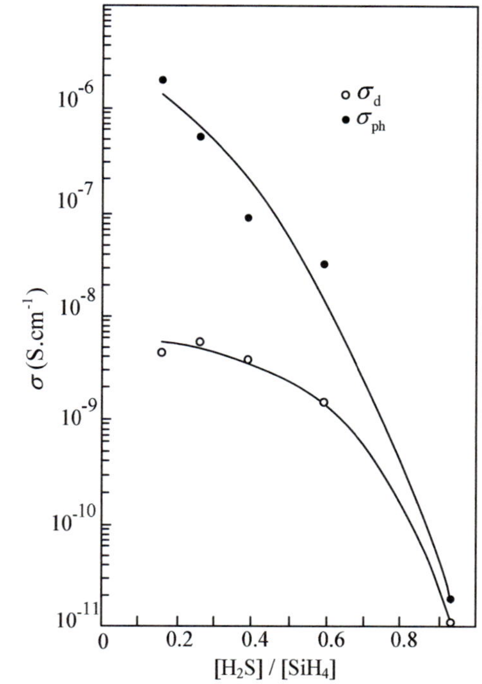

Figure 12. Compositional dependence of dark conductivity (σ_d) and photoconductivity (σ_{ph}) of a-Si_{1-x} S_x alloys. Excitation was performed using a He-Ne laser (10^{15} photon cm^{-2}) (Ref. 15).

Figure (13) shows the variation of σ_d and σ_{ph} for a-Si,Se:H alloys. σ_{ph} was measured with band-pass filtered illumination from a tungsten-halogen lamp providing a uniform carrier generation rate of 10^{21} cm^{-1} s^{-1}. The photoconductivity (σ_{ph}) was found to drop by four orders of magnitude when R_v, and consequently E_{opt}, increases from 1.85 to 2.1 eV. However, the photosensitivity (σ_{ph}/σ_d) remains high, exceeding 10^3 throughout most of the composition range [22].

3.5. Photoluminescence

The photoluminescence (PL) spectra of a-Si,S:H alloy thin films were taken with an excitation energy of 2.41 eV and recorded with a cooled Ge diode [22]. Figure (14) summarizes the variation of integrated intensity (I_{ph}), the full width at half maximum (FWHM) and the photoluminescence peak position as a function of E_{opt}. The PL efficiency η_{PL} drops by a factor of 1/30 when going from unalloyed film to an alloy grown at R_v = 0.93. The PL peak position remains approximately constant at 1.3 eV for all compositions. The FWHM increases from approximately 0.28 eV in the unalloyed material to about 0.60 eV for the samples with the highest sulphur content. These data are evidence of an increase in the width of the band tails with alloying, and in good agreement with earlier results on the variation of Urbach energy with composition. The drop of η_{PL} is a direct consequence of the increase of nonradiative recombination centers as demonstrated by the ESR measurements of N_s [22].

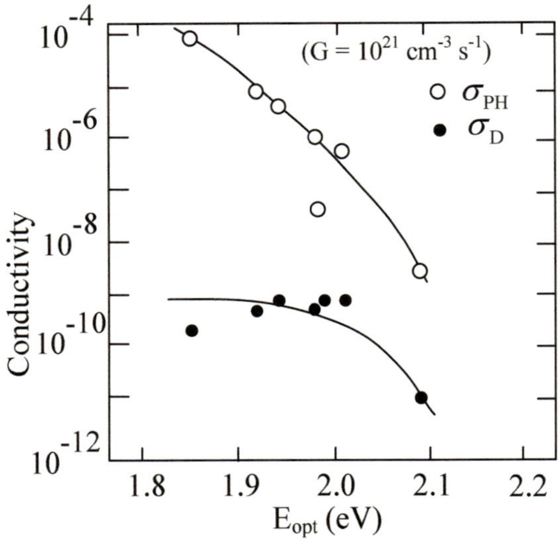

Figure 13. Dark conductivity (σ_d) and photoconductivity (σ_{ph}) versus E_{opt} for a-S,Se:H samples prepared at various compositions (Ref. 17).

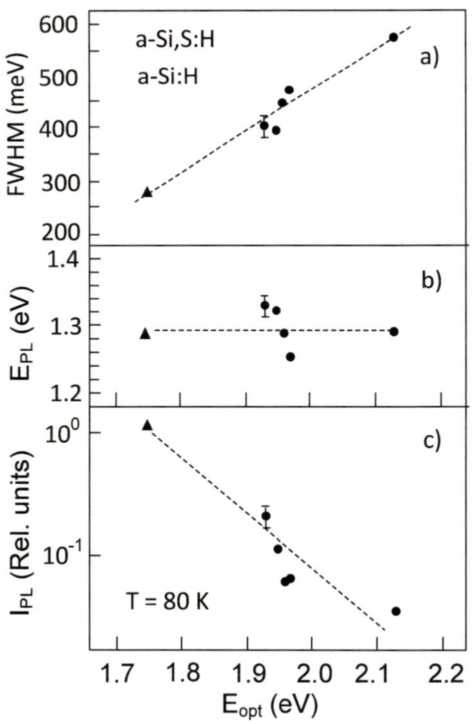

Figure 14. PL peak energy, FWHM, and intensity for various composition of a-Si,S:H alloys (Ref. 18).

4. Conclusions

This work is a review of the structural, spectroscopic, optical, and electronic properties of hydrogenated amorphous silicon-chalcogen alloys. The two alloys considered in this work are a-Si,S:H and a-Si,Se:H. It has been shown that thin films of these materials can be prepared over a relatively wide range of compositions by the glow discharge decomposition of a mixture of either H_2S or H_2Se and SiH_4 gas. Infrared transmission spectroscopy has been employed to probe the bonding structure of these alloys. In a-Si,S:H three sulphur-induced bands at 480, 710, and 760 cm^{-1} have been identified, and several conclusions on their origin have been discussed. The existence of the 480 cm^{-1} band, attributed to Si-S bonds, proves that sulphur can be bonded to the silicon matrix. A shift of the SiH and SiH_2 stretching frequencies have been observed, indicating the presence of significant levels of SiH_xS_y configurations in a-Si,S:H alloys. In a-S-,Se:H, infrared spectroscopy measurements reveal the existence of a spectral feature at 390 cm^{-1}, ascribed to Si-Se bonds. It has been also observed that the SiH and SiH_2 stretching mode frequencies shift to higher wave numbers with increasing selenium content, and thus provide evidence for the presence of significant levels of $(Si)_xSe_{3-x}SiH$ and $(Si)ySe_{2-y}$ SiH_2 configurations in a-Si,Se:H alloys.

Optical absorption measurements carried out on a-Si,S:H and a-Si,Se:H alloys reveal that incorporation of chalcogen atoms in the gaseous phase increases the band gap of these materials. Subgap absorption measurements were carried out to investigate the distribution of defect states in the band gap. It was found that there is a systematic increase in Urbach energy and defect density as the concentration of chalcogen atoms in the gaseous phase increases, and thus indicate a substantial increase in the width of the band tails.

Photoconductivity measurements on these alloys show a decrease in σ_{ph} with increasing E_{opt}. However, the photosensitivity $(\sigma_{ph/}\sigma_d)$ remains high.

Photoluminescence measurements show that incorporation of chalcogen atoms in a-Si:H causes a drop in PL efficiency and an increase in the FWHM, again providing further evidence of an increase in the width of the band tail with alloying.

Author details

Shawqi Al Dallal*

Address all correspondence to: shaldallal@gmail.com

College of Graduate Studies and Research, Ahlia University, Manama, Bahrain

References

[1] J. Chevallier, S. Kalem, S. Al Dallal, and J. Bourneix, *Journal of Non-Crystalline Solids*, 51, p. 277 (1982).

[2] S. Girson, F. Briones, and S. L. Vicent, *Philosophical Magazine*, 56, p. 449 (1987).

[3] A. D. Stewart, and D. I. Jones, *Philosophical Magazine*, B57, p. 431 (1988).

[4] P. Harper, and B. Wherrett, *Nonlinear Optics*, Academic Press, New York (1977).

[5] F. Z. Henari, *Optics Communications*, 281, p. 5894 (2008).

[6] G.M. Ferreira, Chi Chen, R.J. Koval, J.M. Pearce, R. W. Collins, and C. R. Wronski, *Journal of Non-Crystalline Solids*, 338–340, p. 694 (2004).

[7] Ken Zweibel, *Solar Energy Materials & Solar Cells*, 63(4), p. 375, (2000).

[8] J. M. Pearce, N. Podraza, R. W Collins, M. M. Al-Jassim, Jones, K. M., J Deng, and C. R Wronski, *Journal of Applied Physics*, 101 (11): 114301-7 (2007).

[9] R. J. Koval, Chi Chen, G. M. Ferreira, A. S. Ferlauto, J. M. Pearce, P. I. Rovira, C. R. Wronski, and R. W. Collins, *Materials Research Society Symposium Proceedings*. 715, A6.1,(2002).

[10] J. Bullot and M. P. Schmidit, *Physica Status Solidi (b)*, 143(2), fmi–fmi, 345–806, K95–K166 (1987).

[11] A. Carbone, F. Demichelis, G. Kaniadakis, G. Della Mea, F. Freire, and P. Rava. *Journal of Materials Research*, 5(12), pp. 2877–2881(1990).

[12] B. Racine, A. C. Ferrari, N. A. Morrison, I. Hutchings, W. I. Milne, and J. Robertson, *Journal of Applied Physics*, 90, 10, p. 15 (2001).

[13] R. Martins, J. Figueiredo, V. Silva, H. Aguas, F. Soares, A. Marques, I. Ferreira, and E. Fortunato, *Journal of Non-Crystalline Solids*, 299, p. 1283, (2002).

[14] E. Fortunato, L. Pereira, H. Aguas, I. Ferreira, and R. Martins, *Proceedings of IEEE*, 93, p. 1281, (2005).

[15] R. Martins, L. Raniero, L. Pereira, D. Costa, H. Aguas, S. Pereira, L. Silva, A. Gonçalves, I. Ferreira, and E. Fortunato, *Philosophical Magazine*, 89, 28–30, pp. 2699–2721, (2009).

[16] R. Martins, H. Aguas, I. Ferreira, E. Fortunato, S. Lebib, and P. R. I Cabarrocas, *Chemical Vapor Deposition*, 9, 6, pp. 333–337, (2003).

[17] S. Al Dallal, S. Al Alawi, and Mahmood Hammam, *Journal of Non-Crystalline Solids*, 356, pp. 2323–2326, (2010).

[18] S. Al-Dallal, S. M. Al-Alawi, and S. Aljishi, *Proceedings of the 6th International Symposium on Advanced Materials*, (1999).

[19] S. Al-Dallal, M. Hammam, and S. M. Al-Alawi, *Philosophical Magazine B*, 63, p. 839 (1991).

[20] S. Al-Dallal, M. Hammam, and S. Al-Alawi, *Journal of Non-Crystalline Solids*, 114, p. 789 (1989).

[21] S. Al-Dallal, S. Aljishi, M. Hammam, S. M. Al-Alawi, M. Stutzmann, Shu Jin, T. Muschik, and R. Schwarz. *Journal of Applied Physics* 70 (9), p. 4926, (1991).

[22] S. Aljishi, S. Al-Dallal, S. M. Al-Alawi, M. Hammam, H. S. Al-Alawi, Stutzmann, S. Jin, T. Muschik, and R. Schwarz, *Solar Energy Materials*, 23, p. 334 (1991).

[23] S. Al-Dallal, M. Hammam, S. M. Al-Alawi and S. Aljishi, *Thin Solid Films*, 205, p. 89, (1991).

[24] S. Al-Dallal, F.Z. Henari, S. M. Al-Alawi, S. R. Arekat, and H. Manaa, *Journal of Non-Crystalline Solids*, 345–346, p. 302 (2004).

[25] S. Al-Dallal, S. Aljishi, M. Hammam, S. M. Al-Alawi, and R.Schwarz, *Proceedings of the International Conference on Condensed Matter Physics (ICCMPA)*, April 13–16, p. 331, (1992).

[26] G. Lucovsky, R.J. Nemanich, and J.C. Knights, *Physical Review B*, 19, p. 2064 (1979).

[27] M. H. Brodsky, and M. Cardona, and J. J. Cuomo, *Physical Review B*, 16, p. 3556 (1977).

[28] E. C. Freeman, and W. Paul, *Physical Review B*, 18, p. 4288 (1978).

[29] H. Shanks, C. J. Fang, L. Ley, M. Cardona, F. J Demond, and S. Kalbitzer, *Physica Status Solidi (b)*, 100, 43 (1980).

[30] G. Lucovsky, *Solid State Communications*, 29, 571(1979).

[31] Sanderson, R. T., *Chemical Bonds and Bond Energy*. San Diego: Academic Press, p. 41 (1976).

[32] C. J. Fang, L. Ley, H. R. Shanks, K. J. Gruntz, and M. Cardona, *Physical Review B*. 22, p. 6140 (1980).

[33] R.Tsu, M. Izu, S. R Ovshinsky, and F. H, Pollak, *Solid State Communications*, 36, p. 817 (1980).

[34] J. Chevallier, S. Kalem, S. Al Dallal, and J. Bourneix, *Journal of Non-Crystalline Solids*, 51, p. 277 (1982).

[35] L. J. Bellamy, *The Infrared Spectra of Complex Molecules*, vol. 1. London: Chapman & Hall, p. 394 (1975).

[36] S. Aljishi, M. Stutzmann, S. Jin, C. Herrero, S. Al Dallal, M. Hammam, and S. M. Al-Alawi, Proc. 13th ICALS Conf., *Journal of Non-Crystalline Solids* (1989).

[37] E. A. Ebsworth, R. Taylor, and L. A. Woodward, *Transactions of the Faraday Society*, 55, p. 211 (1959).

[38] M. Tenhover, R. S. Henderson, D. Lukco, M. A. Hazle, and R. K. Grasselli, *Solid State Communications*, 51, 455 (1984).

[39] K. M. Guggenheimer, *Proceedings of the Physical Society*, 58, p. 456 (1946).

[40] D. R. Herschback, and V. W. Laurie, *Journal of Chemical Physics*, 35, p. 458 (1961).

[41] R. M. Atkins, and P. L. Timms, *Spectrochemica Acta*, 33A, p. 853 (1977).

[42] N. Saito, T. Yamada, Yamagushi, I. Nakaaki, and N. Tanaka, *Philosophical Magazine* B52 (5) (1985).

[43] S. Al Dallal, S. Al Alawi, and M. Hammam, *Journal of Non-Crystalline Solids*, 356, pp. 2323–2326 (2010).

[44] Z. E. Smith, V. Chu, SK. Shepard, S. Aljishi, J. Kolodzey, D. Slobodin, T. L. Chu, and S. Wagner, Physics Letters, 50, p. 1521 (1987).

Chapter 3

Superalloys for Advanced Ultra-Super-Critical Fossil Power Plant Application

Xishan Xie, Yunsheng Wu, Chengyu Chi and
Maicang Zhang

Additional information is available at the end of the chapter

http://dx.doi.org/10.5772/61139

Abstract

Superalloys are world-wildly used not only for aerospace but also for chemistry, oil & gas and power engineering application. In recent years the 700 °C level Advanced Ultra-Super-Critical (A-USC) technology with high thermal efficiency is developing in the world to reduce the coal consumption and pollution emissions. Any kind of advanced ferritic and austenitic heat-resisting steels can not meet 700 °C A-USC technology requirement. In this case high quality Ni-base superalloys must be adopted for 700 °C A-USC technology. The research and development of Ni-Fe and Ni-base superalloys such as HR6W, GH2984, Haynes 230, Inconel 617/617B, Nimonic 263, Haynes 282, Inconel 740 and 740H are reviewed in this chapter.

Keywords: Ni-base superalloys, A-USC power plant, Long time aging, Stress-rupture strength, Structure stability

1. Introduction

Superalloys are world-wildly used not only for aerospace but also for chemistry, oil & gas and power engineering application.

Electricity is one of the most important pillars for mankind living and economic development. Most of the countries in the world electricity is generated by coal fired power plants. In half of the century steam temperatures of coal fired power plants are going up from 475 °C to 600 °C. In this temperature range heat-resisting steels such as ferritic and austenitic steels are wildly used. Recently for further raising thermal efficiency and reduction of CO_2 emission the 700 °C advanced ultra-super-critical (A-USC) power plant development projects are initiated in Europe, United States, Japan, China and India also. In 700 °C A-USC project the highest temperature parts are boiler super-heater and re-heater tube components. The fire-side metal temperature of these components can reach 750-760 °C even higher. Today's modern power plants are required for very long time service such as 30-40 years. There are very strict requirements of these high temperature tube materials such as:

1. The high temperature stress rupture strength for 10^5 h should be higher than 100 MPa;

2. The fire-side metal oxidation layer after 10^5 h should be less than 1 mm;

3. Good structure stability at long time service and no harmful phase formation;

4. Good process performance such as fabrication and welding ability etc.

At above mentioned serious condition any kind of ferritic and austenitic heat-resisting steels can not fulfill these requirements. The high quality Ni-base superalloys must be used for A-USC power plant application. Figure_1 clearly shows the ratio change of high temperature materials with the steam parameters development for coal fired power plants. The Ni-base superalloys will occupy 29% for a 360 bar/700 °C/720 °C A-USC power plant [1].

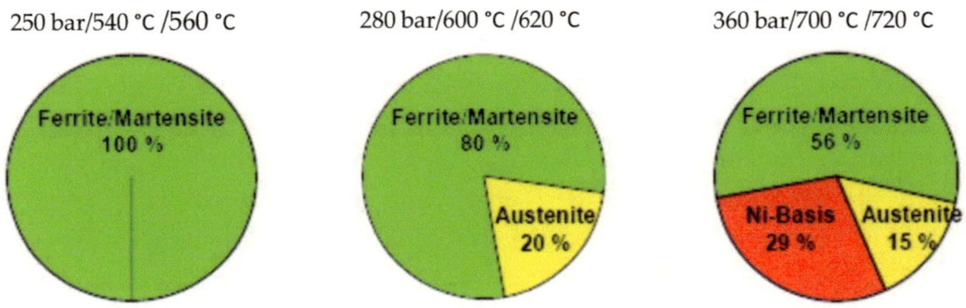

Figure 1. The ratio change of high temperature materials with the steam parameters development for coal fired power plants.

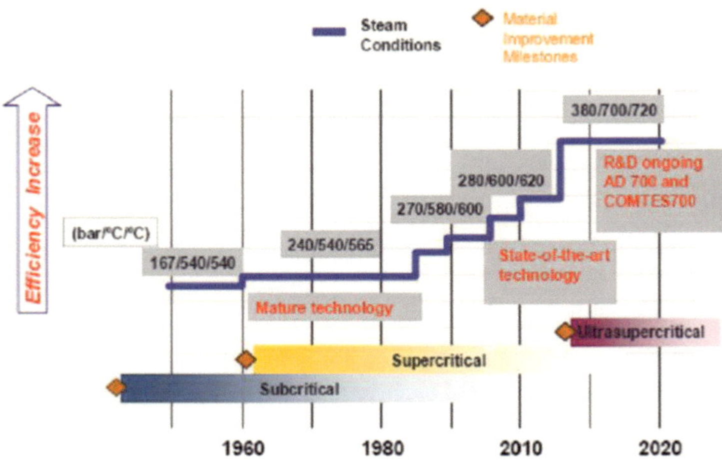

Figure 2. The steam parameters development of coal fired power plants in the world.

Figure 2 shows the steam parameters development of coal fired power plants in the world [2]. Now the "bottle neck" issue is adopting high performance long life superalloys. The European Union initiated their 700 °C A-USC Project from the year of 1998. Up to now they have not got successful results because of the failure of Ni-base alloy weldment at their test bed.

The purpose of this chapter is to give an overview of today's Ni-base superalloys for high temperature (700 °C even higher) A-USC power plant materials selection and application.

2. Superalloys for 700°C A-USC power plant

The nominal chemical compositions of candidate Ni-Fe and Ni-base superalloys for 700 °C A-USC power plant are shown in Table 1. The alloys of HR6W (developed by Sumitomo Metal, Japan) and GH2984 (developed by Institute of Metal, China) are Ni-Fe base superalloys. The others are Ni-base superalloys developed by Haynes and Special Metals Corp. (formerly INCO Alloys International, USA).

Alloy	ASME Code	C	Cr	Co	Mo	Nb	Ti	Al	Si	Mn	Fe	W	Cu	Ni
HR6W	2684	0.08	23	-	-	0.2	0.1	-	0.5	1.0	Bal.	7	-	45
GH2984	No	0.06	19	-	2.2	1.1	1.1	0.4	≤0.5	≤0.5	33	-	-	Bal.
Haynes230	2063	0.10	22	≤5	2.0	-	-	0.3	0.4	0.5	≤3	14	0.02(La)	Bal.
Inconel 617B	2439	0.06	22	12	9.0	-	0.4	1.0	≤1.0	≤1.0	≤1.5	-	-	Bal.
Haynes282	No	0.06	20	10	8.5	-	2.1	1.5	≤0.15	≤0.3	≤1.5	-	-	Bal.
Nimonic263	No	0.06	20	20	6.0	-	2.2	≤0.6	≤0.4	≤0.6	≤0.7	-	≤0.2	Bal.
Inconel740	2702	0.03	25	20	0.5	2.0	1.8	0.9	0.5	0.3	≤0.7	-	-	Bal.
Inconel740H	2702	0.03	25	20	0.5	1.5	1.35	1.35	0.15	0.3	≤0.7	-	-	Bal.

Table 1. Nominal chemical compositions (wt %) of candidate high temperature materials for 700 °C A-USC power plant.

HR6W and GH2984 are based on Ni-Fe-Cr austenite matrix and solid solution strengthened by Mo and W and also precipitation strengthened by Nb, Ti and Al. Most of Ni-base superalloys are based on Ni-Cr-Co austenite matrix with solid solution strengthening by molybdenum (0.5%-9%Mo) and γ' precipitation strengthening by adding Ti, Al and Nb also. Haynes 230 is mainly a solid solution strengthening Ni-base superalloy strengthened by Cr, W and Mo.

The creep strengths for 10^5 h at different temperatures of above mentioned high temperature materials are shown in Figure_3 [3-6]. Inside temperatures of superheater and reheater tube loops for 700 °C/720 °C A-USC boiler are 700 and 720 °C respectively. However the fireside temperature of these tubes can reach 760 °C even higher. The arrow indication at the 100MPa margin is critical for 700 °C/720 °C A-USC boiler tube materials selection. From point of view on this critical margin only Inconel 740 Ni-base superalloy is acceptable. It seems to us that

Haynes 282 may characterize with the highest creep strength among these candidate materials. However Haynes 282 is still under the developing for superheater and reheater tube products and also have not been issued ASME code.

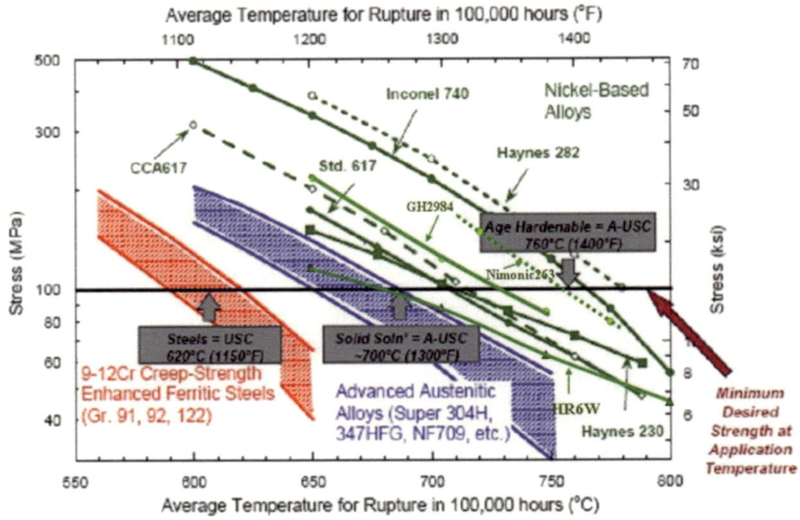

(R. Romanosky, 2011; J.T. Guo, 2005; H. Semba, 2008; X.S. Xie, 2013)

Figure 3. Creep strength comparison among some candidate materials.

The fireside coal ash corrosion is also a critical criterium for 700 °C A-USC boiler superheater and reheater tube selection. Figure_4 shows a complete series results of different high temperature materials [7]. It is clearly shown that high contents of Mo and W are harmful for coal ash corrosion. Inconel 740 with very low content of Mo (0.5% Mo only) characterizes with excellent coal ash resistance (as the star indicated in Figure_4).

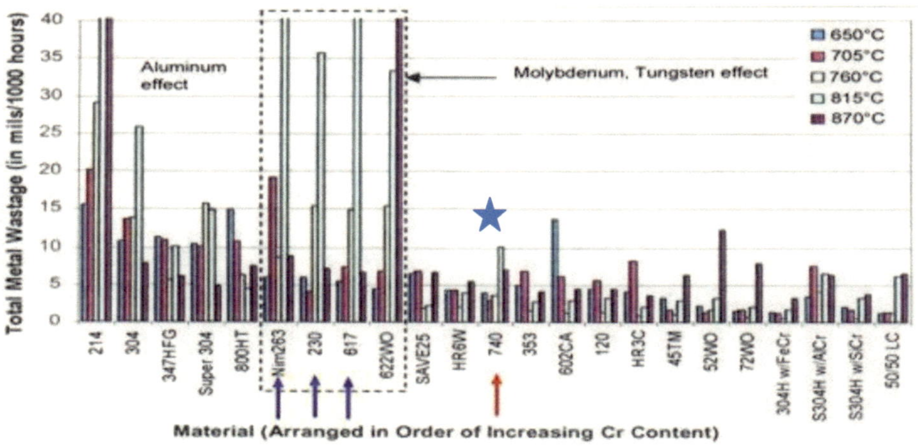

Figure 4. Coal ash corrosion of some candidate materials.

http://dx.doi.org/10.5772/61139

2.1. HR6W (45Ni-Fe-23Cr-7W-0.2Nb-0.1Ti-0.08C)

HR6W is a Ni-Fe base superalloy mainly strengthening by 7%W which was developed by Sumitomo Metals, Japan in 1986. The large amount of W is not only for solid solution strengthening but also for precipitation strengthening of Laves phase. According the thermodynamic calculated phase diagram (Figure_5) the main precipitated phase is Laves phase and also $M_{23}C_6$, M(C, N) and α–Cr formation at 700 °C. A small amount of B addition in HR6W is for grain boundary strengthening [8].

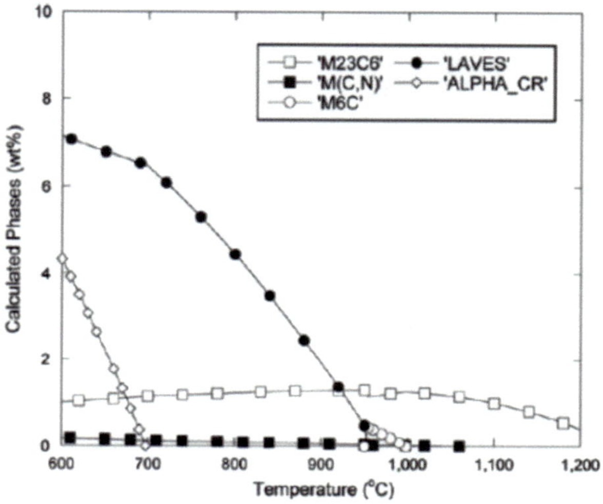

Figure 5. The calculated thermodynamic equilibrium phase diagram of HR6W alloy.

The fraction of main strengthening phase Laves is dramatically increased with the W content (from 3% to 8%), but there is almost no influence on $M_{23}C_6$ and MX phases (Figure_6a). However the fraction of α–Cr is dramatically increased when the Cr is higher than 23.4% (Figure_6b) [8].

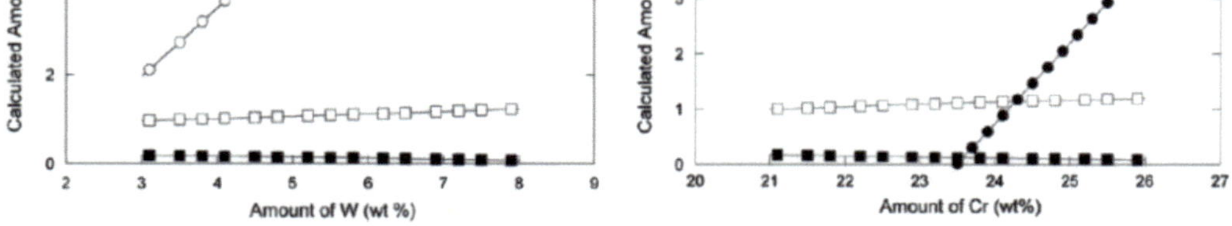

Figure 6. The calculated amount of equilibrium phases at 700 °C in HR6W alloy change with W (a) and Cr (b).

Tungsten is concentrated in Laves phase and the Cr diffuse away from Laves phase to the interphase of Laves/γ matrix. In result of that the concentration of Cr promotes α–Cr formation. Figure_7 shows α–Cr particles with Laves phase at grain boundaries and also in the grains at the vicinity of Laves phase [8].

Figure 7. The EBSD picture of HR6W alloy crept at 700°C/100 MPa, 7998.6 h.

TEM images (Figure_8) clearly shows the precipitated Laves phase morphology after long time stress rupture tests at 700, 750 and 800 °C [9]. It can be seen that the growth rate of Laves phase is very sensitive to the test temperature. The dispersive precipitated Laves phase is already grown to a certain size at 750 °C after long time stress rupture test.

The stress-rupture strength of HR6W in the temperature range from 650-800 °C are shown in Figure_9 [10]. The extrapolated stress-rupture strength for 10^5 h at 700 °C is 88 MPa only. It is difficult to fulfil 700 °C A-USC bolier superheater and reheater tube requirement.

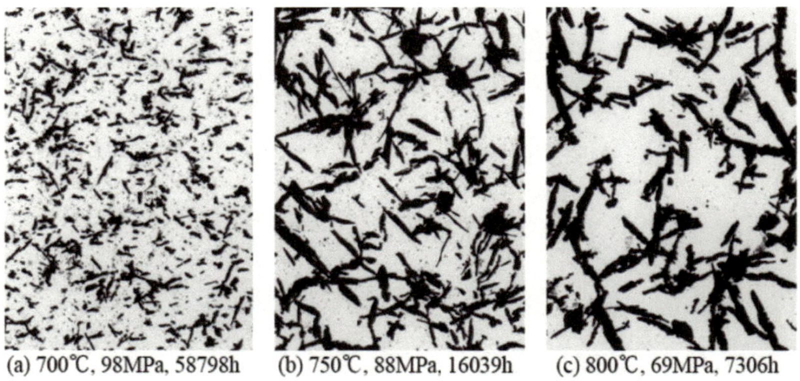

Figure 8. The microstructure of HR6W alloy after creep test at 700 °C (a), 750 °C (b)and 800 °C (c) respectively.

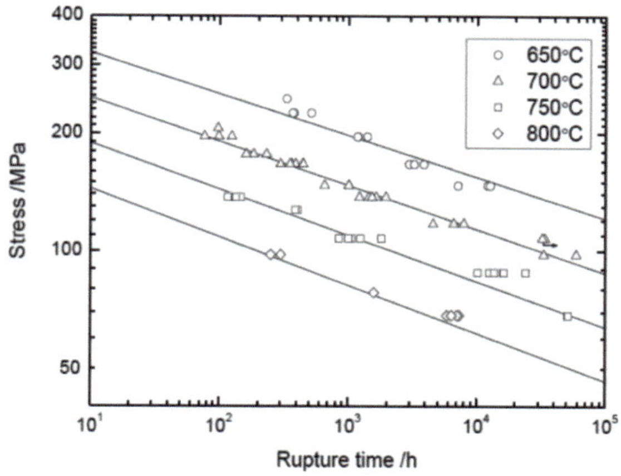

Figure 9. Stress rupture strength curves of HR6W at different temperature.

2.2. GH2984 (Ni-33Fe-20Cr-2Mo-1Nb-1Ti-0.35Al-0.06C)

GH2984 is a Ni-Fe base superalloy with the solid solution strengthening of big amount of Cr (20%) and small amount of Mo (2%). Certain amount of Nb, Ti, Al addition is for Ni_3(Nb, Ti, Al) type γ' phase precipication strengthening. GH2984 was developed by the Institute of Metal Research, China in the year of 1970'. GH2984 contains 33% Fe, which is an economic superalloy in comparison with other Ni-base superalloys. GH2984 has been used as a marine ship boiler superheater components for long term service more than 10 years [11]. It is recommended for tube components in the temperature range of 650-700 °C.

The main precipication strengthening phase in GH2984 is γ'-Ni_3(Nb, Ti, Al). GH2984 contains 5.74% γ' with the average size of 23 nm after standard heat treatment. However the brittle σ phase formation (about 3%) happens after long time aging at 700 °C. Except this the γ' growth rate is also high in the temperature range of 700-750 °C [11]. GH2984 contains 20%Cr and show good oxidation resistance. Figure_10 shows its oxidation dynamic curves at 650 and 700 °C [12]. The average oxidation rate at 700 °C for 100h is 0.0058 g/m^2 h only.

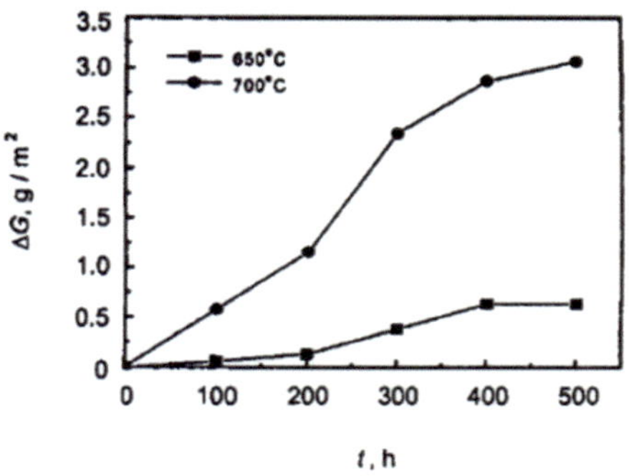

Figure 10. The oxidation curves of GH2984 alloy at 650 and 700 °C.

Careful control of the main γ' strengthening elements Nb, Ti and Al is important for GH2984. The optimum contents of these γ' strengthening elements are recommended as 1.35%Nb, 1.24%Ti and 0.65%Al [13]. The stress rupture strength curves of GH2984 in the temperature range of 650-800 °C are shown in Figure_11 [5]. Because of the low fraction ofγ' strengthening phase the stress rupture strength of GH2984 still can not fulfill 700 °C A-USC boiler superheater and reheater tube components requirement. Recently Institute of Metal Research, China is making the modification of GH2984 for developing GH2984G and try to fulfill the 700 °C A-USC project in China.

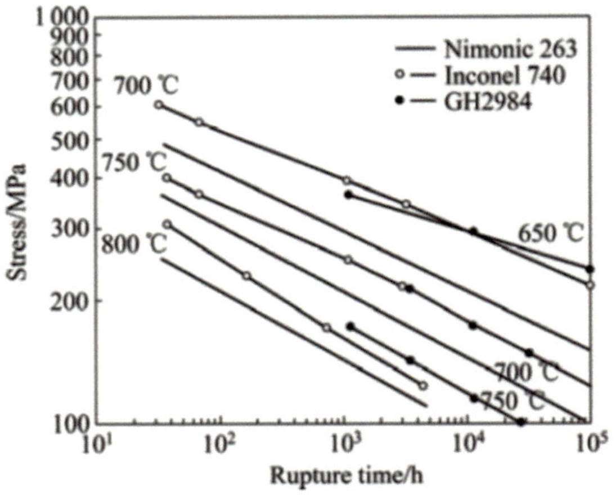

Figure 11. Stress rupture curves of GH2984 alloy in comparison with Nimonic263 and Inconel740

2.3. Haynes 230 (Ni-22Cr-2Mo-14W-0.3Al-0.1C-La)

Hayne 230 developed by Haynes International in 1984, is a typical solid solution strengthening Ni-base alloy with high contents of Cr (22%) and W (14%) and partially Mo (2%) [14]. Haynes 230 characterizes with good workability for hot deformation in a wide temperature range from 925 °C to 1175 °C. There are no harmful phase (such as σ and μ phase) formation in the temperature range of 649-871 °C for long time aging till 16,000 h [15].

Figure_12a shows a typical austenitic structure with the grain boundary precipitated M_6C carbide at mill annealed condition [16]. The zig-zag grain boundaries can be formed by grain boundary carbide precipitation as shown in Figure_12b. It can block grain boundary sliding for improvement of creep resistance at high temperature.

A detail study on the strengthening mechanism of Haynes 230 shows that a fine intermetallic phase $Ni_2(Cr, W)$ can precipitate in γ-matrix for improvement of creep resistance. Figure_13 shows the precipitation of $Ni_2(Cr, W)$ phase after 1000 h creep test at 750 °C, 28 MPa [17]. Because of the very low fraction of $Ni_2(Cr, W)$ phase the strengthening effect is also not strong. In result of that Haynes 230 still can not meet the stress rupture strength higher than 100 MPa at 750 °C for 10^5 h.

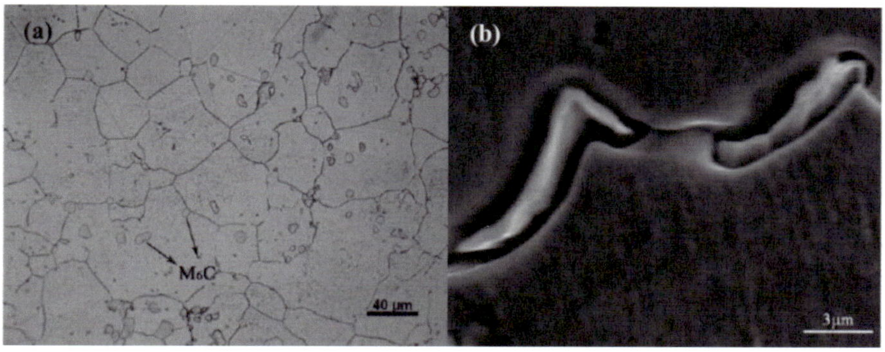

Figure 12. The microstructure of Haynes230 alloy at supplied condition (a) carbide precipitation at grain boundaries (b).

2.4. Inconel 617/617B (Ni-22Cr-12Co-9Mo-0.4Ti-1Al-0.06C)

Inconel 617 was originally developed by INCO Alloys International in the year of 1975. It is a mainly solid solution strengthening superalloy based on Ni-Cr-Co γ-matrix by adding high content of Mo (9%) and also a small amount of Al and Ti for γ' precipitation strengthening [18]. For creep resistant improvement Alloy 617B was further developed by Tyssen-Krupp VDM, Germany for strictly control of Al and Ti contents and also boron addition for grain-boundary strengthening. Alloy 617B sometimes is also designated as 617CCA or 617Mod.

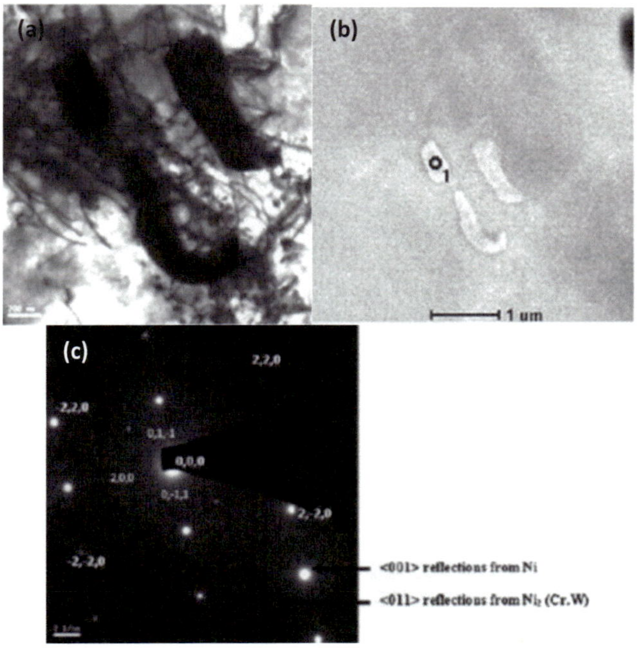

Figure 13. TEM (a), SEM (b) and diffraction pattens (c) of $Ni_2(Cr,W)$ in Haynes230 alloy after creep test at 750 °C /28 MPa, 1000h.

Phase diagram calculation by Thermo-Calc can help us to understand the phase stability in the different temperature ranges. As the equilibrium phase diagram of Inconel 617B (Figure_14) shows that the σ and μ phase can be formed in different temperature ranges.

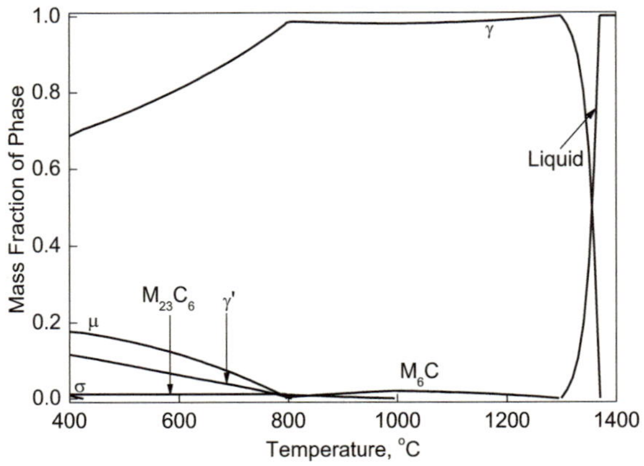

Figure 14. Phase diagram of Inconel 617B calculated by Thermo-Calc.

Quanyan Wu [19, 20] has systematically studied its microstructure behaviour in the temperature range of 482-871 °C for long time aging till 65,000 h. A group of TTT curves are shown in Figure_15. It can be seen from this diagram that there are no σ and μ phase formation. The main precipitated phases are γ', MX, M_6C and $M_{23}C_6$ [19].

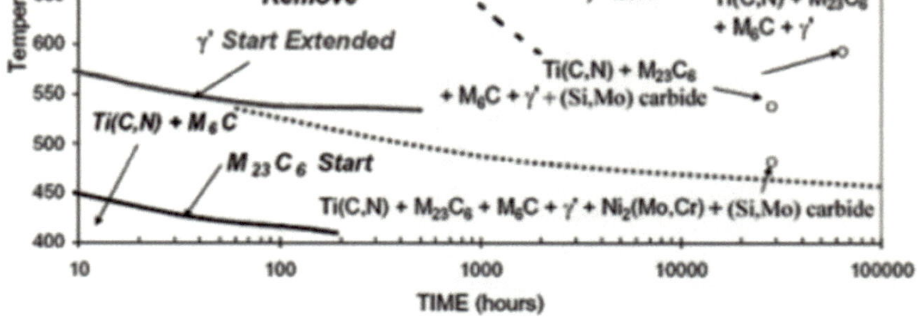

Figure 15. TTT curves of modified Inconel617 based on observation of TEM and SEM results of long-term aging samples

According Wu's study the average size of γ' after long time aging at 704 °C till 43,100 h is about 40-60 nm, but its volume fraction is low about 5% only. However the size of γ ' grows rapidly at 704 °C till 65,000h (about 200 nm) and γ' volume fraction decreases to a lower level.

The γ' precipitation and growth behavior can be clearly seen from Figure_16 at long time aging for 1000 and 3000 h at 700, 750 and 800 °C. The strengthening phase γ' grows rapidly with increasing temperature and simultaneously γ' dispersion degree and its volume fraction both decrease rapidly [20].

In result of that Inconel 617/617B (617CCA or 617Mod) can not meet 700 °C A-USC project requirement as Figure_3 indicated.

2.5. Nimonic 263 (Ni-20Cr-20Co-6Mo-2Ti-0.6Al-0.06C)

Nimonic 263 was originally developed by Rolls-Roycc in 1971 as a high stress rupture strength and corrosion resistant Ni-base superalloy. It is based on Ni-Cr-Co austenitic matrix with high content of Mo (6%) for further solid solution strengthening. Especially 2%Ti and 0.6%Al addition to this alloy for forming about 10% $Ni_3(Al, Ti)$ type γ' creates good solid solution with γ' precipication strengthening and aso boron addition for grain boundary strengthening in a good combination for the improvement of high temperature creep resistance. As we can see in Figure_3 that Nimonic 263 can just reach the 10^5 h stress rupture strength at 750 °C for 700 °C A-USC boiler superheater and reheater tubing requirement.

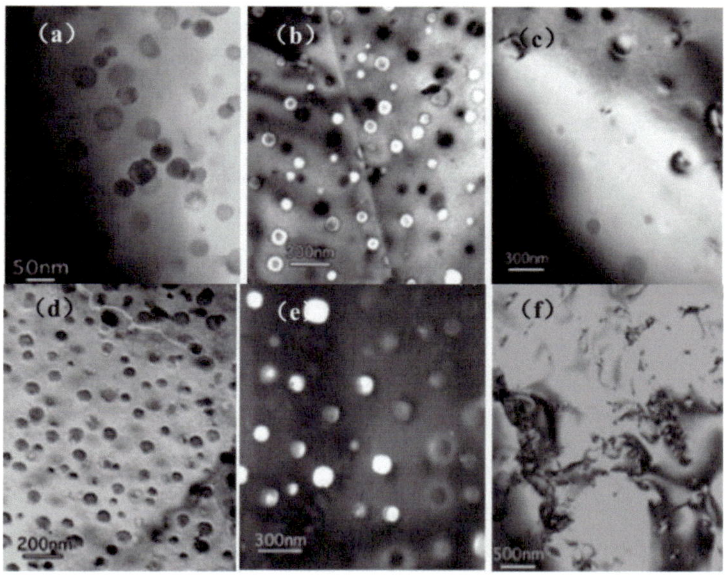

Figure 16. TEM images of CCA617 alloy aged at 700 °C/1000 h (a), 750 °C/1000 h (b), 800 °C/1000 h (c), 700 °C/3000 h (d), 750 °C/3000 h (e) and 800 °C/3000 h (f).

Figure_17a shows the dispersively precipitated fine γ' (23±4 nm) at 750 °C, 50 h aging. These dispersively precipitated fine γ ' particales coagulate to a larger size (68±20 nm) at 850 °C for

50h aging as shown in Figure_17b. The TTT diagram of Nimonic 263 is clearly shown in Figure_17c. It can be learned from this diagram that the basic structure for Nimonic 263 is mainly fine γ' precipitation in Ni-Cr-Co-Mo γ-matrix and carbide $M_{23}C_6$ destributed at grain boundaries. It can be seen also from this TTT diagram that the Ni_3Ti type η phase can be formed at high temperature long time aging. The Ti-rich γ'- Ni_3 (Al, Ti) in Nimonic 263 is a meta-stable precipitated phase for good strengthening effect. However the meta-stable Ti-rich type γ' phase will transform to a stable Ni_3Ti type η phase for long time aging at high temperature [21, 22].

Figure_18 shows the very large plate-like η phase formation in Nimonic 263 after long time aging at high temperature [21].

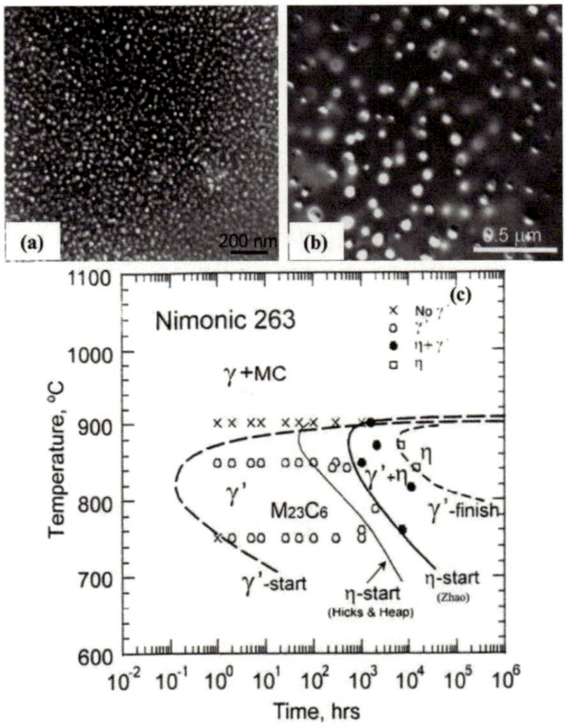

Figure 17. TEM dark field images of Nimonic263 aged at 750 °C (a) and 850 °C (b) for 50 h and TTT curve (c).

Nimonic 263 has been adopted as a good high stress rupture strength and good corrosion resistant Ni-base superalloy in aero-engines for not very long service time (~10^3-10^4 h). However the fossil power plant should be put in service for a very very long time such as 30-40 years. The long time structure stability is a critical issue for superheater and reheater tubing application in 700 °C A-USC boiler.

Phase diagram calculation by Thermo-Calc can help us to understand the phase stability in the different temperature ranges. As the equilibrium phase diagram of Nimonic 263 (Figure_19) shows that the η, σ and μ phase can be formed in different temperature ranges [23].

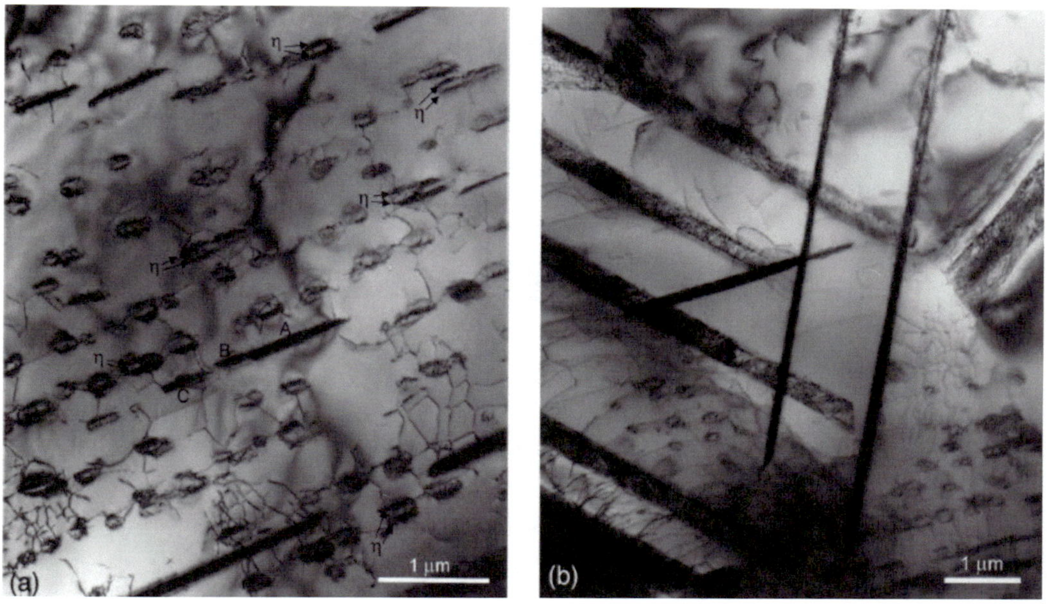

Figure 18. The TEM images of arranged γ' phase transformed to η phase after heat treatment at 800 °C for 8 h then aged at 900 °C for 1250 h (a) and dislocations distribution at η/γ' boundaries (b).

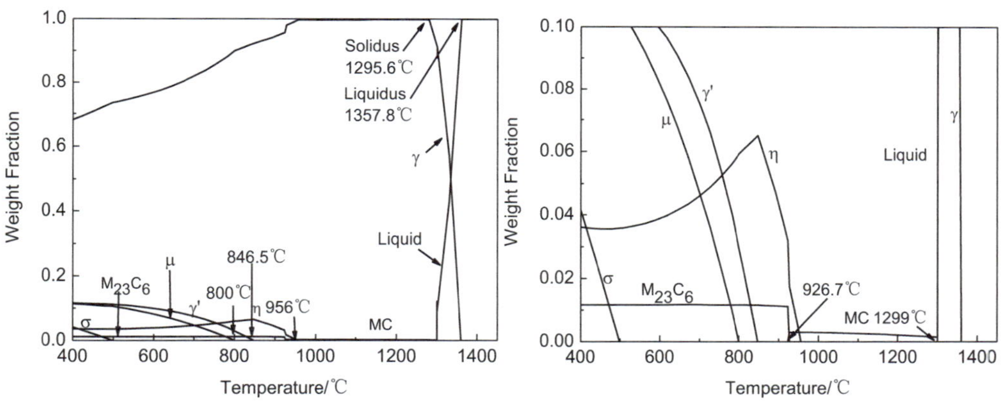

Figure 19. Phase diagram of Nimonic 263 calculated by Thermo-Calc.

It is clearly seen that except γ', η and μ phase can be formed in a wide range of high temperatures. Figure_20 is an example that a big amount of large plate-like Ni_3Ti type η phase formed after stress rupture test at 775 °C, 115 MPa for 12601 h [23]. It is clearly that strengthening effect of γ' is decreased dramaticly by the transformation from γ' to η. It means that the structure stability of Nimonic 263 should be improved for 700 °C A-USC boiler superheater and reheater tubing components at very long time service.

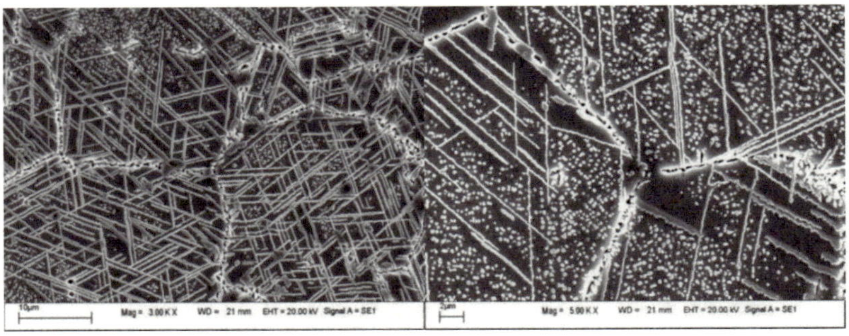

Figure 20. Microstructure instability of Nimonic 263 after the stress rupture test at 775 °C, 115 MPa of 12601 h.

2.6. Haynes 282 (Ni-20Cr-10Co-8.5Mo-2.1Ti-1.5Al-0.06C)

Haynes 282 was developed by Haynes International, USA in 2005. It is a corrosion resistant and high creep strength Ni-base superalloy. The original purpose for Haynes 282 is to develop a weldable Ni-base superalloy to replace Waspaloy for making different components in aero-engine industry. As today's knowledge understanding the Waspaloy is a good Ni-base superalloy in the intermediate temperature range for aero-engine application. However it is difficult for flat product and especially is almost cannot be welded for required components [24].

With 57%Ni, 20%Cr, 10%Co and 8.5%Mo for solid solution strengthening and 2.1%Ti, 1.5%Al for forming γ' precipitation strengthening. Haynes 282 falls into a subcategory within the larger category of the familiar γ' strengthened superalloys.

Phase diagram calculation result of Haynes 282 by Thermo-Calc is shown in Figure_21 [25]. It can be seen that except the γ' precipitation possibly μ phase and occasionally σ phase can be also formed in a certain temperature range. The carbides MC, $M_{23}C_6$ and M_6C with a small amount should be existed in Haynes 282. The basic structure of Haynes 282 at as heat treated condition is about 17% γ' dispersively distributed in γ-matrix with grain boundaries precipitated carbides $M_{23}C_6$ (0.67%), M_6C (0.05%) and also MC (0.16%) randomly distributed in γ-matrix [26].

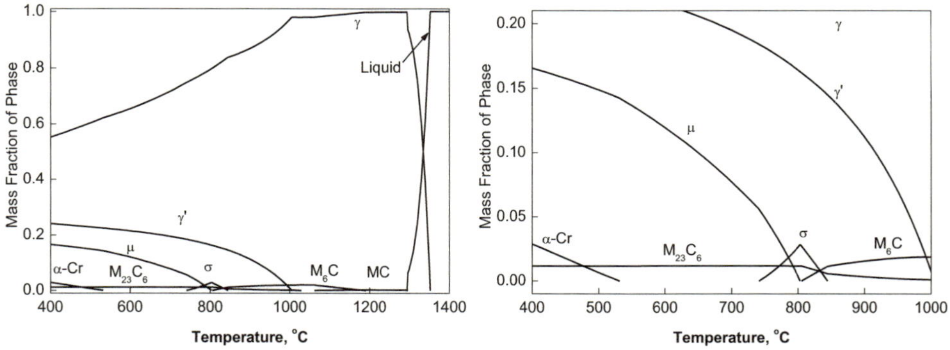

Figure 21. Phase diagram of Haynes 282 calculated by Thermo-Calc.

Figure_22 shows the typical structure after long time aging at 649 °C, 760 °C and 816 °C for 1000 h [27]. These SEM images show that Hayne 282 characterizes with good structure stability. It still keep a structure with fine γ' precipitation in γ-matrix and carbides ($M_{23}C_6$ and M_6C) distribute at grain boundaries. The harmful phases such as μ and σ have not been found yet.

Figure_23 clearly shows that the tensile yield strengths of Haynes 282 after very long time aging at high temperatures are much higher than Haynes 230 Ni-base alloy [27].

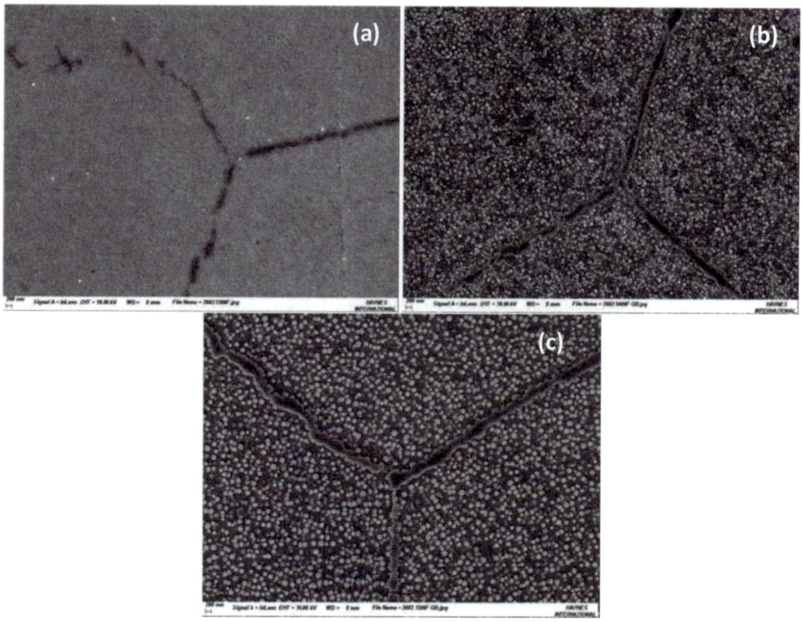

Figure 22. SEM images of Haynes282 aged at 649 °C (a), 760 °C (b) and 816 °C (c) for 1000 h.

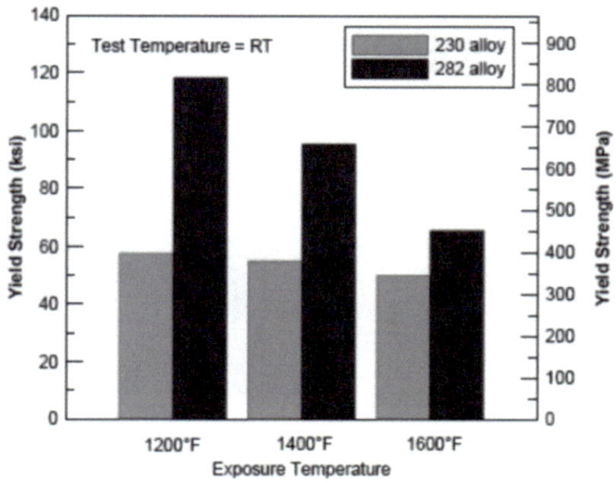

Figure 23. Comparison of room temperature tensile properties between Haynes282 (after 16,000 h aging) and Haynes230 alloy (after 20,000 h aging)

Furthermore the high temperature stress rupture strength of Haynes282 is also a little higher than Inconel 740, as the dash line indicated in Figure_3. However Haynes International are just developing 282Alloy for 700 °C A-USC boiler superheater and reheater tubing application. Haynes will accumulate the important data together from seamless tube product for asking ASME Code in the future.

2.7. Inconel 740 (Ni-25Cr-20Co-0.5Mo-2Nb-1.9Ti-0.8Al-0.03C)

As above mentioned all Ni-Fe and Ni-base superalloys such as HR6W, GH2984, Haynes 230, Inconel 617/617B, Nimonic 263 (see Table 1) can not meet 700 °C A-USC boiler superheater and reheater tubing requirement (see Figure_3 and 4).

Inconel 740, a new Ni-Cr-Co-Mo-Nb-Ti-Al superalloy, is developed in the year of 2000 by Special Metal Corp.(SMC, Huntington) USA for European THERMIE AD700 A-USC project with steam parameters of 700 °C, 35 MPa [28]. At this condition the superheater and reheater fire-side metal temperature can be 750-760 °C even higher. As we mentioned that the superheater and reheater require stress rupture strength(100 MPa, 10^5h) at temperatures 750-760 °C, together with the high corrosion resistance(≤1 mm cross-section loss in 10^5 h). Inconel740 characterizes with the highest stress rupture strength and corrosion/oxidation resistance among today's commercial available candidate materials and can meet above mentioned strict requirements.

Typical microstructure and phase fraction of Inconel740 at standard heat treatment are shown in Figure_24. The main strengthening phase γ' (12.98%) homogeneously distributes in Ni-Cr-Co γ-matrix, $M_{23}C_6$ carbide (0.115%) and a very small amount(0.054%) of high Si-containing G-phase mainly precipitated at grain boundaries and MC-(Nb.Ti)C (0.183%) carbide formed at solidification process randomly distributes in the alloy.

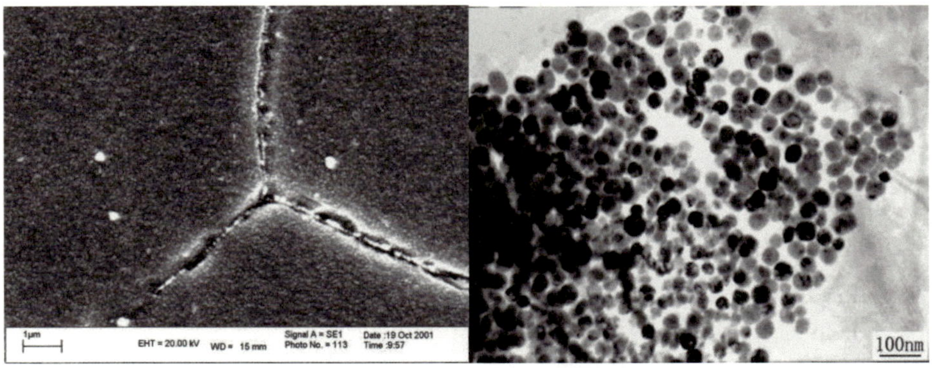

Phase	γ'	MC	$M_{23}C_6$	G
wt%	12.980	0.183	0.115	0.054

Figure 24. Typical microstructure and phase fractions of Inconel 740 after standard heat treatment.

The microstructure of Inconel 740 is quite stable at 700 °C long time aging as shown in Figure_25. However, it is quite different at 760 °C long time aging. There are a lot of plate-like η-phase nearby grain boundaries and rapidly grow to the grains. Moreover the large globular high Si-containing G phase distributed at grain boundaries as determined by EDAX(see Figure_26) [29,30].

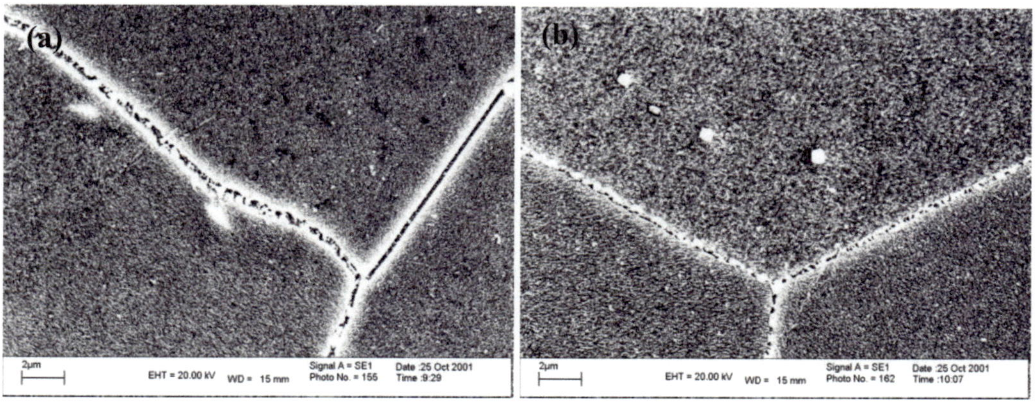

Figure 25. Stable microstructure of Inconel 740 at 704 °C long time aging: (a) 704 °C/1000 h and (b) 704 °C/2000 h.

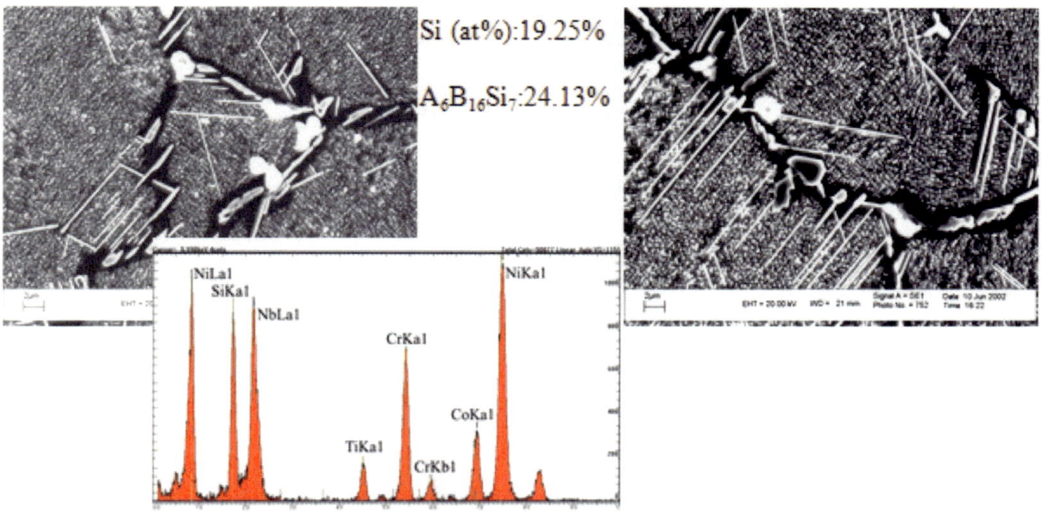

Figure 26. The large amount of η-phase and G-phase in Inconel 740 after 760 °C long time aging.

Detail quantitative determination of brittle high Si-containing G-phase after 704-760 °C long time aging is shown in Figure_27. It is very harmful for Inconel 740 that the amount of G phase(0.471%) after 2000 h aging at 760 °C is almost 10 times than the amount of G-phase(0.054%) as at standard heat treatment condition. Moreover the unstability of γ' strengthening phase and the formation of η-phase have been also detail studied as shown in Figure_28 [31].

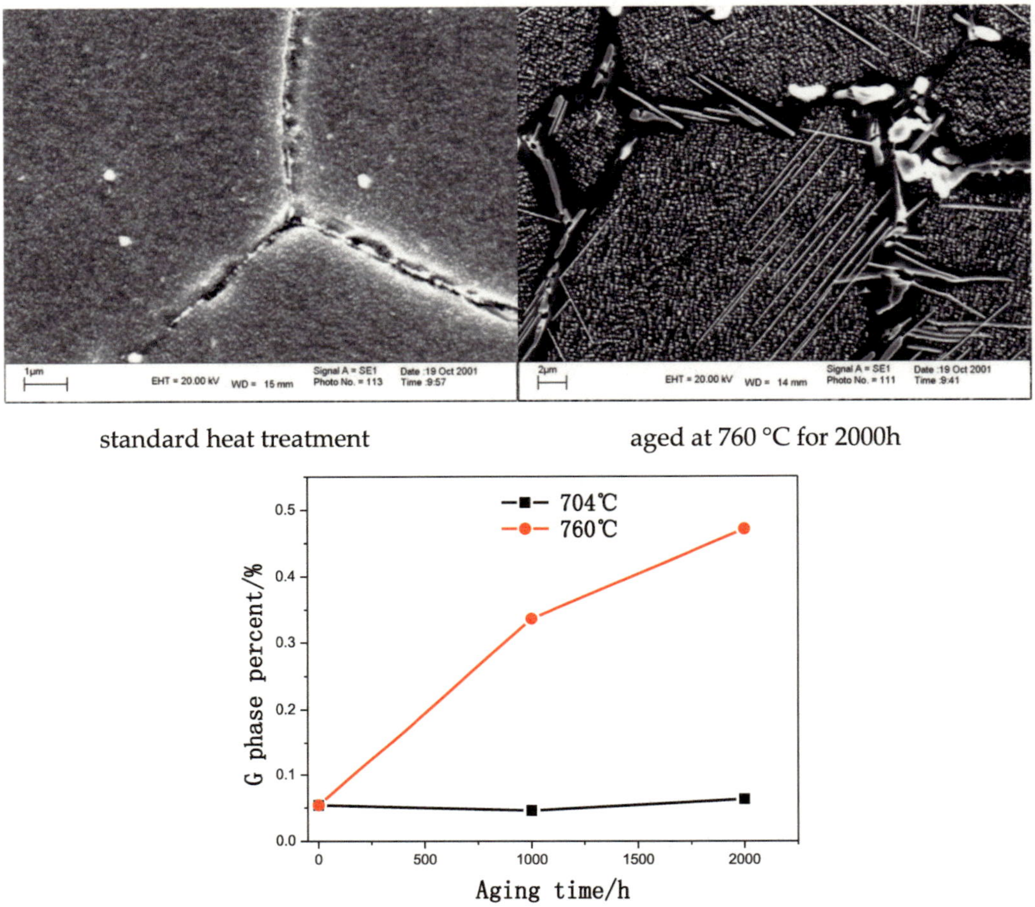

Figure 27. High Si-containing G-phase formation and plate-like η-phase precipitation in Inconel 740 at 760 °C long time aging.

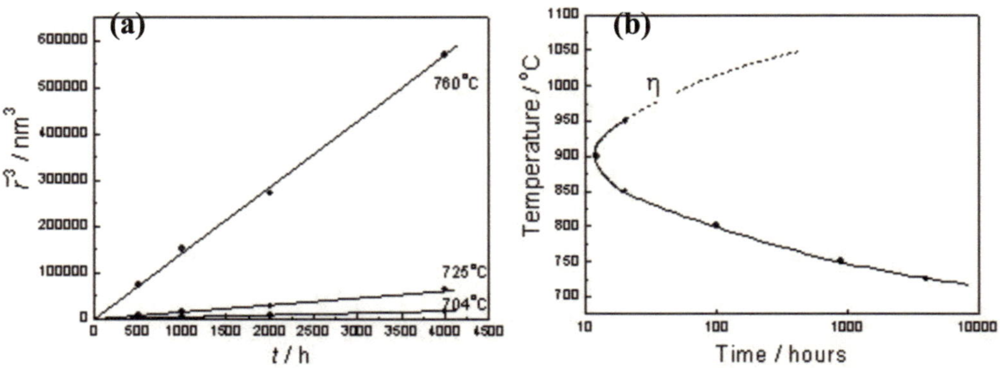

Figure 28. Coarsening of γ' phase(a) and the precipitation of η-phase(b) in Inconel 740.

It can be recognized that the structure stability is good for Inconel 740 during long time aging at the temperatures under 725 °C. The unstability of Inconel 740 at 760 °C has been confirmed by 3 important factors: high coarsening rate of γ' phase, plate-like η-phase formation and high Si-containing brittle G-phase formation at grain boundaries. It will clearly develop degradation of strength and ductility both. Just for these reasons the structure stability improvement and modification of Inconel 740 for developing Inconel 740H should be done to fulfill the superheater/reheater requirements at the temperature 750 °C and above [32].

2.8. Inconel 740H (Ni-25Cr-20Co-0.5Mo-1.5Nb-1.35Ti-1.35Al-0.03C)

Inconel 740H is a modification of Inconel 740 for improvement of structure stability by the adjustment of Nb, Ti, Al and Si [33]. Chemical composition and phase fraction comparison between Inconel 740H and Inconel 740 is shown in Table 2 and 3 respectively. Figure_29 shows the comparison of calculated phase diagrams of Inconel 740 and Inconel 740H. It can be seen from these results that the γ' strengthening phase is more stable in Inconel 740H than Inconel 740. The η-phase is eliminated in Inconel 740H and there is not existence of brittle high Si-containing G phase in Inconel 740H.

Alloy	C	Cr	Ni	Co	Mo	Nb	Ti	Al	Mn	Fe	Si
740	0.03	25	bal	20	0.5	2	1.8	0.9	0.3	0.7	0.5
740H	0.03	25	bal	20	0.5	1.5	1.35	1.35	0.3	0.7	0.15

Table 2. Chemical composition (wt%) comparison between Inconel 740H and Inconel 740.

Alloy	γ'	MC	$M_{23}C_6$	G
740	12.980	0.183	0.115	0.054
740H	14.623	0.220	0.202	0

Table 3. Phase fraction (wt%) comparison between Inconel 740H and 740.

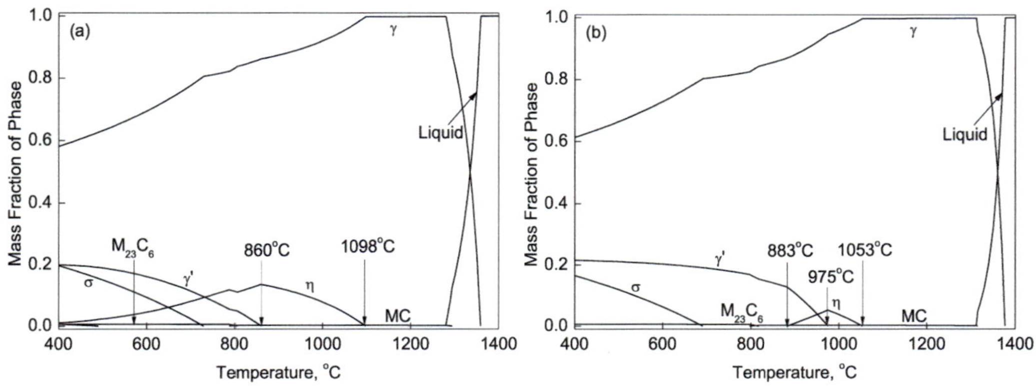

Figure 29. Phase diagram comparison between Inconel 740 (a) and Inconel 740H (b).

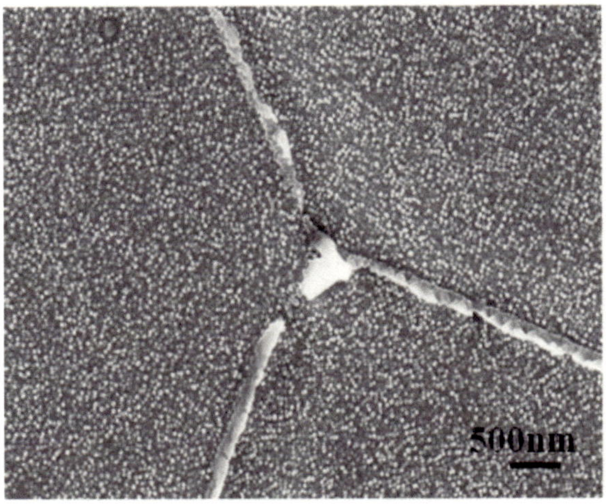

Figure 30. Typical microstructure of Inconel 740H after standard heat treatment.

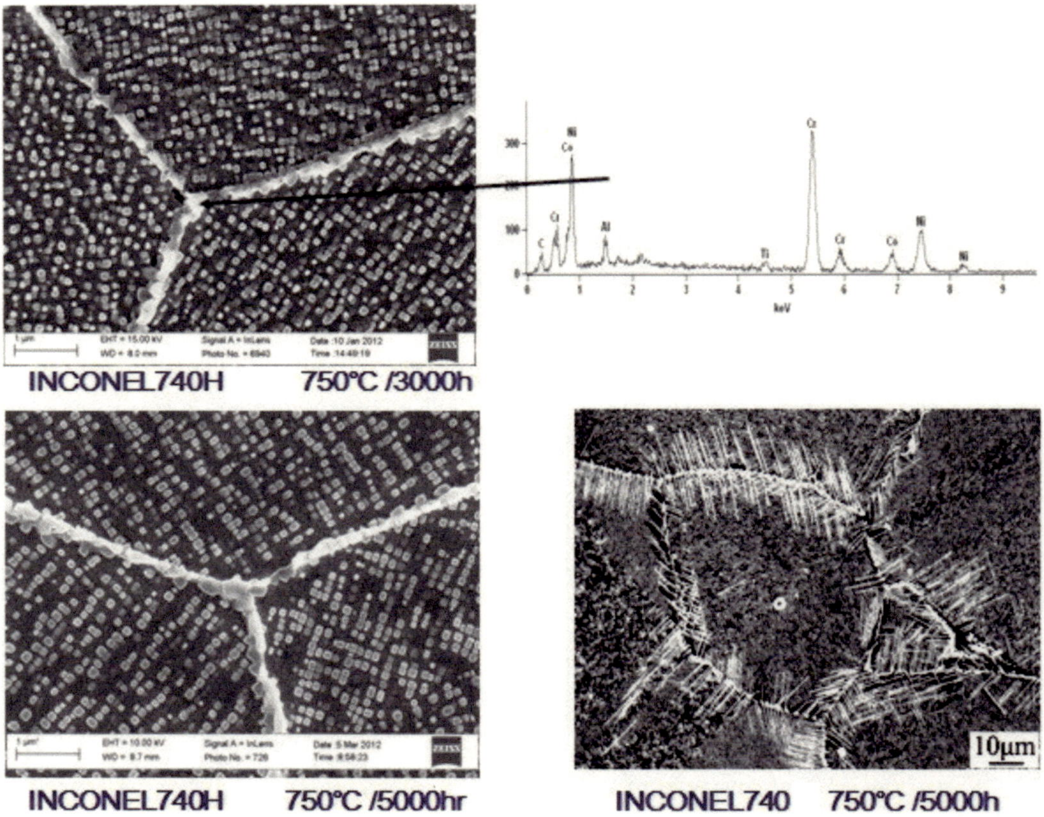

Figure 31. Long time structure stability comparison between Inconel 740H and 740 at 750 °C.

Typical microstructure of Inconel 740H after standard heat treatment as shown in Figure_30 is very similar to the structure of Inconel 740 in comparison with Figure_24. The main strengthening phase γ' (14.623%) homogeneously distributes in Ni-Cr-Co γ-matrix, $M_{23}C_6$ carbide(0.202%) and also MC-(Nb,Ti)C carbide(0.220%) formed at solidification process randomly distributes in this alloy. However, the comparison of structure stability of Inconel 740H and Inconel 740 at 750 °C and 800 °C long time aging is quite different(see Figure_31 and Figure_32) [34]. Inconel 740H keeps stable γ' strengthening in Ni-Cr-Co γ-matrix and $M_{23}C_6$ carbide at grain boundaries. However, plate-like η-phase formation and high Si-containing G-phase precipitation obviously happen in Inconel740. Moreover the γ' growth rate comparison between Inconel 740H and Inconel 740(see Figure_33) clearly shows that Inconel 740H characterizes very good structure stability at 750 °C long time aging.

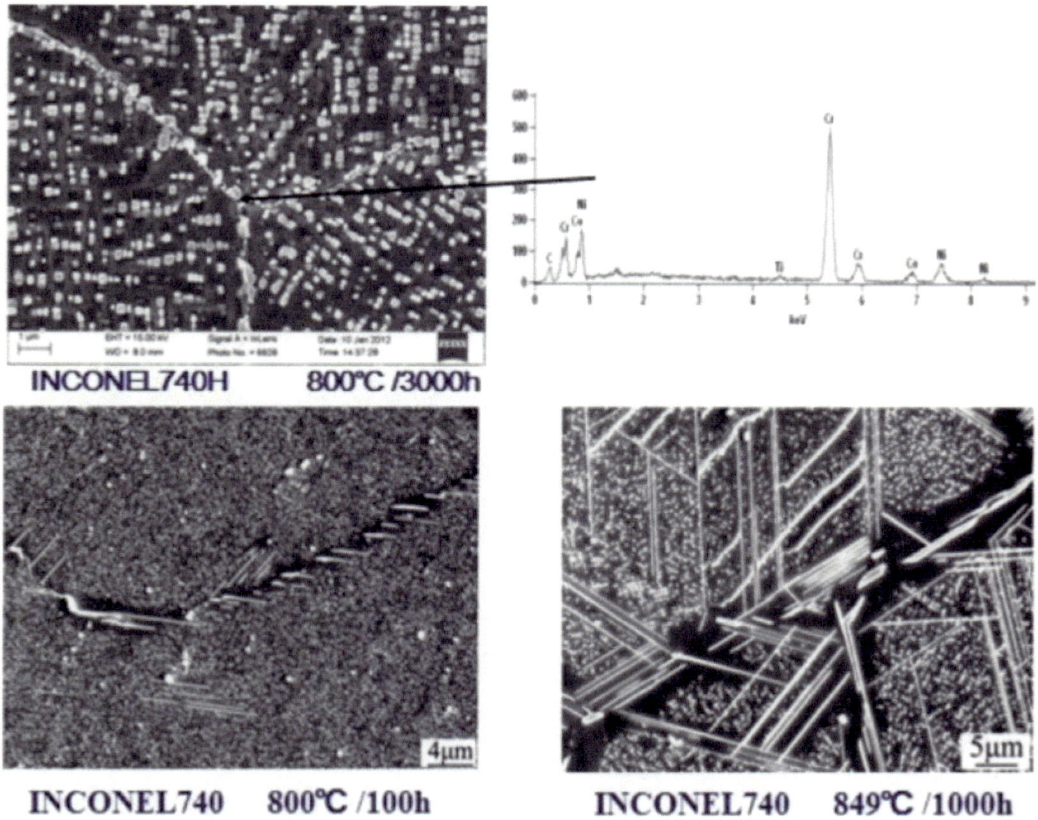

Figure 32. Long time structure stabiltiy comparison between Inconel 740H and 740 at 800 °C aging.

The structure stability study has been also conducted after high temperature stress rupture tests [30]. The structure of Inconel 740 sample tested at 775 °C, 170 MPa, after 2779 h shows a large amount of η-phase near grain boundaries and G phase formation at grain boundaries(see Figure_34a). However, the structure of Inconel740H sample tested at 750 °C, 280 MPa, after

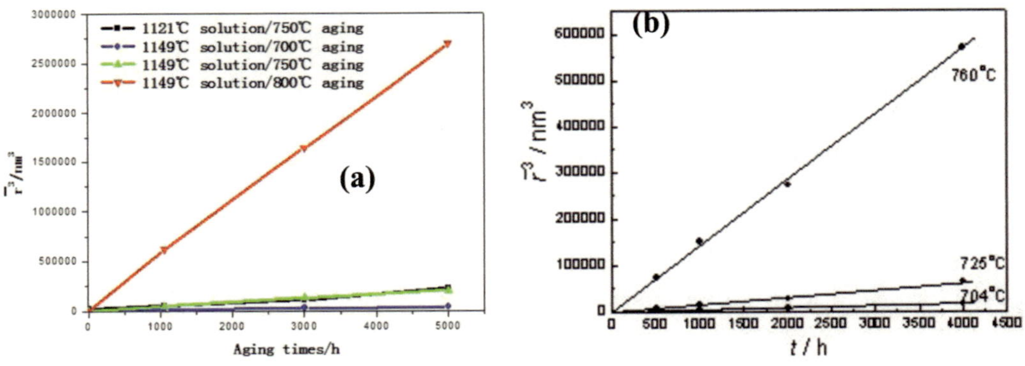

Figure 33. γ' growth rate comparison between Inconel 740H(a) and 740(b) at high temperatures.

1087 h still keeps stable γ' precipitation in γ-matrix and $M_{23}C_6$ carbide distributed at grain boundaries and there are no existence of plate-like η-phase and brittle G-phase at grain boundaries(see Figure_34b).

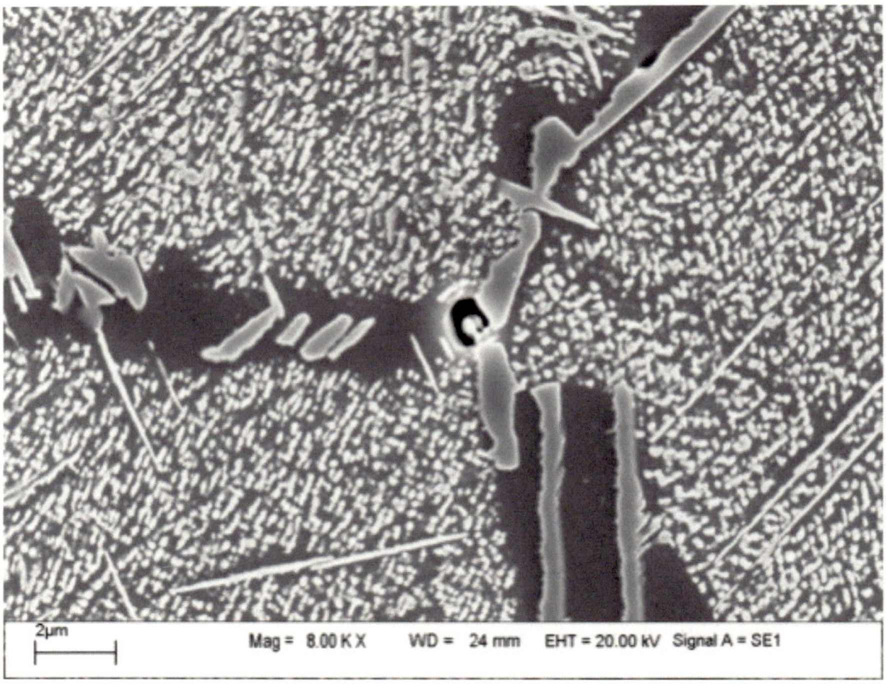

Figure 34. Microstructure comparison of Inconel740(a) and Inconel740H(b) stress rupture tested samples.

The impact toughness of Inconel 740H after high temperature long time aging is much higher than Inconel 740 as shown in Table 4.

Alloy	Exposure	CVN J/cm^2
740	0 h	45
	750 °C-2000h	13.6
740H	0h	90.6
	700 °C-1050 h	29.5
	750 °C-1000 h	35.7
	800 °C-1000 h	50.5
	700 °C-3042 h	23.3
	750 °C-3000 h	29.8
	800 °C-3000 h	46.2
	700 °C-5000 h	34.7
	750 °C-5000 h	23.8
	800 °C-5000 h	45.3
	solution treatment at 1121 °C	93.7
	750 °C-1000 h(solution treatment at 1121 °C)	51.2
	750 °C-3000 h(solution treatment at 1121 °C	44.4
	750 °C-5000 h(solution treatment at 1121 °C)	45.2

Table 4. Impact toughness comparison of Inconel 740H and Inconel 740 after high temperature long time aging.

3. Summary

A brief review of some candidate Ni-Fe and Ni-base superalloys for 700 °C A-USC fossil power plant application have been discussed. However up to now still have no success to build a 700 °C A-USC model plant, because high temperature materials test has not successfully passed at any 700 °C A-USC test beds.

One more issue must be mentioned that welding is very important for boiler tubing component making. Furthermore the thick wall pipe welding for header and main steam piping is more critical for building a power plant.

From the view point of high temperature stress rupture strength and corrosion/oxidation resistance among above mentioned Ni-Fe and Ni-base superalloys with ASME Code as indicated in Table 1. Inconel 740H may be a perspective high temperature material for 700 °C A-USC Project. However the weldability of Inconel 740H is still an important factor for real application.

Except today's commercial available Ni-Fe and Ni-base superalloys as listed in Table 1,some new Ni-Fe base and Ni-base superalloys are still developing in the world for 700 °C A-USC

Project. We hope that in close cooperation of world materials scientists and engineers the high temperature materials for 700 °C A-USC technology will be developed and put in a model power plant for long time service in the future.

Author details

Xishan Xie[1*], Yunsheng Wu[1], Chengyu Chi[2] and Maicang Zhang[1]

*Address all correspondence to: xishanxie@mater.ustb.edu.cn

1 High Temperature Materials Research Laboratories, University of Science &Technology Beijing, Beijing, China

2 CPI Nuclear Power Institute, China Power Investment Corporation, Beijing, China

References

[1] Chi C Y, Yu H Y, Xie X S: Research and development of austenitic heat-resistant steels for 600 °C superheat/reheater tubes of USC power plant boilers. World Iron & Steel. 2012; 04: 50-65. DOI:10. 3969/j. issn. 1672-9587. 2012. 04. 008

[2] Viswanathan R, Henry J F, Tanzosh J: US program on materials technology for ultra-supercritical coal power plants. Journal of Materials Engineering and Performance. 2005; 14(3): 281-292. DOI: 10.1361/10599490524039

[3] Robert Romanosky. Global Advanced Fossil-Gen Summit. 2011, Shanghai, China.

[4] Guo J T, Du X K: A superheater tube superalloy GH2984 with excellent properties. Acta Metallurgica Sinica. 2005; 11: 1221-1227.

[5] H. Semba, H. Okada, M. Yonemura: 34th MPA-Seminar Materials and Components Behaviour in Energy & Plant Technology, 2008. Stuttgart, Germany, 14.1-14.8.

[6] Xie X S, Chi C Y, Zhao S Q: Superalloys and the development of advanced ultra-supercritical power plants. Materials Science Forum. 2013; 747: 594-603. DOI: 10.4028/www.scientific.net/MSF_747-748.594

[7] Baker B A, Smith G D: Corrosion resistance of Alloy 740 as superheater tubing in coal fired ultra supercritical boilers. CORROSION 2004. NACE International, 2004.

[8] Shingledecker J P, Evans N D: Creep-rupture performance of 0.07 C–23Cr–45Ni–6W–Ti, Nb austenitic alloy (HR6W) tubes. International Journal of Pressure Vessels and Piping. 2010; 87(6): 345-350. DOI: 10.1016/j.ijpvp.2010.03.011

[9] Igarashi M, Semba H, Yonemura M: Advances in materials technology for A-USC power plant boilers. Advances in Materials Technology for Fossil Power Plants: Pro-

ceedings from the Sixth International Conference, August 31-September 3, 2010, Santa Fe, New Mexico, USA. ASM International, 2011, 1022300: 72.

[10] SembaH, OkadaH, YonemuraM: In Proceeding of 34th MPA-Seminar. Stuttgart, 2008.

[11] Guo J T, Zhou L Z, Yuan C: Microstructure and properties of several originally invented and unique superalloys in China. The Chinese Journal of Nonferrous Metals. 2011; 02: 237-250.

[12] Wang S H, Du X K, Guo J T: Anticorrosion performance of GH984 alloy under various environmental conditions. Corrosion Science and Protection Technology. 2002; 05: 284-287.

[13] Wang S H, Du X K, Zhao J F: Study of Al, Ti, Nb contents of GH984 al1oy used in superheater tube of naval vessels. Materials for Mechanical Engineering. 1999; 02: 36-39.

[14] Twancy H M: High-temperature oxidation behaviour of a wrought Ni-Cr-W-Mn-Si-La alloy. Oxidation of Metals. 1996; 45: 323-348.

[15] Boehlert C J, Longanbach S C; A comparison of the microstructure and creep behaviour of cold rolled Haynes 230 alloy and Haynes 282 alloy. Materials Science and Engineering A. 2011; 528: 4888-4898.

[16] Li J, Rui H, Hong C K: The effect of M23C6 carbides on the formation of grain boundary serrations in a wrought Ni-based superalloy. Materials Science and Engineering A. 2012; 536: 37-44. DOI:10.1016/j.msea.2011.11.060

[17] Chatterjee S, Roy A K: Mechanism of creep deformation of Alloy 230 based on microstructural analyses. Materials Science and Engineering A. 2010; 527(29): 7893-7900. DOI: 10.1016/j.msea.2010.08.069

[18] Lin F S, Xie X S, Zhao S Q: Selection of superalloys for super heater tubes of domestic 700°C A-USC boilers. Journal of Chinese Society of Power Engineering. 2011; 12: 960-968.

[19] Wu Q, Song H, Swindeman R W: Microstructure of long-term aged IN617 Ni-base superalloy. Metallurgical and materials transactions A. 2008; 39(11): 2569-2585. DOI: 10.1007/s11661-008-9618-y

[20] Wu Q Y. Microstructural evolution in advanced boiler materials for Ultra-Supercritical Coal power plants [thesis]. Cincinnati : University of Cincinnati; 2006.

[21] Zhao J C, Ravikumar V, Beltran A M: Phase precipitation and phase stability in Nimonic 263. Metallurgical and Materials Transactions A. 2001; 32(6): 1271-1282. DOI: 10.1007/s11661-001-0217-4

[22] Hicks B, Heap M: Report No, B48749, Lucas Gas Turbine Equipment Ltd., Materials Laboratory, Burnley, Sept. 25, 1968.

[23] Xie X S Dong J X, Wang J: Structure Stability Analyses on Nimonic 263 after Long Time Stress Rupture Tests (internal report). High Temperature Materials Research Laboratories, University of Science & Technology Beijing, 2011.

[24] Pike L M: Development of a fabricable gamma-prime strengthened superalloy. Haynes International. 1020: 46904-9013.

[25] Liu J Q, Zeng Y P, Xie X S: Thermodynamic simulation calculation of precipitation phase in a new-type Ni-based superalloy. Materials Review. 2007; 09: 119-122.

[26] Zeng Y P, Liu J Q, Xie X S: Microstructure and properties analysis of a new wrought superalloy. Acta Metallurgica Sinica. 2008; 05: 540-546.

[27] Pike L M: Long term thermal exposure of Haynes 282 alloy. Superalloy 718 and Derivatives, 2010: 644-660. DOI: 10.1002/9781118495223.ch50

[28] Special Metals Corporation. Inconel740 bulletin. Special Metals Corporation, Huntington, WV, USA, 2005.

[29] Zhao S Q, Xie X S: Properties and microstructure after long term aging at different temperatures for a new nickel base superalloy. Acta Metallurgica Sinica. 2003; 04: 399-404.

[30] Zhao S Q, Xie X S, Smith G D: Microstructural stability and mechanical properties of a new nickel-based superalloy. Materials Science and Engineering A. 2003; 355(1): 96-105. DOI: 10.1016/S0921-5093(03)00051-0

[31] Zhao S Q, Dong J X, Xie X S: Thermal stability study on a new Ni-Cr-Co-Mo-Nb-Ti-Al superalloy. Proceedings from the 10th International Symposium on Superalloy. 2004: 19-23.

[32] Xie X S, Zhao S Q, Dong J X: An investigation of structure stability and its improvement on new developed Ni-Cr-Co-Mo-Nb-Ti-Al superalloy. Materials Science Forum. 2004; 475: 613-516.

[33] Xie X S, Zhao S Q, Dong J X: A new improvement of inconel alloy 740 for USC power plants. Proc of Advances in Materials Technology for Fossil Power Plants, ASM, 2008.

[34] Chi C Y, Zhao S Q, Lin F S: High Temperature Long-term Structure Stability Study on Inconel740/740H Ni-Base Superalloy. Keynote Lecture at Proc of The 5th Symposium on Heat Resistant Steels and Alloys for High Efficiency USC/A-USC Power Plants 2013, Seoul, Korea, 2013.

Chapter 4

Precipitation Process in Fe-Ni-Al-based Alloys

Hector J. Dorantes-Rosales, Victor M. Lopez-Hirata,
Jorge L. Gonzalez-Velazquez,
Nicolas Cayetano-Castro and
Maribel L. Saucedo-Muñoz

Additional information is available at the end of the chapter

http://dx.doi.org/10.5772/61070

Abstract

This chapter covers first the precipitation and coarsening processes in Fe-Ni-Al alloys aged artificially at high temperatures, as well as their effect on the mechanical properties. These results show the precipitation evolution, morphology of precipitates, coarsening kinetics and mechanical properties such as hardness. Additionally, the effect of alloying elements such as copper and chromium is also studied on the precipitation and coarsening processes. The main results of this section are concerning on the coarsening kinetics and its effect on hardness. Besides, the diffusion couple method is employed to study the precipitation and coarsening process in different Fe-Ni-Al alloy compositions, as well as its effect on the hardness. All the above aspects of precipitation and coarsening are also supported with Thermo-Calc calculations.

Keywords: Fe-Ni-Al alloys, Beta prime phase, precipitation, aging, coarsening, hardness

1. Introduction

The Fe-Ni-Al system is very important for the design and development of several Fe-based alloys. For instance, the precipitation process of the β'phase is relevant for the alloy strengthening at high temperatures in different engineering alloys such as PH stainless steels and Fe-Cr-Ni-Al-based alloys, etc. [1]. They are applied to fabricate different industrial components which usually require a combination of good mechanical strength and oxidation resistance at elevated temperatures. The β′phase consists of an ordered phase of a NiAl type with a B2 (CsCl) crystalline structure. These alloys show excellent mechanical properties and corrosion resistance at high temperatures around 1000°C with a melting point of about 1638°C. They also have a density of about 5.7 g/cm^3 which is lower than that presented by the Ni-based superalloys, about 8 g/cm^3 [2].

In a similar way to Ni-superalloys, the Fe-based superalloys support its mechanical strengthening on the formation of coherent precipitates in a soft matrix. Likewise, the mechanical properties of precipitation hardened alloys are closely related to the morphology, spatial distribution, volumetric fraction and average radius of the precipitated particles in the phase matrix. These microstructural characteristics can be controlled by means of heat treatments; however, they are also modified during the operation of industrial components at high temperatures for prolonged times [3–6].

The coarsening process is a metallurgical phenomenon which takes place at high temperatures and consists of the dissolution of small precipitates and subsequent mass transfer to the larger precipitates during the heating of alloys. This phenomenon usually has a strong effect on the mechanical properties of alloys. The Lifshitz-Slyozov-Wagner (LSW) theory for diffusion-controlled coarsening predicts a coarsening growth kinetics with a time dependent of $t^{1/3}$ considering the spherical precipitates without elastic interaction with the phase matrix and precipitate volume fraction close to zero [7,8]. There are several modified LSW theories for diffusion-controlled coarsening which incorporated higher volume fractions and the precipitate morphology different from spheres, as well as multicomponent alloys. Nevertheless, the temporal power law of $t^{1/3}$ is sustained in all the above cases, but the time-independent precipitate size distribution of LSW theory becomes broader and more symmetric with the increase in volume fraction [9–13]. The coarsening resistance of precipitates is a very important factor to retain a good strength at high temperatures in these alloys. An alternative to have a better coarsening resistance, it can be obtained either by a low value of lattice misfit which maintains a coherent interface with low interfacial energy between the precipitate and the matrix, low solute solubility or slow atomic diffusion in the matrix phase [13]. The addition of alloying elements is a good alternative to control the coarsening kinetics because of its effect on the interfacial energy, solubility of precipitates or atomic diffusion. For instance, the addition of Cu is expected to have, at least, effect on the solubility since Cu has low solubility in bcc Fe and it has a good solubility in Ni and more than 4 at.% in Al [14]. A small addition of Cu to the Fe-Ni-Al alloys is expected to modify the coarsening kinetics of β′ precipitates. Another possibility is the chromium addition which has a higher solubility in the ferrite matrix and thus it could also influence on the coarsening kinetics [15].

The precipitation and coarsening process are usually analyzed by artificial aging of a given composition alloy. Nevertheless, an alternative way for studying the precipitation reactions in ternary alloys is using diffusion couples [16–17] which permit to analyze the precipitation process for different alloy compositions in the same specimen. This method has been used to determine solubility limits and precipitation evolution which is based on the microstructure observation of different composition aged alloys produced by a continuous composition gradient in the specimen. There are different methods to create the macroscopic composition gradient in the specimen such as diffusion couples, imperfect arc melting of sandwiched metals, imperfect homogenization of coarse precipitates, etc. This method has also been applied to study the precipitation and coarsening processes in Fe-Ni-Al alloys [18].

Thus the studies of precipitation process in Fe-Ni-Al alloys can be summarized as follows: The precipitation of the NiAl type β′ phase in an Fe-rich phase matrix has been analyzed in different works [5,6]. The effect of alloying elements on the precipitation process and its mechanical properties has been also studied by different authors [3,4,14,15]. The application of diffusion couples to analyze the precipitation process in Fe-Ni-Al alloys has been reported in the literature [15].

Therefore, the present chapter covers first the precipitation and coarsening processes in Fe-Ni-Al alloys aged artificially at high temperatures, as well as their effect on the mechanical properties. These results show the precipitation evolution, morphology of precipitates, coarsening kinetics and mechanical properties such as hardness. Additionally, the effect of alloying elements such as copper and chromium is also studied on the precipitation and coarsening processes. The main results of this section are concerning on the coarsening kinetics and its effect on hardness. Besides, the diffusion couple method is employed to study the precipitation and coarsening process in different Fe-Ni-Al alloy compositions, as well as its effect on the hardness. All the above aspects of precipitation and coarsening are also supported with Thermo-Calc calculations.

2. Phase diagrams of Fe-Ni-Al system

Figures 1 (a–d) show the calculated isothermal ternary phase diagrams of the Fe-Ni-Al system corresponding to 750, 850, 950 and 1100°C, respectively [19]. In general, it can be seen that the liquid and ferrite phase regions decrease with the decrease in temperature. The β′ phase region increases as temperature increases. The austenite phase region also shows an increase as temperature increases. The γ′ phase also extends its presence as temperature decreases. These diagrams show good agreement with the experimental ones reported in the literature [5,20].

If chromium is added to the above ternary system, some changes may occur since this has an extended solubility in both the austenite and the ferrite phases. The calculated pseudo-ternary Fe-Ni-Al phase diagrams with the addition of 1 wt.% Cr are shown in Figures 2 (a–d) for 750, 850, 950 and 1100°C, respectively. It can be noticed, in general, that the β′ phase region increases as temperature decreases from 1100°C to 950°C, and then it decreases from 850°C to 750 °C. The ferrite phase region also increases with temperature, but it is slightly wider than that of the ternary system. The β′ and liquid region are not present for the four temperatures. The austenite phase region also increases with temperature and the γ′ phase region increases as temperature decreases. There are no experimental data reported in the literature for comparison.

3. Precipitation process

According to the ternary phase diagram, shown in Figure 1, it is possible to carry out a hardening precipitation treatment in the Fe-rich region. Even more, the precipitation reaction can take place in two ways [20]:

$$\beta' \rightarrow \beta' + \alpha \tag{1}$$

$$\alpha \rightarrow \alpha + \beta' \tag{2}$$

That is, the supersaturated solid solution of either the β′ or α phase can produce a mixture of the ferrite α phase and the ordered intermetallic β′ phase which can promote an increase in the mechanical strength and thus a better performance in creep conditions. Additionally, it is expected to have a coherent interface between these two phases since the lattice parameter of these is very similar [3]. Additionally, this fact gives better creep resistance in this type of alloys.

For instance, the X-ray pattern for the Fe-10wt.%Ni-15wt.%Al alloy solution treated at 1100 °C and then aged at 850 °C for 5 h is shown in Figure 3. The XRD pattern for the solution treated specimen only indicates the peaks from the bcc ferritic α phase. On the other hand, the XRD pattern corresponding to the aged specimens shows the main reflections corresponding to the ordered NiAl-type β′ phase. The principal morphology characteristics of precipitation in Fe-rich Fe-Ni-Al alloys is the presence of cuboids of the β′ phase dispersed in the ferrite matrix showing a preferential alignment on the <100> directions of the ferritic α phase due to the lowest elastic-strain energy [13]. This alignment can be clearly observed in the DF-TEM micrograph of the Fe-10wt.%Ni-15wt.%Al alloy aged at 850 °C for different times and its corresponding electron diffraction pattern with a zone axis [100] shown in Figure 4.

4. Microstructure evolution during coarsening

An important characteristic for an alloy with good creep resistance is that the interparticle distance between the β′ precipitates should be as short as possible. Thus, it is important to know the microstructure evolution of precipitates during the coarsening stage, as well as its kinetics and the effect on the mechanical properties of creep resistance alloys.

As an example, the precipitation evolution during the coarsening process for the Fe-10wt.%Ni-15wt.%Al, Fe-10wt.%Ni-15wt.%Al-1wt.%Cr and Fe-10wt.%Ni-15wt.%Al-1wt.%Cu alloys is shown in Figures 5, 6 and 7 (a–f), respectively. The morphology of the β′precipitates is rounded cuboids in the three aged alloys from the early stages of aging at all temperatures, Figures 5–7 (a–c).The straight sides of precipitates suggest the presence of a coherent interface between the precipitates and the matrix.

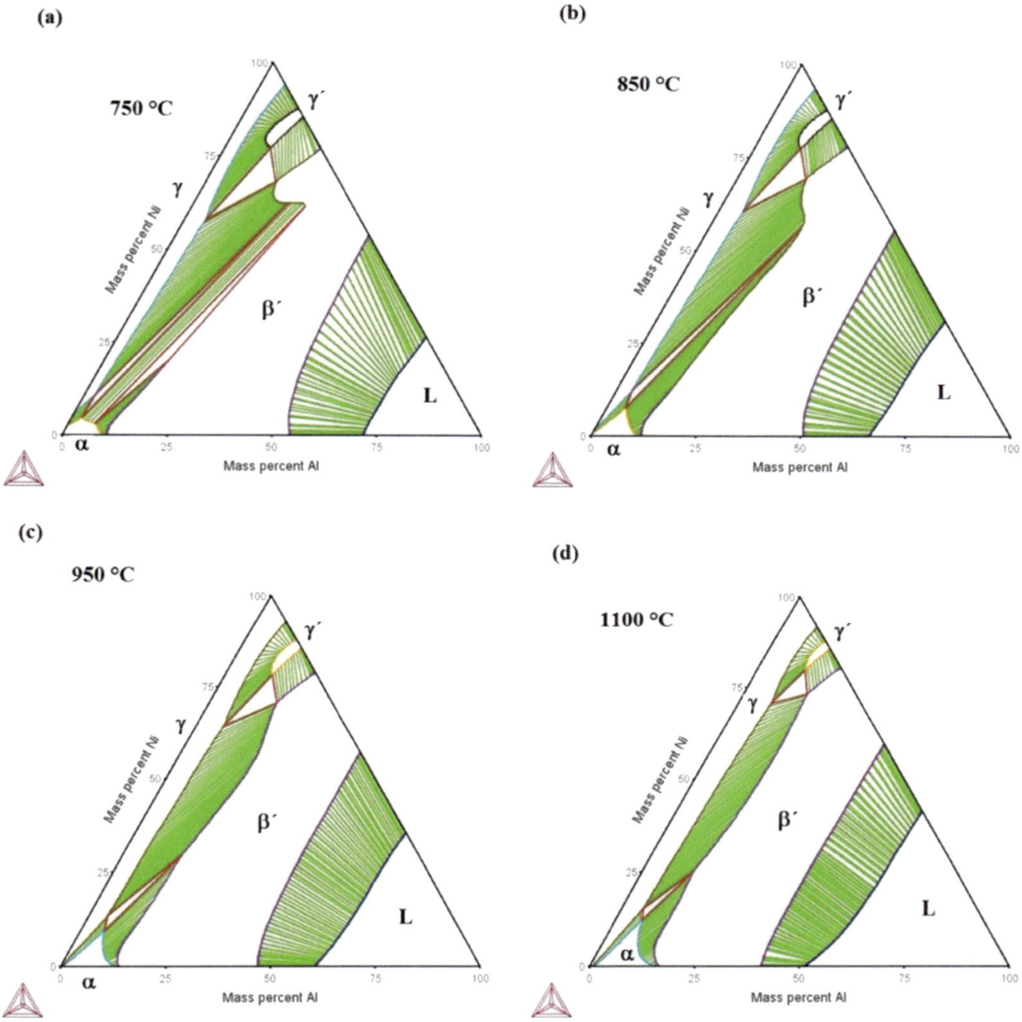

Figure 1. Calculated isothermal phase diagrams of the ternary Fe-Ni-Al system at (a) 750°C, (b) 850°C, (c) 950°C and (d) 1100°C.

The precipitates become aligned with the ferritic matrix over the course of time, Figures 5 (c–d).This alignment has been reported [13] to occur in the <100> crystallographic direction of the ferrite matrix since it corresponds to the softest one in the bcc crystalline structure. This suggests that the alloy specimens have the <100> orientation in the aged specimens. A further aging promotes the increase in size of precipitates in both of the aged alloys. The increase in precipitate size is higher as the aging temperature increases. Aging for longer times has a tendency to form square or rectangular arrays of cuboid precipitates in both alloys; however, this fact seems to be higher in the ternary alloy aged at 750°C and 850°C than that in the other one, Figure 5 (e,f).

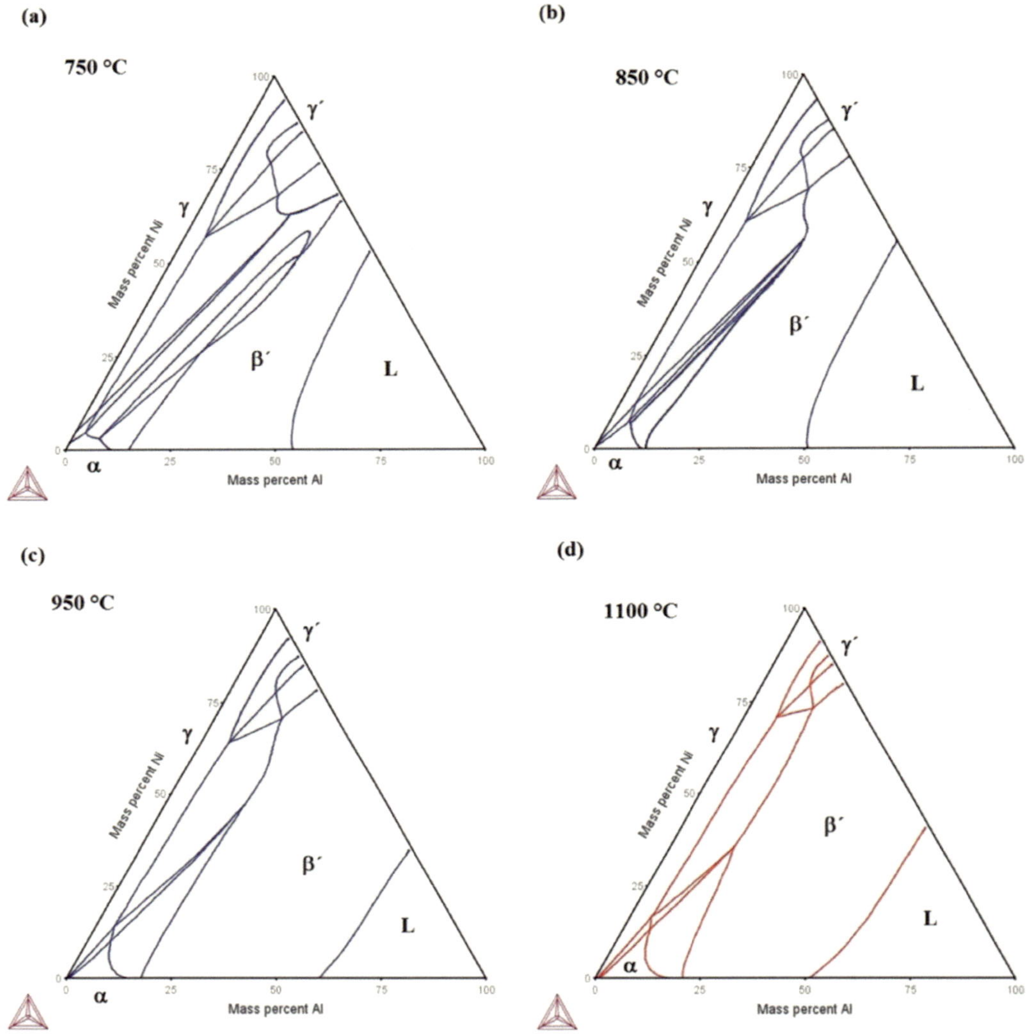

Figure 2. Isothermal phase diagrams of the pseudo-ternary Fe-Ni-Al-1wt.%Cr system at (a) 750°C, (b) 850°C, (c) 950°C and (d) 1100°C.

At aging at 950°C, elongated arrays of precipitates are aligned with respect to the ferritic matrix in the aged alloys, Figures 5, 6 and 7 (f–i). Some coalescence of precipitates is observed in these arrays and the straight sides of some precipitates become curved. This suggests the loss of coherency between the precipitates and the matrix. This characteristic is more notorious in the case of the aged ternary alloy. This behavior seems to be related to a higher elastic-strain effect in this alloy [13]. No splitting of precipitates was observed to occur in the aged alloys. The volume fraction is, in general, higher for the ternary alloy; however, the volume fractions are very close in alloys aged at 750°C. It is interesting to notice that the end of plates is like a rounded-tip for the aged Fe-10wt.%Ni-15wt.%Al-1wt.%Cu alloy, while it is more or less flat

http://dx.doi.org/10.5772/61070

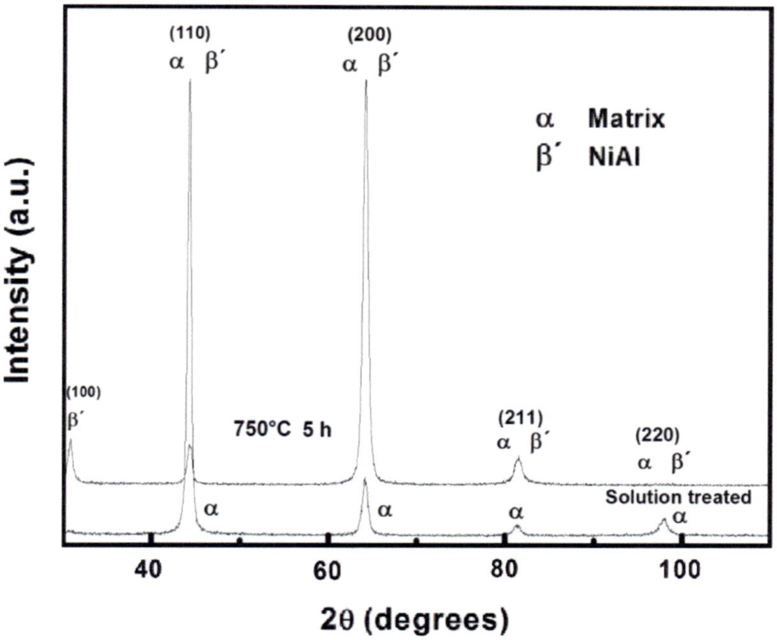

Figure 3. XRD pattern of the Fe-10wt.%Ni-15wt.%Al alloy solution treated at 1100°C and then aged at 850°C for 5 h.

for the other aged alloys. This fact suggests that the coherency between matrix and precipitates is lost in the former alloy and thus the plate grows more easily in that direction. This causes a larger length of plates in this case. The alignment of precipitates on elastically softest directions occurred faster in the Cu-containing alloy. This can be attributed to a higher coherency-strain energy which may be caused by a larger lattice misfit because of the Cu addition.

5. Coarsening kinetics

The coarsening kinetics of precipitates is usually an important parameter to analyze the creep resistance in the heat-resistant alloys, as well as to know the effect of alloying elements on the coarsening process of precipitates, which enables us to design better creep resistance alloys. The conventional way for analyzing the coarsening kinetics is to plot the equivalent radius, determined from the area of a precipitate, against

the aging time. The variation of radius r with time t obeys usually a power law, $r = kt^n$, and the time exponent usually indicates the coarsening mechanism. The Lyfshitz-Slyozov-Wagner (LSW) theory for diffusion-controlled coarsening rules the coarsening behavior of precipitates in different alloys [7–13].

Furthermore, several modifications of this theory have been reported [13] to consider the shape, volume fraction and coherency effects on the growth kinetics.

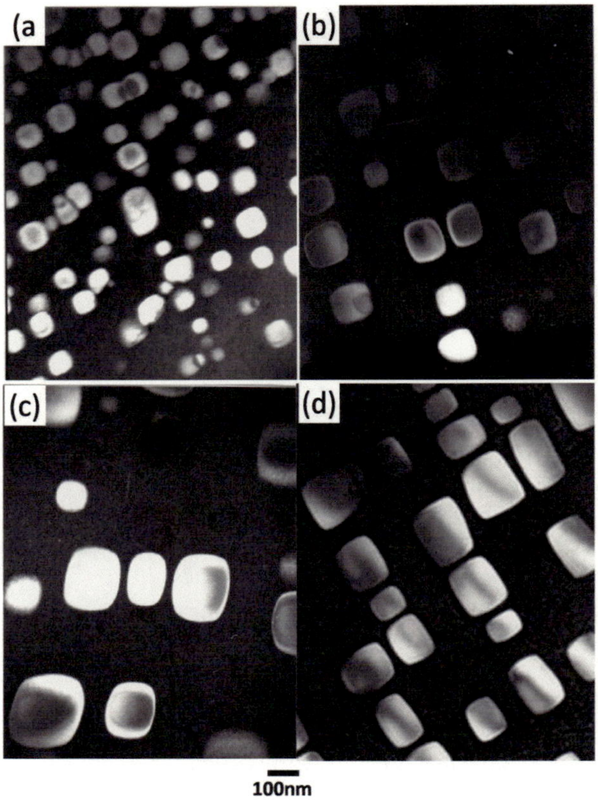

Figure 4. DF-TEM micrograph and electron diffraction pattern for the Fe-10wt.%Ni-15wt.%Al alloy aged at 850°C for (a) 1 h, (b) 5 h, (c) 25 h and (d)75 h.

To show the coarsening kinetics in Fe-Ni-Al alloys, the average equivalent circus radius, *r*, of β′ precipitates, expressed as r^3- r_o^3, is plotted as a function of aging time in Figure 8 (a–c)for the aged Fe-10wt.%Ni-15wt.%Al, Fe-10wt.%Ni-15wt.%Al-1wt.%Cr and Fe-10wt.%Ni-15wt.%Al-1wt.%Cu alloys, respectively. These figures show straight line curves for the different cases which suggest that the LSW theory for diffusion-controlled coarsening is followed in this study. The LSW theory and modified LSW theories [7,13] express mathematically the variation of particle radius with time as follows:

$$r^3 - r_o^3 = kt \tag{3}$$

where r_o and r are the average radius of precipitates at the onset of coarsening and time t, respectively, and k is a rate constant which can be determined from the slope of straight lines in Figure 8. The r_o value was determined from the linear regression analysis for each case. Figure 8 also shows the values of rate constant k for each case and they indicate that the coarsening process takes place more rapidly for the Fe-10wt.%Ni-15wt.%Al alloy aged at the three temperatures than that of the other two alloys. In general, the slowest growth kinetics can be observed for the aged Fe-10wt.%Ni-15wt.%Al-1wt.%Cr alloy. This behavior suggests the effect of chromium and copper on the Ostwald ripening process in this alloy.

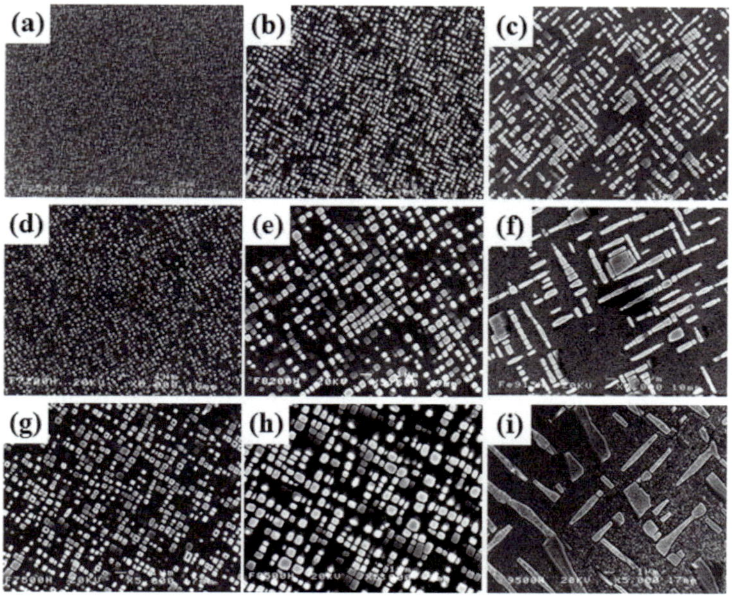

Figure 5. SEM micrographs of the precipitation evolution for the Fe-10wt.%Ni-15wt.%Al alloy aged at (a,d.g) 750°C, (b,e,h) 850°C and (c,f,i) 950°C for 25, 200 and 500 h, respectively.

A further analysis of the rate constant *k* can be conducted by obtaining the Arrhenius plot of this, as shown in Figure 9. The slope from the straight lines permits to determine the activation energy for the coarsening of β′precipitates in each alloy. It can be noticed that the lowest value corresponds to the ternary alloy and the highest energy is for the aged Fe-10wt.%Ni-15wt.%Al-1wt.%Cu alloy. This is in agreement with the fastest and lowest coarsening kinetics, respectively, observed in these two alloys. The activation energy is in the range of 200–240 kJ/mol, which is close to the value of 188 and 234 kJ/mol reported for the bcc Fe-rich Fe-Al alloys [21]. In order to understand the effect of alloying element on the coarsening process in these alloys, Figure 10 (a) illustrates the HAADF-STEM image of the Fe-10wt.%Ni-15wt.%Al-1wt.%Cr alloy aged at 950°C for 50 h. Some β′ precipitates and the ferrite matrix can be observed in the micrograph. The STEM intensity profile corresponding to Fe, Ni, Al and Cr elements, following the line A to B indicated in Figure 10 (a), are shown in Figure 10 (b). The presence of chromium is slightly higher in the ferrite matrix than that observed in the β′ precipitates. The iron content of precipitates is higher than that of nickel and aluminum. In order to show more clearly the effect of alloying elements, a Thermo-Calc equilibrium analysis [19] was carried out for the two alloy compositions at temperatures between 750°C and 950°C. Thermo-Calc analysis indicates that the ferrite matrix is richer in chromium, about 1.1 at.% for all temperatures, in comparison with that in the β′ precipitates. This detail is in good agreement with the chemical behavior reported in the literature for Fe-Ni-Al-Cr alloys [3]. The mole fraction of alloying elements is shown in Figure 11 for the equilibrium β′ phase for the Fe-10wt.%Ni-15wt.%Al and Fe-10wt.%Ni-15wt.%Al-1wt.%Cr alloys at temperatures between 750°C and 950°C. These precipitates are expected [2] to be an (Fe,Ni)Al intermetallic compound. Thermo-Calc analysis shows that the iron content in the β′ precipitates increases with the aging temperature, from about 10 to 50 at.%, while the nickel and aluminum contents decrease with temperature, from about 45 to 20 at.% in both alloys. However, the content of iron in the β′ precipitates is slightly lower in the Fe-10wt.%Ni-15wt.%Al-1wt.%Cr alloy than that corresponding to the other alloy, while the nickel and aluminum contents of the former alloy are slightly higher than that of the latter alloy. In the case of the alloy with chromium, the chromium contents show a slight increase with temperature, from 0.04 to 0.22 at.%. This chemical behavior suggests that the β′ precipitates are an intermetallic compound closer to the NiAl compound in both alloys aged at 750°C since the Fe content is low. In contrast, the β′ precipitates are a compound of (Fe,Ni)Al type in both alloys aged at 850°C and 950°C showing a considerable amount of iron, which is in good agreement with Figure 5. This behavior may be favorable for having a higher coarsening resistance in these precipitates since the atomic diffusion process would be more complex which may cause its retardation [13]. As mentioned above, the coarsening resistance in the alloy with chromium is higher than that observed in the other alloy. This fact can be attributed mainly to the presence of chromium in both the ferrite matrix and the β′ precipitates which cause a slower diffusion process during the coarsening of precipitates.

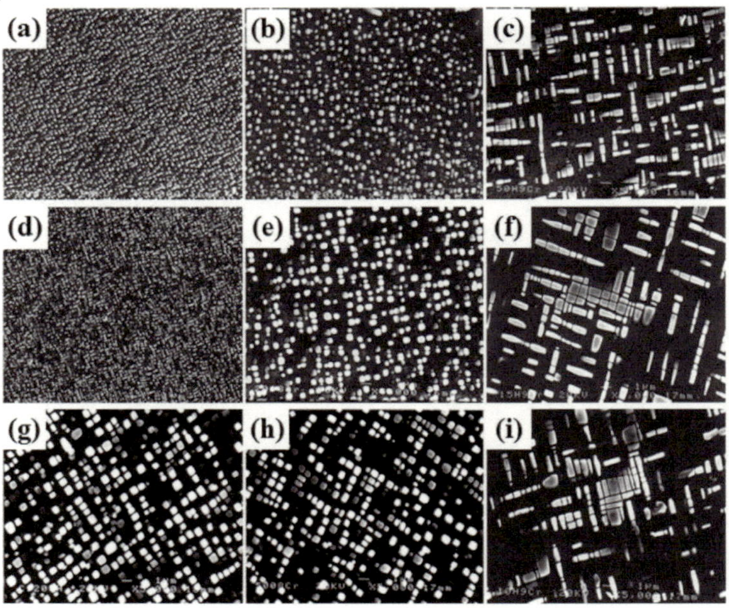

Figure 6. SEM micrographs of the precipitation evolution for the Fe-10wt.%Ni-15wt.%Al-1wt.%Cr alloy aged at (a,d.g) 750°C, (b,e,h) 850°C and (c,f,i) 950°C for 25, 200 and 500 h, respectively.

http://dx.doi.org/10.5772/61070

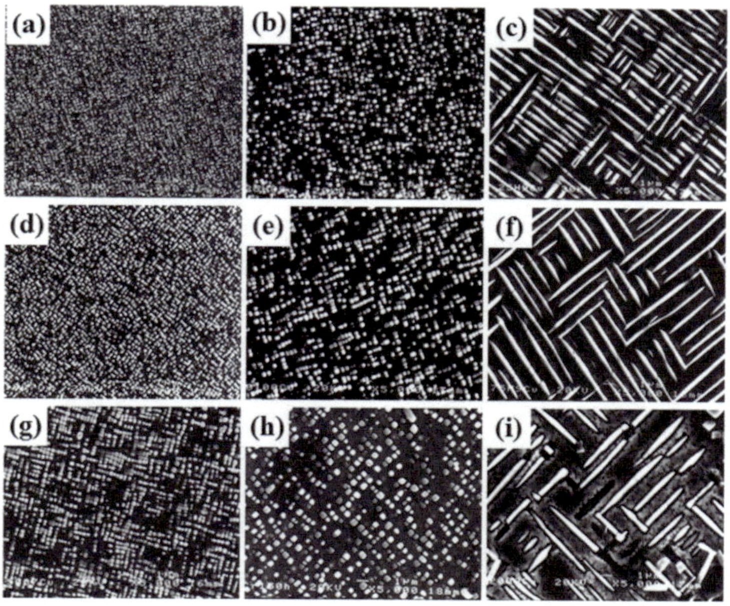

Figure 7. SEM micrographs of the precipitation evolution for the Fe-10wt.%Ni-15wt.%Al-1wt.%Cu alloy aged at aged at (a,d.g) 750, (b,e,h) 850 and (c,f,i) 950 °C for 25, 75 and 200 h, respectively.

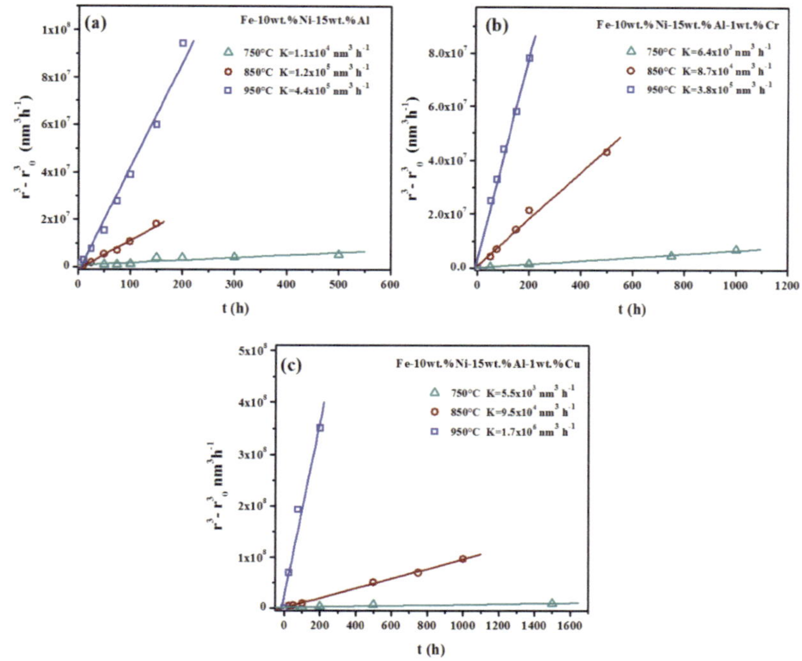

Figure 8. Plot of the cube equivalent radius as a function of aging time.

Figure 12 (a) shows a HAADF-STEM image of the Fe-10wt.%Ni-15wt.%Al-1wt.%Cu alloys aged at 750°C for 100 h. The corresponding HAADF-STEM EDS line scan profile is also shown in Figure 12 (b) for this specimen. The Fe-rich content of the ferritic matrix is evident. In contrast, the β′ precipitates are composed of Fe, Ni, Al and Cu. Furthermore, most of the Cu content is located within the β′ precipitates. This fact suggests a delay in the volume diffusion and thus, the coarsening resistance seems to be improved.

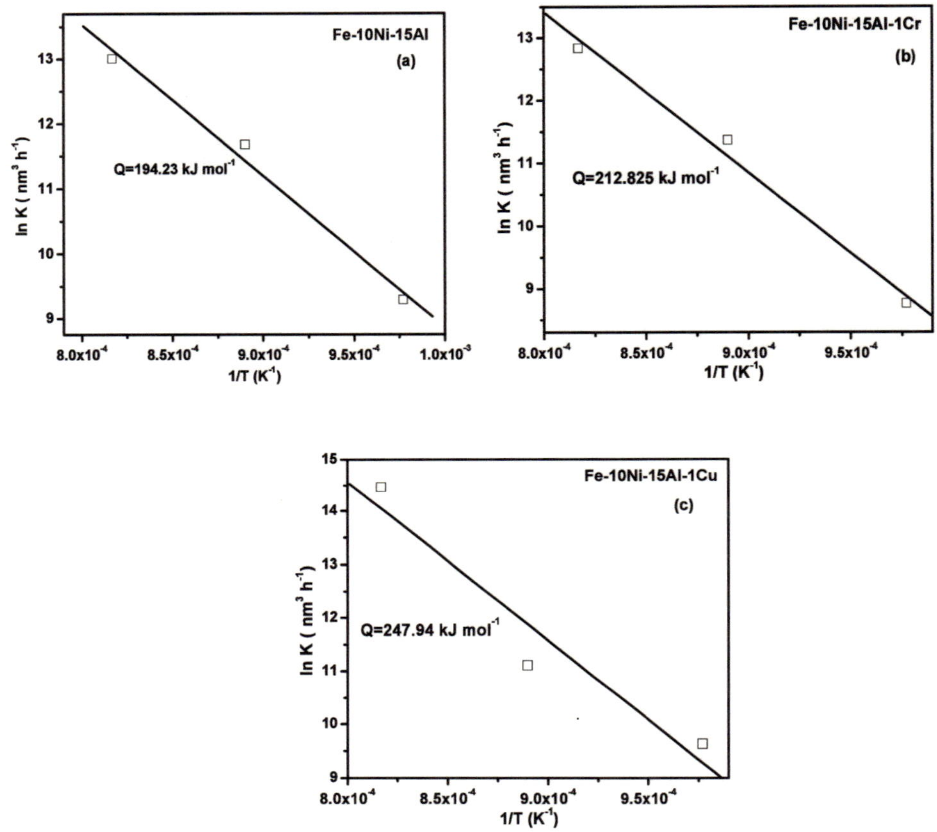

Figure 9. Arrhenius plot of the rate constant *k*.

The probability density $\rho^2 h(\rho)$ was determined with the following equation [14,15]:

$$\rho^2 f(\rho) = \frac{N_{(r,r+\Delta r)}}{\sum N_{(r,r+\Delta r)}} \frac{\bar{r}}{\Delta r} \frac{9}{4} \quad (4)$$

where r is the average radius of particles and $N_i(r+\Delta r)$ indicates the particle number in a given radius interval Δr. The normalized radius is defined as the ratio of $r/\bar{r}$. It can be noticed that the particle size distribution is broader and lower than the size distribution predicted by the LSW theory in the alloys aged for prolonged times, while it is more similar to that predicted by the LSW theory for short aging times.

The precipitate size distribution, plot of the probability density versus the normalized radius ϱ, is shown in Figures 13 (a–f)for the Fe-10wt.%Ni-15wt.%Al, Fe-10wt.%Ni-15wt.%Al-1wt.%Cr and Fe-10wt.%Ni-15wt.%Al-1wt.%Cu alloys, respectively, aged at 750°C for 50 and at 950°C for 150 h. The size distribution of the LSW theory for diffusion-controlled coarsening is also shown in these figures. The size distribution of both alloys, aged at 750°C for 50 h, is more symmetrical and closer to that one of LSW theory. In contrast, the size distribution of both alloys, aged at 950°C for 150 h, is broader and more symmetrical, which is a coarsening characteristic observed in different binary and ternary alloys, either with large volume fraction of precipitates or with high coherency-elastic strain effect [12,13]. This behavior is in complete agreement with the size distribution predicted using the modified LSW theories [11].

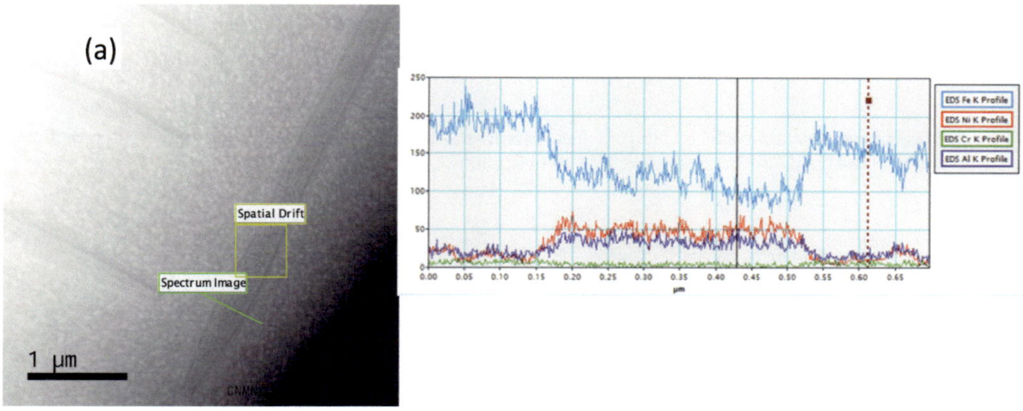

Figure 10. (a) HAADF-STEM image of (Fe,Ni)Al precipitates in a Fe matrix and (b) intensity profile of Fe, Ni, Al and Cr elements.

6. Hardness of aged alloys

The aging curves are shown in Figure 14 for the three alloys aged at 750, 850 and 950°C. The highest hardness corresponds to the aged Fe-10wt.%Ni-15wt.%Al-1wt.%Cu alloy, while the lowest one is observed to occur in the aged Fe-10wt.%Ni-15wt.%Al alloy. This hardness can be related to the presence of a small amount of copper in the β′ precipitates, Figure 11. In the case of Fe-10wt.%Ni-15wt.%Al-1wt.%Cr alloy, the increase in Fe content for the β′precipitates has been reported [3,4] to cause an increase in strength, while its decrease tends to improve its ductility. The highest resistance to the overaging in these alloys occurs during the aging at 750°C, which shows the slowest coarsening kinetics in both alloys. In contrast, the lowest coarsening resistance was observed to take place in the highest aging temperature because of the fastest Ostwald ripening process [13].

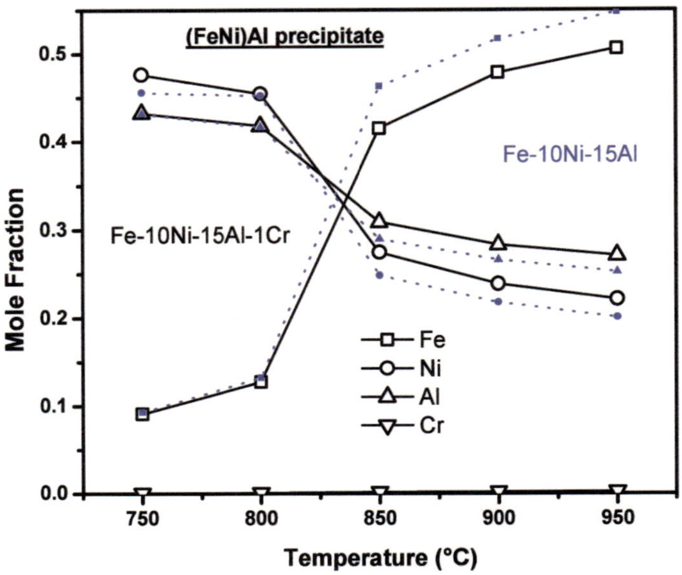

Figure 11. Calculated fraction mole of alloying elements in the β′precipitates against aging temperature for the Fe-10wt.%Ni-15wt.%Al and Fe-10wt.%Ni-15wt.%Al-1wt.%Cr alloys.

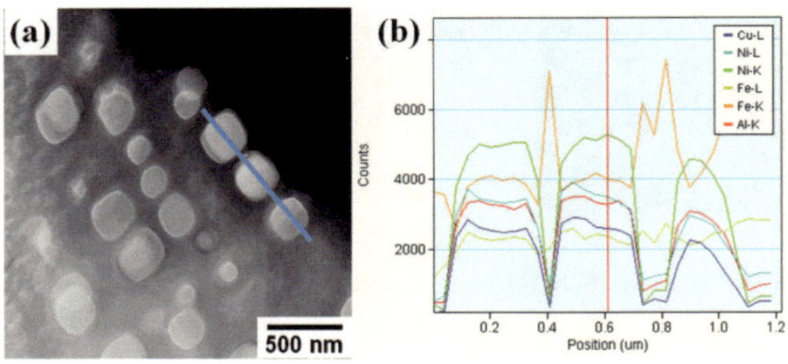

Figure 12. (a) HAADF-STEM image of (Fe,Ni)Al precipitates in a Fe matrix and (b) intensity profile of Fe, Ni, Al and Cu elements.

7. Application of diffusion couples

An alternative way for studying the precipitation reactions in ternary alloys is using diffusion couples which permit to analyze the precipitation process for different alloy compositions in the same specimen [16,17]. Furthermore, the diffusion couples have been used widely to study the precipitation process in binary alloys. This technique is called macroscopic composition gradient method. This procedure permits to determine solubility limits and precipitation evolution and it is based on the microstructure analysis of different composition alloys formed

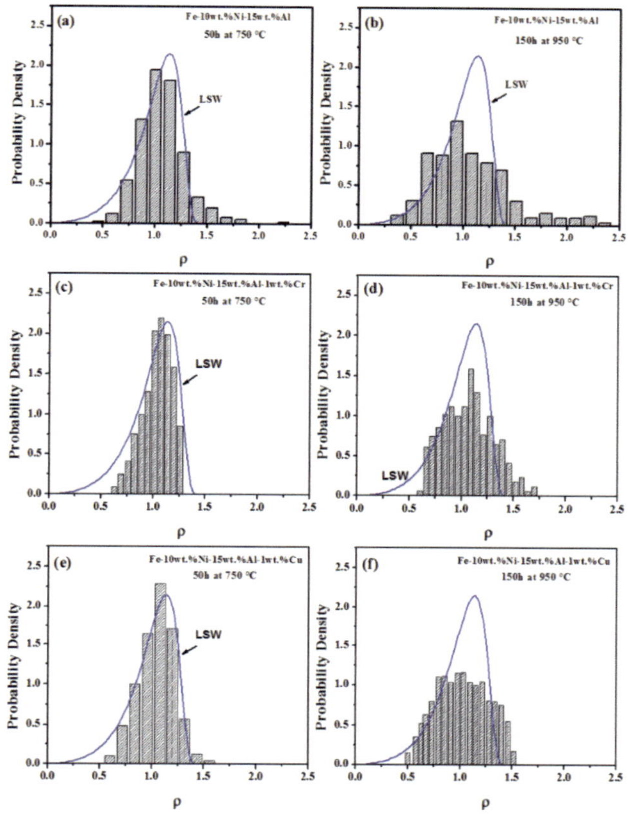

Figure 13. Precipitate size distributions for the (a–b) Fe-10wt.%Ni-15wt.%Al, (c-d) Fe-10wt.%Ni-15wt.%Al-1wt.%Cu and (e–f) Fe-10wt.%Ni-15wt.%Al-1wt.%Cr alloys aged at 750°C and 950°C for 50 and 150 h, respectively.

by a continuous composition gradient. The macroscopic composition gradient can be created in a specimen by different methods as stated above.

Figure 15 shows the Fe-25at.%Ni-25at.% Al alloy /Fe diffusion couple after diffusion heat treatment at 1100°C for 96 h. The black arrows indicate the interface of diffusion couple while dotted arrow indicates the path of linear chemical analysis with EDS. The concentration profiles of Fe, Ni and Al are shown in Figure 15.

The zero value in the x-axis is the position of the diffusion couple interface. The concentration profiles show that a composition gradient is present from the interface on both sides of the diffusion couple. That is, the Ni and Al compositions decrease toward Fe side, while the iron content decreases towards the alloy side. This profile behavior indicates that interdiffusion process occurs from high to low concentration regions.

The precipitation evolution is shown in Figure 16 for the diffusion couple solution treated and then aged at 900°C for different times. The SEM micrographs correspond to six different zones in the diffusion couple, designated as C_1–C_6. The Al and Ni contents increase from position

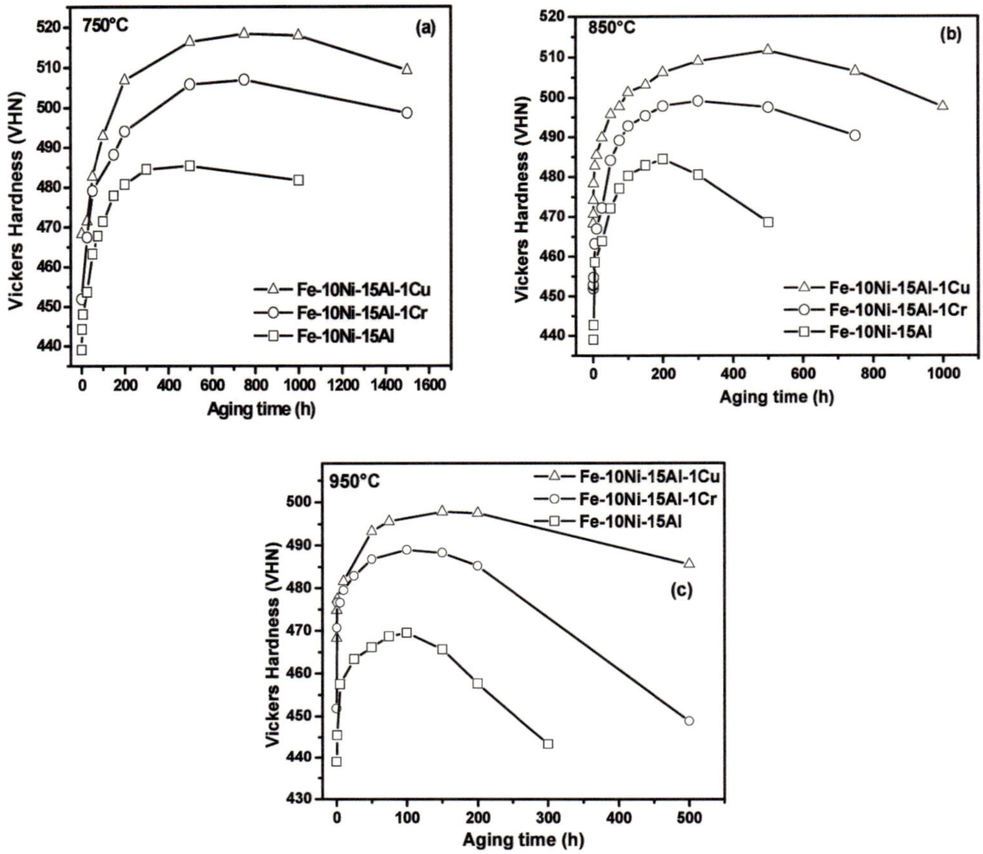

Figure 14. Aging curves for the Fe-10wt.%Ni-15wt.%Al, Fe-10wt.%Ni-15wt.%Al-1wt.%Cu and Fe-10wt.%Ni-15wt.%Al-1wt.%Cr alloys aged at 750, 850 and 950°C.

C_1 to positionC_6, as shown in Table 1. It is important to mention that the solution-treated diffusion couple indicated almost no precipitation. The precipitate morphology is rounded cuboids at the initial stages of aging. The cuboid precipitates seem to be aligned in a given crystallographic direction of the ferrite phase matrix. According to the literature [15], this direction correspond to the <100> direction of the ferrite matrix with the intention of decreasing the coherency-strain energy.

The X-ray diffraction patterns for the solution-treated and then aged alloy side in the diffusion couple are shown in Figure 17. A single-phase, the ferrite phase, is confirmed in the solution-treated specimen, while the presence of the XRD peaks corresponding to the β phase is clearly detected in the XRD pattern of the diffusion couple aged at 900°C for 50 h. No other phases were detected.

The volume fraction of precipitates increases with the increase in Ni and Al contents. As the aging progresses, the precipitate coarsening for all composition positions can be noted.

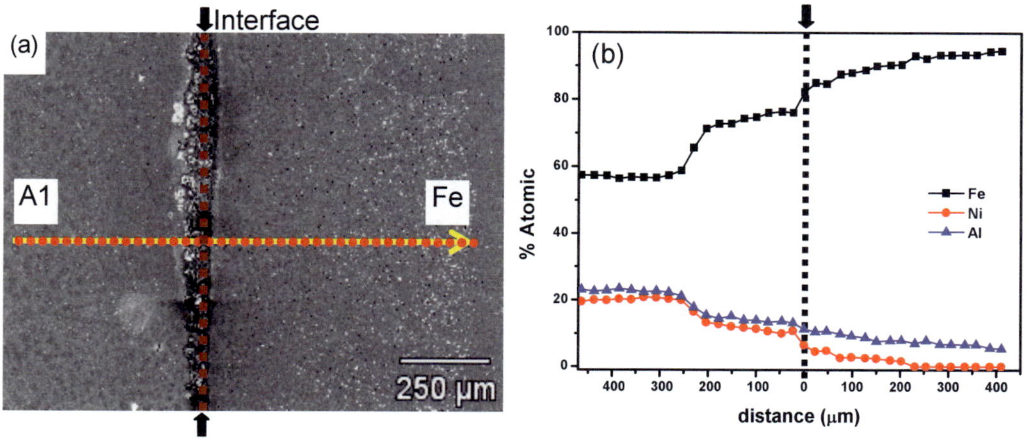

Figure 15. SEM micrograph of the (a) diffusion couple and (b) EDS chemical composition.

As a result of prolonged aging, the precipitate morphology becomes elongated and irregular. The coarsening process is observed to take place firstly in the zones which have higher content of solutes. This fact can be associated with the higher precipitate volume fraction which facilitates the coarsening process [13]. It is important to notice that the precipitate faces are curved in this stage of aging. This could indicate the coherency between the precipitates and the matrix. This fact can be adopted as the reason for the faster coarsening kinetics of the elongated precipitates.

For the first four compositions C_1–C_4, the precipitate radii r were measured, taking into account only precipitates with rounded cuboid shape, at any aging time t. The plot of ln r vs. ln t is shown in Figure 18. The coarsening growth kinetics is expected to follow this equation [7, 13]:

$$r = k_r t^n \tag{5}$$

where r and t are the mean radius of precipitates and time t, respectively, k_r is a rate constant and n the time exponent. The slope value n can be obtained from Figure 18 and these were 0.29, 0.29, 0.30 and 0.31 for the compositions C_1 to C_4, respectively. These values are close to 1/3 which is predicted by the diffusion-controlled coarsening LSW theory [8] and the modified LSW theory for ternary alloys [10]:

$$r^3 - r_o^3 = k_r t \tag{6}$$

where r_o and r are the mean radius of precipitates at the onset of coarsening and time t, respectively, and k_r is a rate constant.

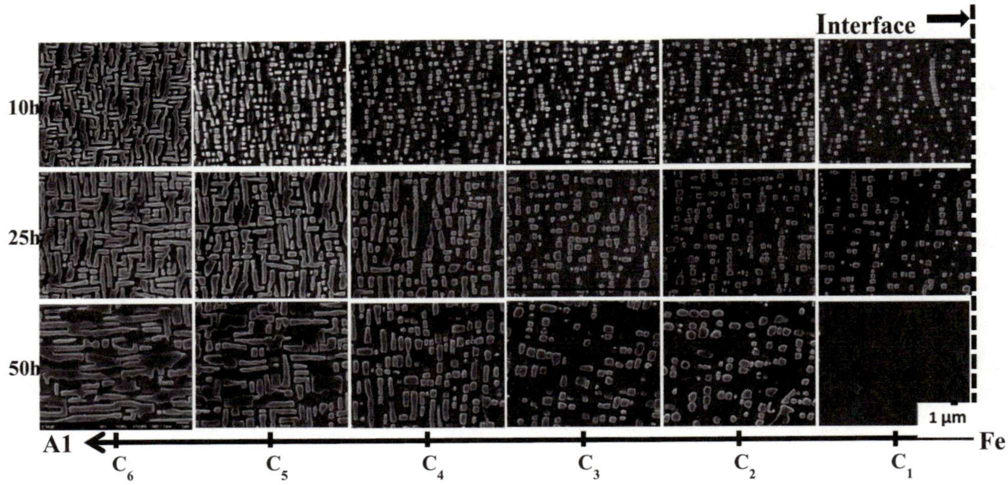

Figure 16. SEM micrographs of different positions in the diffusion couple.

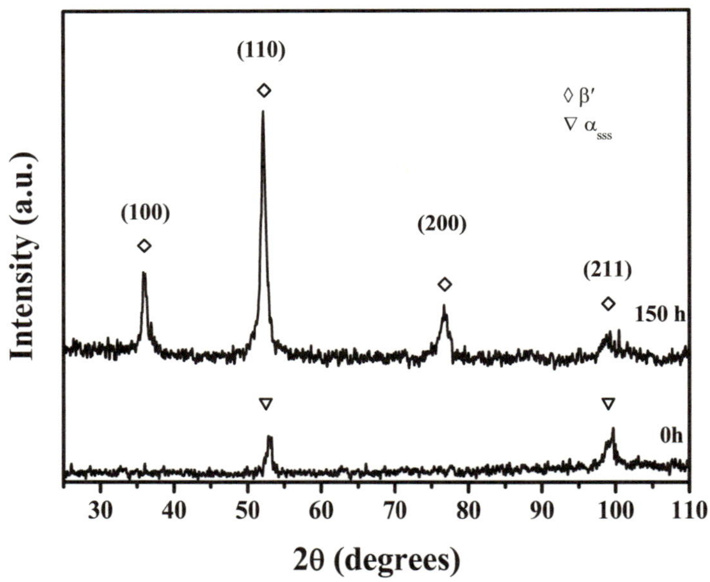

Figure 17. X-ray diffraction pattern of the diffusion couple.

Figure 19 shows the graph of r^3–r_o^3 against time t. The rate constant k_r for the coarsening process of β' phase in the ferritic matrix were determined from the slope of the straight lines and they are 2.4, 2.5, 3.0 and 3.90× 10^{-5} nm^3h^{-1} for the compositions C_1–C_4, respectively. These values are similar to those reported in the literature [14,15].

Position	Al (at.%)	Fe (at.%)	Ni (at.%)
C_1	11.16	76.87	11.98
C_2	13.91	72.71	13.39
C_3	15.09	70.12	14.88
C_4	20.0	61.67	18.33
C_5	23.24	54.26	22.5

Table 1. Chemical composition of selected positions

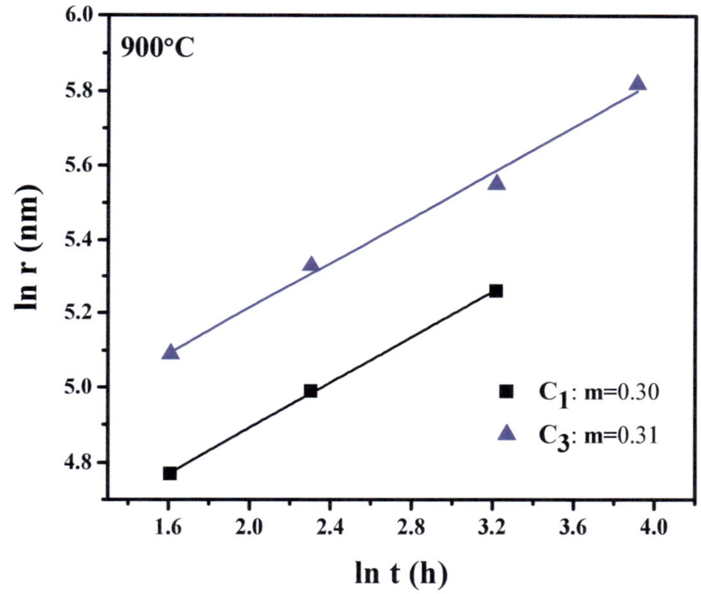

Figure 18. Plot of ln *r* versus ln *t*.

Figure 20 shows the precipitate size distributions for C_1 and C_4 compositions at different aging time along with the theoretical distribution functions predicted from the LSW theory. In these figures, the ordinates are $\varrho^2 f(\varrho)$ where ϱ represents the normalized particle size as defined by the LSW theory.

The particle size distributions are symmetric and sharp for short aging times, but they become broader and more asymmetric for prolonged aging times. This behavior has been described early [10,13] occurring in alloy systems with high volume fractions due to the presence of elastic interactions in alloy systems. Therefore, this fact is associated with the periodic formation of precipitate groups and the possible precipitate coalescence. Figure 21 shows the variation of the average radius *r* with the precipitate volume fraction for the diffusion couple after aging for 25 h. It is manifest the increase in the precipitate radius as the precipitate volume fraction increases because of the faster coarsening kinetics.

The plot of Vickers hardness versus Fe composition is shown in Figure 22 for the diffusion couple specimen aged at 900°C for different times. It can be noted that the hardness increases with the decrease in Fe content since the volume fraction of β' precipitates also shows an increase with the decrease in iron content or the increase in Ni and Al contents. The hardness of diffusion couple also increased with the aging time because of the coherent precipitates formed during the aging process.

In summary, the above results indicate that the hardness of the aged Fe-Ni-Al alloys show an increase with the increase in Al and Ni contents due to the increase in the volume fraction of β′ precipitates; nevertheless, the coarsening resistance decreases rapidly with the increase in Al and Ni contents due to the increase in the volume fraction of β′ precipitates. With the aim of having good coarsening resistance and the highest hardness in the Fe-rich aged Fe-Ni-Al alloys, the Ni and Al contents have to be lower than about 15 at.% Ni and 15 at.% Al, as stated by Figures 15 and 16. If higher contents of Al and Ni were necessary, the coarsening kinetics will be faster. In this case, it would be better to consider Ni-rich Fe-Ni-Al alloys with an austenite matrix and β′ precipitates which are expected to have higher coarsening resistances due to the lower atomic diffusion process in the fcc austenite phase than that in the bcc ferrite phase [2].

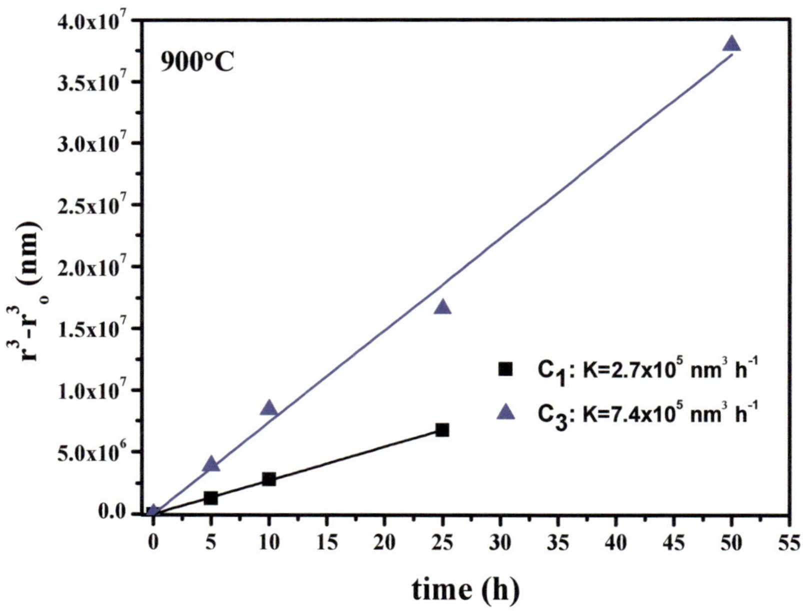

Figure 19. Plot of the cube equivalent radius as a function of aging time.

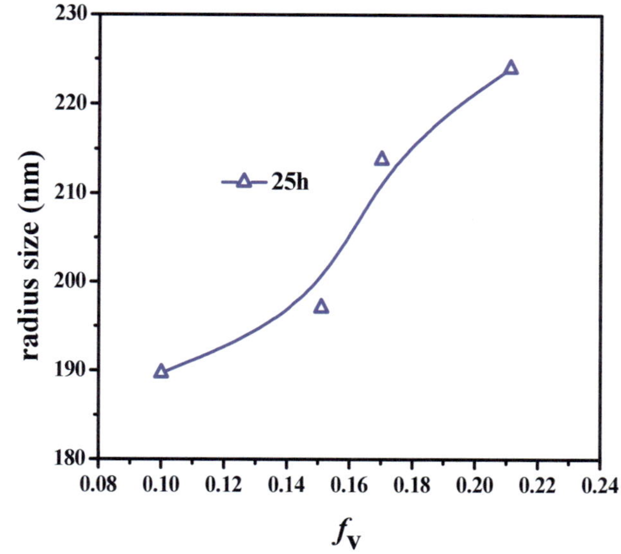

Figure 20. Radius as a function of volume fraction of precipitates.

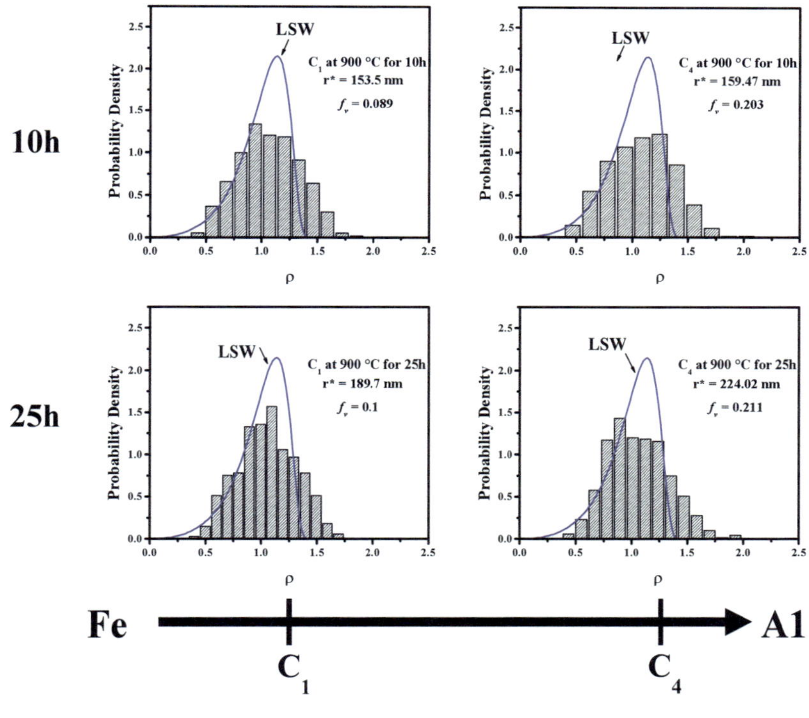

Figure 21. Precipitates size distribution for different compositions.

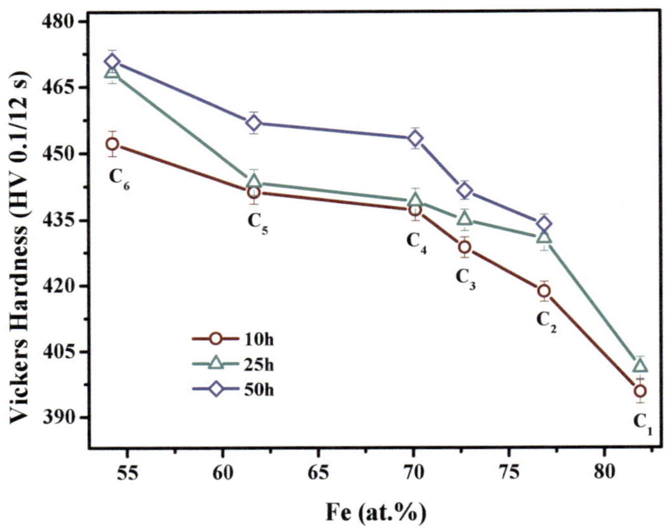

Figure 22. Plot of Vickers hardness versus at.% Fe.

8. Conclusions

An analysis of the precipitation evolution in the Fe-rich Fe-Ni-Al-based alloys was carried out using two different methods, the traditional way using one alloy composition and the diffusion couples which enables to analyze the precipitation in different alloy composition in the same specimen. This study indicates clearly the precipitation hardening of these alloys by the presence of the β′ phase precipitates in the ferrite phase matrix. The morphology of the β′ precipitates is rounded cuboids and it changes to elongated plates aligned in the <100> crystallographic direction. The coarsening growth kinetics of the β′ precipitates, in general, followed the modified LSW theories for diffusion-controlled coarsening. The addition of alloying elements such as copper or chromium promotes the increase in coarsening resistance and aging peak hardness by either its dissolution into the ferrite matrix or precipitates.

Acknowledgements

The authors wish to acknowledge financial support from, GAID, SIP-IPN and Conacyt.

Author details

Hector J. Dorantes-Rosales, Victor M. Lopez-Hirata*, Jorge L. Gonzalez-Velazquez, Nicolas Cayetano-Castro and Maribel L. Saucedo-Muñoz

*Address all correspondence to: vlopezhi@prodigy.net.mx

Instituto Politecnico Nacional (ESIQIE-CNMN), Department of Metallurgy and Materials, Mexico City, Mexico

References

[1] Guo Z, Sha W, Vaumosse D. Microstructural evolution in a PH13-8 stainless steel after ageing. Acta Mater 2003;51(1):101–16.

[2] Sauthoff G. Intermetallics. VCH 1995.

[3] Stalybrass C, Sauthof G. Ferritic Fe-Al-Ni-Cr alloys with coherent precipitates for high temperature applications. Mater Sci Eng 2004;A 387:985 –90.

[4] Stalybrass C, Schneider A, Sauthof G. The strengthening effect of (Ni,Fe)Al precipitates on the mechanical properties at high temperatures of ferritic FeNiAlCr alloys. Intermetallics 2005;13:1263 –8.

[5] Muñoz-Morris MA, Morris DG. Microstructure and mechanical behavior of a Fe-Ni-Al alloy. Mater Sci Eng 2007;444A:236 –41.

[6] Soriano-Vargas O, Saucedo-Muñoz ML, Lopez-Hirata VM, Paniagua-Mercado AM. Coarsening of β′ precipitates in an isothermally-aged Fe_{75}-Ni_{10}-Al_{15} alloy. Mater Trans JIM 2010;51:442 –6.

[7] Lifshitz IM, Slyozov VV. The kinetics of precipitation from supersaturated solid solution. J Phys Chem Solids 1961;19:35 –50.

[8] Wagner C. Theorie der alterung von niederssclagen durch umlosen. Z Electrochem 1961;65:581 –91.

[9] Baldan A. Progress in Ostwald ripening theories and their applications to nickel-base superalloys. J Mater Sci 2002;37:2171 –202.

[10] Kuehmann CJ, Voorhees PW. Ostwald ripening in ternary alloys. Met Mater Trans 1996;27A:937 –43.

[11] Thornton K, Akaiwa N, Voorhees PW. Large-scale simulations of Ostwald ripening in elastically stressed solids: I. Development of microstructure. Acta Mater 2004;52:1365 –78.

[12] Ardell AJ. The effect of volume fraction on particle coarsening: theoretical considerations. Acta Metall 1972;30:61 –71.

[13] Kostorz G. Phase transformations in materials. Wiley-VCH; 2001.

[14] Rosales-Dorantes HJ, Cayetano-Castro N, Lopez-Hirata VM, Saucedo-Muñoz ML, Villegas-Cardenas JD, Hernández-Santiago F. Coarsening process of coherent β′ precipitates in Fe-10wt.%Ni-15wt.%Al and Fe-10wt.%Ni-15wt.%Al-1wt.%Cu alloys. Mater Sci Technol.2013;29:1492 –8.

[15] Cayetano-Castro N, Saucedo-Muñoz ML, Dorantes-Rosales HJ, Gonzalez-Velazquez JL, Villegas-Cardenas JD, Lopez-Hirata VM. Ostwald ripening process of coherent β′ precipitates during aging in $Fe_{0.75}Ni_{0.10}Al_{0.15}$ and $Fe_{0.74}Ni_{0.10}Al_{0.15}Cr_{0.01}$ alloys. Adv Mater Sci Eng 5;1–7.

[16] Miyazaki T, Kobayashi S, Koyama T. Determination of the critical nucleus size of precipitates by utilizing the macroscopic composition gradient method. Met Trans 1999;30A:2783 –9.

[17] Miyazaki T, Koyama T, Kobayashi S. A new characterization method of the microstructure using the macroscopic composition gradient in alloys. Met Mater Trans 1996;26A:945 –9.

[18] Contreras-Piedras E, Dorantes-Rosales HJ, Lopez-Hirata VM, Hernandez-Santiago F, Gonzalez-Velazquez JL, Lopez –Monroy I. Analysis of precipitation in Fe-rich Fe-Ni-Al alloys by diffusion couples. Mater Sci Eng A 2012;558:366 –70.

[19] Thermo-Cal software 2008 v. 3.1/ Niv5.TDB.

[20] Hao S, Takayama T, Ishida K, Nishizawa T. Miscibility gap in Fe-Ni-Al and Fe-Ni-Al-Co systems. Met Trans 1984;15A:1819 –28.

[21] Mehrer H. Diffusion in Solid Metals and Alloys. Springer-Verlag, 1990.

Chapter 5

Welding Metallurgy of Corrosion-Resistant Superalloy C-276

Manikandan Manoharan, Arivazhagan Natarajan and Nageswara Rao Muktinutalapati

Additional information is available at the end of the chapter

http://dx.doi.org/10.5772/61104

Abstract

Microsegregation occurs during solidification of fusion zone in alloy C-276. The concomitant precipitation of topologically close-packed phases P and μ has been reported to be responsible for the hot cracking observed in this alloy during welding. The clue to preventing hot cracking hence lies in suppressing microsegregation in the fusion zone. An important avenue towards this is the introduction of current pulsing during gas tungsten arc welding. Current pulsing was found to be effective in mitigating microsegregation; it was also found to refine the microstructure in the weld zone and improve the mechanical behavior of weld joints. Judicious choice of filler wire is of paramount importance to get weld joints free from segregation and with a good combination of mechanical properties. Joints made using arc welding methods were found to be highly resistant to corrosion in salt spray tests. Non-arc-based methods—laser welding and electron beam welding—were found to be effective in largely keeping the microsegregation at bay. This chapter elaborates on these issues.

Keywords: Superalloy C-276, Fusion weld joints, Microsegregation, Microstructure, Mechanical behavior, Corrosion resistance

1. Introduction

Alloy C-276 is a highly corrosion-resistant Ni-based superalloy. It has good resistance to corrosion in both oxidizing and reducing conditions. The alloy exhibits excellent resistance to corrosion in various environments encountered during chemical processing and allied industries and, as such, finds widespread application in these industries [1]. It has also been a highly attractive material for naval/marine applications owing to its excellent resistance to corrosion in seawater environments, especially under crevice corrosion conditions [1]. Testing for crevice corrosion in both quiescent seawater and flowing seawater showed that C-276 did not corrode. Critical crevice and critical pitting temperatures in tests conducted as per ASTM G48 methods C and D showed higher resistance of the alloy to crevice and pitting corrosion compared to stainless steel 304 SS, Incoloy 825 and 925, Inconel 625 and 725 [1].

Alloy C-276 is a nickel-based single phase superalloy. The nominal chemical composition of the alloy is given in Table 1. The main alloying elements are Cr, Mo, Fe and W. The alloy is designed based on solid solution strengthening; there is no precipitation hardening operating in this alloy [2].

Base/Filler Metal	Chemical Composition (% Wt.)								
	Ni	Mo	Cr	W	Co	Mn	Fe	Nb	Others
Hastelloy C-276	Bal	16.36	15.83	3.45	0.05	0.41	6.06	—	0.17 (V), 0.005(P), 0.002 (S), 0.02 (Si), 0.005(C)
ERNiCrMo-3	Bal	10.0	22.0	—	—	0.5	1.0	4.5	0.015 (P), 0.015 (S), 0.5 (Si), 0.5 (Cu), 0.4 (Al), 0.4 (Ti), 0.1 (C)
ERNiCrMo-4	Bal	17.00	16.5	4.5	2.50	1.0	7.0	—	0.04 (P), 0.03 (S), 0.5 (Cu), 0.02(C)

Table 1. Chemical Composition of Base Metal and Filler Wires

Chromium facilitates the formation of passive films in a wide range of oxygen-bearing environments, thereby contributing to alloy's corrosion resistance. Chromium also contributes to solid solution strengthening. Molybdenum imparts alloy resistance to reducing chemicals. In addition, by virtue of its large atomic size compared to that of nickel, it has a strong solid solution strengthening effect [3]. The behavior of W is similar to that of Mo. It is, in fact, an even more effective solid solution strengthener than Mo.

Alloy C-276 is generally supplied in the solution annealed condition. Solution annealing leads to the dissolution of the alloying elements in the austenitic matrix. Standard solution annealing treatment is comprised of soaking at 1120°C followed by rapid quenching to room temperature to prevent the formation of carbides and/or embrittling phases. Representative microstructure of the alloy in the solution treated condition is shown in Fig. 1.

Welding is a very important process which is used for the manufacturing of various products out of this alloy. Arc welding is a commonly used technique. The most common arc welding methods are gas tungsten arc welding (GTAW) and gas metal arc welding (GMAW). The choice of filler wire is an important part of the weld design process. Two types of filler wire have been employed: (i) filler wire with composition matching with that of base metal and (ii) overalloyed filler wire where alloying elements are present in the filler wire material at a level higher than in base metal. For a majority of applications, the filler wire with base metal composition is adequate [4]. For situations where the weld joints are expected to serve in highly aggressive environments, overalloyed filler wire is selected [4].

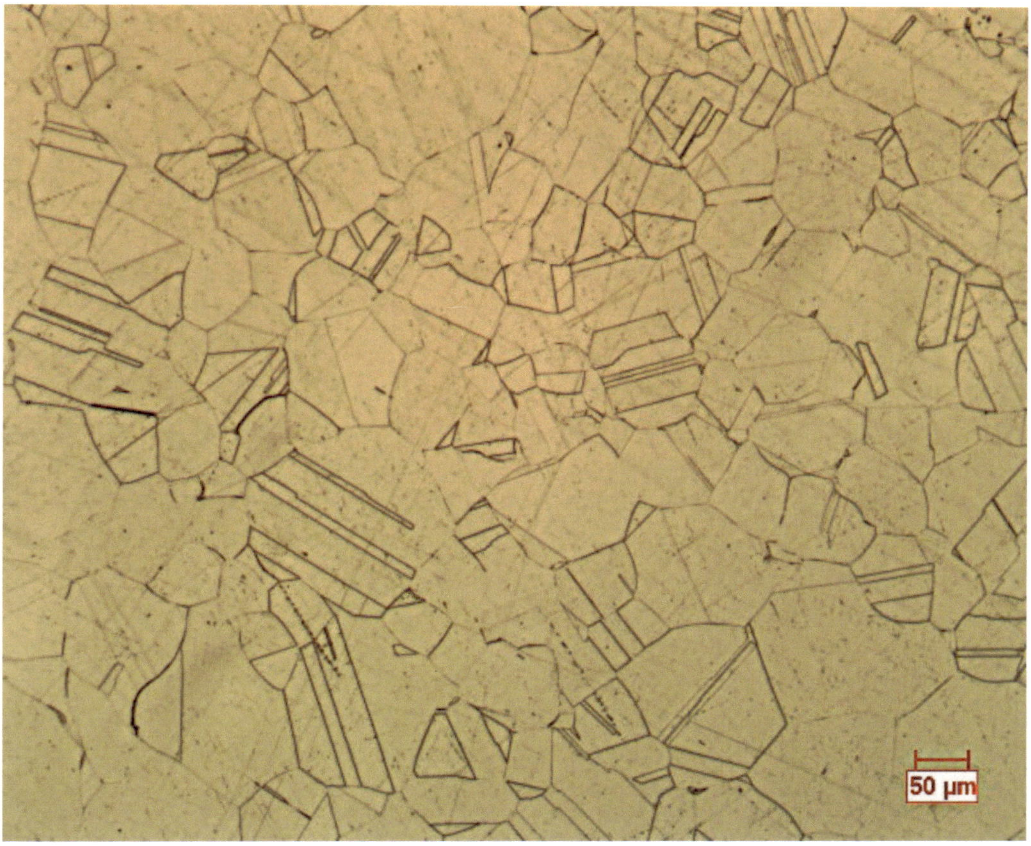

Figure 1. Microstructure of Hastelloy C-276 in as-received condition.

The performance of alloy C-276 can become adversely affected under certain circumstances and these were documented in the literature by different workers. It was emphasized that deleterious phases appear when the material is exposed to high temperatures. Tawancy [5] reported the precipitation of M_2C and μ phases in the form of a continuous layer at the grain boundary when the alloy is aged at 537°C for a long time, leading to a reduced tensile elongation and higher corrosion rate in boiling sulfuric–ferric sulfate solution. Akhtar et al. [6] reported the precipitation of molybdenum-rich μ phase after aging at 850°C, leading to a drop in impact energy; a completely intercrystalline brittle fracture occurred after aging for 120 h. Raghavan et al. [7] observed the formation of three distinct phases when the alloy was aged in the temperature range 650°C–900°C. Most abundant was the molybdenum-rich μ phase. The next most abundant phase was molybdenum-rich M_6C carbide. The third phase was tentatively identified as P phase, with a composition remarkably similar to that of μ phase.

Both μ and P phases belong to the group of topologically close-packed (TCP) phases. In general, the TCP phases in superalloys have a detrimental effect on several properties. Their rupture strength, tensile ductility and impact toughness at room temperature and corrosion resistance suffer a reduction when TCP phases appear in the microstructure [8].

The welding metallurgy of alloy C-276 gets complicated by the appearance of these TCP phases in fusion zone during welding. Cieslak et al. [9] carried out arc welding studies on alloys C-4, C-22 and C-276. All these grades are highly corrosion-resistant Ni-based alloys derived from Ni-Cr-Mo ternary system. Cieslak et al. [9] found that elemental segregation occurs during welding, leading to the formation of brittle TCP phases P and μ in alloys C-22 and C-276. The authors reported that alloy C-276 shows the highest susceptibility to hot cracking among the three alloys. They established that weld metal hot cracks are associated with intermetallic secondary solidification constituents P and μ. Perricone and Dupont [10] reviewed the subject of TCP phases in superalloys based on Ni-Cr-Mo system and summarized that the occurrence of TCP phases μ and P in alloy C-276 adversely affects weldability and other properties of interest.

The TCP phases P and μ are both rich in Mo and W, i.e., these elements preferentially partition into P and μ phases, depleting the gamma matrix of these elements to that extent. The higher the level at which Mo and W are present in the Ni-Cr-Mo-based alloy, the higher is the propensity for the formation of these brittle phases. For example, Perricone and DuPont [10] found no evidence for the occurrence of these phases in the weldments of a 12% Mo containing Ni-Cr-Mo alloy. However, upon increasing the Mo level to 24% at the expense of Ni, P and μ phases appear in the weld microstructure. Zheng [11] also reported that high concentration of Mo and W in Ni-Cr-Mo alloys leads to the occurrence of TCP phases during solidification. These phases find a place in the ternary equilibrium diagram for the Ni-Cr-Mo system. P phase forms at relatively high temperatures compared to μ phase. The latter phase is believed to be the product of a solid state transformation from the former.

2. Problem statement and critical review of the work done

Microsegregation in the weldment is the root cause of the problem and if it can be minimized by judicious selection of the welding process and parameters, the problem of hot cracking gets mitigated. Different approaches have been taken to suppress microsegregation in C-276 weldments.

Researches have been carried out to examine if pulsing of current during GTAW can be adopted to mitigate the problem of microsegregation in alloy C-276 weldments. In case of autogenous GTAW, there is significant segregation of Mo and Ni in the weld zone. The subgrain boundary shows higher levels of Mo and lower levels of Ni compared to subgrain interior (Figs. 2 a and 2b). In the case of autogenous pulsed current gas tungsten arc welding (PCGTAW), such segregation is essentially absent (Figs. 3 a and 3b).

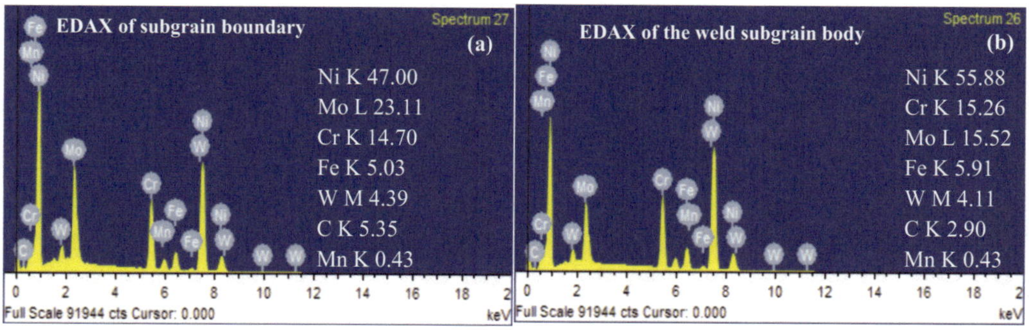

Figure 2. EDAX analysis of autogenous GTA welded alloy C-276: (a) subgrain boundary and (b) subgrain body.

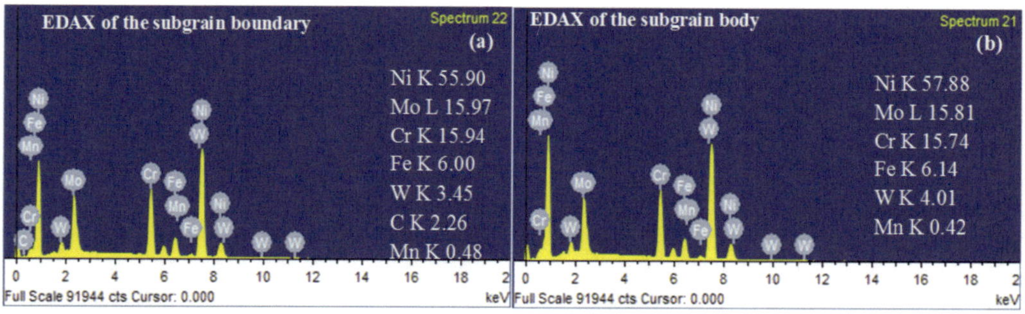

Figure 3. EDAX analysis of autogenous PCGTA welded alloy C-276: (a) subgrain boundary and (b) subgrain body.

When welding was done using matching filler wire ERNiCrMo-4, the situation was the same. The chemical composition of the filler is also included in Table 1. In case of GTAW, it is seen that the subgrain boundary is enriched in Mo and impoverished in Ni and W compared to subgrain body (Figs. 4 a and 4b). In case of PCGTAW, in contrast to the case of GTAW, there is no significant difference between the subgrain boundary and subgrain body with reference to levels of any of the elements (Figs. 5 a and 5b). The microsegregation that was noticed in GTA fusion zone is thus essentially absent in the PCGTA fusion zone. It is believed that the occurrence of secondary constituents P and μ, deleterious in the context of hot cracking susceptibility of the alloy, also comes down to that extent. There is indeed some evidence to this effect. Welds made with GTAW, both autogenous and with filler wire, show indications suggestive of the occurrence of secondary phase(s) in weld interface regions. Welds made with PCGTA, in contrast, did not show any secondary phases. PCGTAW also resulted in refinement of microstructure in fusion zone. Figure 6 shows scanning electron microscope (SEM) images of weld zones of C-276 produced by autogenous welding using GTAW and PCGTAW, respectively. The lower heat inputs coupled with higher instantaneous cooling rates associated with PCGTAW are considered responsible for the refinement of microstructure in the fusion zone. PCGTAW also caused breakage of dendrite arms contributing to the refinement. Further,

PCGTAW led to a better combination of strength and toughness of weld joints. Table 2 shows the results of tensile and impact tests of autogenous and ERNiCrMo-4 weldments using GTAW and PCGTAW. The refined grain structure with reduced severity of microsegregation in PCGTAW is believed to be responsible for the improved mechanical properties of the weld joint.

There have been other process/parameter changes attempted for welding of alloy C-276 to bring down the extent of microsegregation in the fusion zone of C-276 weld joints. For example, electron beam welding and pulsed laser welding have been shown to improve the quality of weldments. Guangyi et al. [12] reported that microsegregation in C-276 joints produced by pulsed laser beam welding is relatively less compared to the situation obtained with arc welding process. Manikandan et al. [13] studied laser-welded C-276 joints and found that microsegregation was within acceptable limits. The higher cooling rates and shorter solidification times associated with laser welding are believed to be responsible for the reduced elemental segregation.

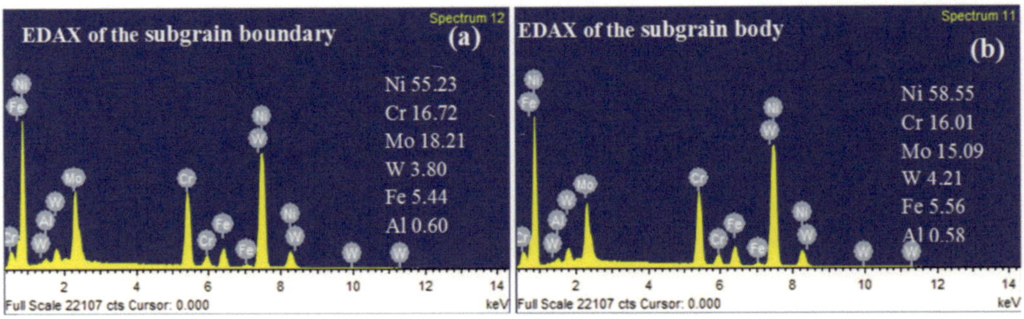

Figure 4. EDAX analysis of GTA welded alloy C-276 with ERNiCrMo-4 filler wire: (a) subgrain boundary and (b) subgrain body.

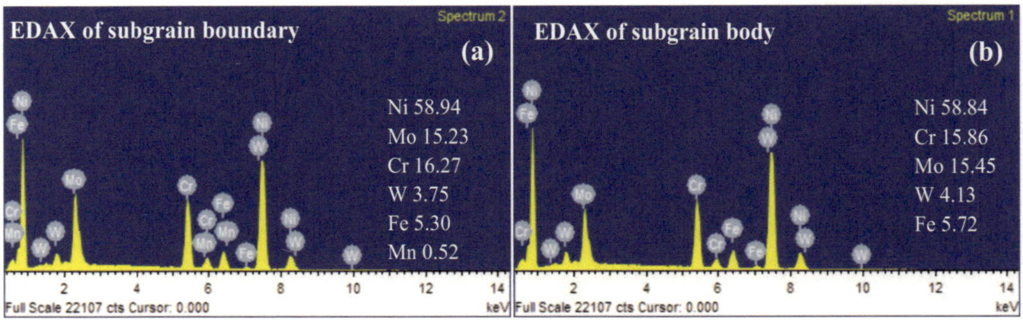

Figure 5. EDAX analysis of PCGTA welded C-276 alloy with ERNiCrMo-4 filler wire: (a) subgrain boundary and (b) subgrain body.

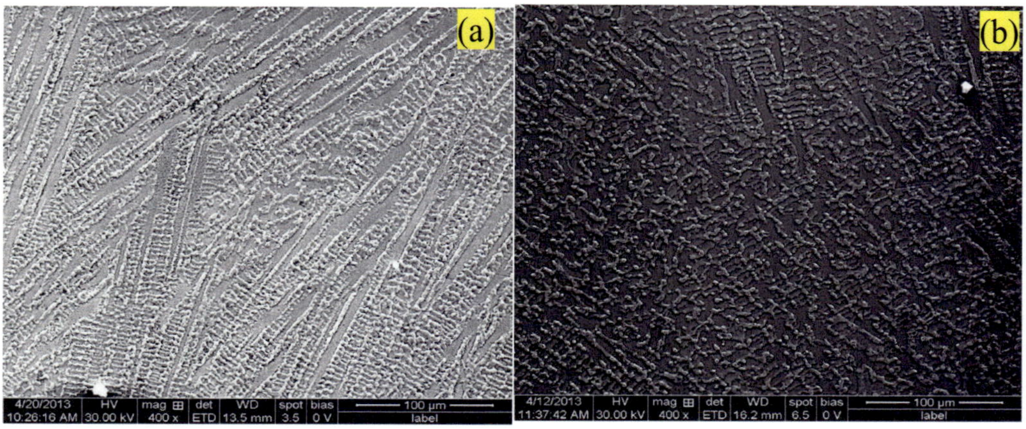

Figure 6. SEM micrograph of autogenous welded alloy C-276 (a) GTA and (b) PCGTA.

Welding Process	Results of Tensile Testing			Impact Toughness
	Average UTS (MPa)	Average Yield Strength (MPa)	Average % Elongation	Average Energy (J)
Autogenous GTA	746	394	53	54
Autogenous PCGTA	763	405	66	56
GTA ERNiCrMo-4	766	403	34	53
PCGTA ERNiCrMo-4	815	413	41	56

Table 2. Results of Tensile and Impact Tests of Autogenous and ERNiCrMo-4 Weldments of Alloy C-276 Produced by GTAW and PCGTAW

The choice of the appropriate filler wire composition is also very important in restricting the elemental segregation in fusion zone of C-276 weld joints. Manikandan et al. [14, 15] carried out welding of C-276 using two different filler wire compositions: ERNiCrMo-3 with a nonmatching composition (lower Mo and Fe, higher Cr and Nb than the base metal) and ERNiCrMo-4 with a matching composition. There was much higher segregation of alloying elements in the fusion zone in case of the former filler wire, as shown in Table 3. Both the GTA and PCGTA weldments made with the filler wire ERNiCrMo-4 showed distinctly superior strength-toughness combinations than those made with filler wire ERNiCrMo-3. Table 4 gives the results of tensile testing and impact testing of ERNiCrMo-3 and ERNiCrMo-4 weldments produced using GTAW and PCGTAW. The relatively high degree of segregation in the fusion zone of welds made with the nonmatching filler wire are believed to be contributing to the poor mechanical behavior of weld joints.

Type of Welding	Zone	Ni	Cr	Mo	W
GTAW ERNiCrMo-3	Weld subgrain boundary	40.81	16.50	29.87	3.79
	Weld subgrain body	55.25	19.17	15.31	1.89
GTA ERNiCrMo-4	Weld subgrain boundary	55.23	16.72	18.21	3.80
	Weld subgrain body	58.55	16.01	15.09	4.21

Table 3. Element Levels in Subgrain Boundary and Body of ERNiCrMo-3 and ERNiCrMo-4 GTA Weldments, Determined by EDAX

Welding Process	Results of Tensile Testing		Impact Toughness
	Average UTS (MPa)	Average Yield Strength (MPa)	Average Energy (J)
GTA ERNiCrMo-3	688	380	48
GTA ERNiCrMo-4	766	403	53
PCGTAW ERNiCrMo-3	706	388	50
PCGTA ERNiCrMo-4	815	413	56

Table 4. Results of Tensile and Impact Tests of ERNiCrMo-3 and ERNiCrMo-4 Weldment of Alloy C-276 Produced Using GTAW and PCGTAW

The excellent resistance of alloy C-276 to marine corrosion has been highlighted in the Introduction. The weld joints made of C-276 showed no weight loss when subjected to salt spray testing for 120 h as per ASTM B117. This was true whether it was autogenous welding or welding with filler wire ERNiCrMo-3 or ERNICrMo-4 and whether it was GTAW or PCGTAW. Prima facie, there appears no need for adopting special welding methods or overalloyed filler wire for producing welded structures of C-276 for marine applications [16].

3. Conclusions

1. PCGTAW of alloy C-276 resulted in reduced severity of microsegregation compared to GTAW. It is believed that the occurrence of secondary constituents P and μ, deleterious in the context of hot cracking susceptibility of the alloy, also comes down to that extent.

2. PCGTAW gave rise to superior mechanical properties—higher strength and at the same time better toughness—compared to GTAW. The refined grain structure with reduced severity of microsegregation is believed to be responsible for the improved mechanical properties of the PCGTA weld joint.

3. Laser welding of alloy C-276 can be done to produce weld joints with acceptable level of microsegregation and a good combination of mechanical properties.

4. Appropriate choice of filler wire composition is important in the context of restraining the microsegregation in C-276 weld joints and realizing a good combination of strength and toughness of the weld joints.

5. Weld joints produced in alloy C-276 showed good corrosion resistance when subjected to salt spray testing as per ASTM B117.

Acknowledgements

Some of the research work presented here was supported by Defence Research and Development Organization (No. ERIP/ER/1103952/M/01/1403). We also thank the Aeronautics Research and Development Board for the funding received (No. DARO/08/2031674/M/I) and for procuring the KEMPI DWE welding machine used in the present study. The authors also acknowledge funding from Department of Science and Technology, Instron makers of the universal testing machine used to generate some of the results presented here. The authors also thank the management of VIT University for agreeing to publish this chapter.

Author details

Manikandan Manoharan[1], Arivazhagan Natarajan[2] and Nageswara Rao Muktinutalapati[2*]

*Address all correspondence to: muktinutala@gmail.com

1 Department of Mechanical Engineering, KPR Institute of Engineering and Technology, Coimbatore, India

2 School of Mechanical and Building Sciences, VIT University, Vellore, India

References

[1] Special Metals Corporation. Inconel alloy C-276 Technical Data Sheet. http://www.specialmetals.com/documents/Inconel%20alloy%20C-276.pdf (accessed 29 October 2013).

[2] Cieslak MJ, Knorovsky GA, Headley TJ, Romig AD. The use of new PHACOMP in understanding the solidification microstructure of nickel base alloy weld metal. Metallurgical Transactions A 1986;17A 2107–2116.

[3] Stephen D, Bernard S, editors, Corrosion: Materials. USA: ASM International; 2005.

[4] Haynes International. Technical Data sheet: http://www.haynesintl.com/pdf/h2010.pdf (accessed 29 October 2013).

[5] Tawancy HM. Long-term ageing characteristics of some commercial nickel-chromium-molybdenum alloys. Journal of Materials Science 1981;16 2883–2889.

[6] Akhter JI, Shaikh MA, Ahmad M, Iqbal M, Shoaib KA, Ahmad WJ. Effect of aging on the hardness and impact properties of Hastelloy C-276. Journal of Materials Science Letters 2001; 20 333–335.

[7] Raghavan M, Berkowitz BJ, Scanlon JC. Electron microscopic analysis of heterogeneous precipitates in Hastelloy C-276. Metallurgical Transactions A 1982;13A 979–984.

[8] Sims CT, Hagel WC, editors, The Superalloys. USA: John Wiley & Sons; 1972.

[9] Cieslak MJ, Headley TJ, Romig AD. The welding metallurgy of Hastelloy alloys C4, C22, C276. Metallurgical Transactions A 1986;17A 2035–2047.

[10] Perricone MJ, Dupont JN. Effect of composition on the solidification behavior of several Ni-Cr-Mo and Fe-Ni-Cr-Mo alloys. Metallurgical and Materials Transactions A 2006;37 1267–1280.

[11] Zheng YR. Acta Metallurgica Sinica 1999;35 1242.

[12] Guangyi MA, Dongjiang WU, Dongming GUO. Segregation characteristics of pulsed laser butt welding of Hastelloy C-276. Metallurgical and Materials Transactions A 2011;A42 3853–3857.

[13] Manikandan M, Hari PR, Vishnu G, Arivarasu M, Devendranath Ramkumar K, Arivazhagan N, Nageswara Rao M, Reddy GM. Investigation of microstructure and mechanical properties of super alloy C-276 by continuous Nd:YAG laser welding. Procedia Materials Science 2014;5 2233–2241.

[14] Manikandan M, Arivazhangan N, Nageswara Rao M, Reddy GM. Microstructure and mechanical properties of alloy C-276 weldments fabricated by continuous and pulsed current gas tungsten arc welding techniques. Journal of Manufacturing Processes 2014; 16 563–572.

[15] Manikandan M, Arivazhagan N, Nageswara Rao M, Reddy GM. Improvement of microstructure and mechanical behavior of gas tungsten arc weldments of alloy C-276 by current pulsing. Acta Metallurgica Sinica (English Letters) 2015;28(2) 208–215.

[16] Manikandan M. Development of Technology Based on Pulsed Current Gas Tungsten Arc Welding for Improving the Weldability of Ni-Based alloy C-276. PhD Thesis. VIT University India; 2015.

Chapter 6

Analysis of the Precipitation and Growth Processes of the Intermetallic Phases in an Fe-Ni Superalloy

Kazimierz J. Ducki

Additional information is available at the end of the chapter

http://dx.doi.org/10.5772/61159

Abstract

The chapter characterizes wrought iron-base superalloys and comprises two main parts. The first describes the chemical composition, microstructure, and precipitation reactions in Fe-Ni, Ni-Fe, and Fe-Cr superalloys. The second part presents the influence of prolonged aging on the precipitation and growth processes in an Fe-Ni superalloy of A-286 type. The prepared specimens, after solution heat treatment at 980°C/2 h/water, were aged at temperatures of 715°C, 750°C, and 780°C with the holding time of 0.5-500 h. Transmission electron microscopy (TEM) and X-ray diffraction were used to examine their structures. It was found, that application of a single-stage aging causes precipitation processes of γ' - $Ni_3(Al,Ti)$, η - Ni_3Ti, β - NiTi, G - $Ni_{16}Ti_6Si_7$, and σ - $Cr_{0.46}Mo_{0.40}Si_{0.14}$ intermetallic phases, as well as the carbide $M_{23}C_6$ and boride M_3B_2. The main phase precipitating during alloy aging was the γ'-type intermetallic phase. It was found that the mean diameter of γ' phase precipitates increases as a function of the cube root of aging time, which is consistent with the predictions based on the Lifshitz-Slyozow-Wagner (LSW) theory. The determined value of activation energy for the process of γ' phase coagulation in the examined alloy was E = 297 kJ/mole.

Keywords: A-286 alloy, Precipitation and growth, Intermetallic phases, LSW theory

1. Introduction

High-temperature Fe-Ni superalloys after precipitation strengthening with intermetallic phases have a number of characteristic properties, including excellent mechanical properties, good creep and heat resistance at elevated and high temperatures. They also show very good corrosion resistance, as well as high ductility at low temperatures, and are non-magnetic. The temperature at which it is possible to apply these alloys ranges from the liquid helium temperature (-269°C) to 540–815°C. At such temperatures, the superalloys can be used in conventional and nuclear power generation, aviation technology, chemical and petrochemical industries, electromachinery industry, cryogenics, and the manufacture of tools for the processing of non-ferrous metals and alloys.

Fe-Ni alloys are widely applied for operation at elevated and high temperatures, where they can be used as a structural material intermediate between types of martensitic steel used for operation at temperatures up to 600°C, and types of creep-resistant nickel superalloys intended for operation at temperatures above 700°C. Creep-resistant Fe-Ni alloys precipitation strengthened, compared to Ni- and Co-based alloys, have lower resistance to oxidation and gas corrosion, but have a high creep resistance at an intermediate range of temperatures, i.e. 540÷700°C. It may be expected that along with the continuous intensification of technological processes, increases in the operating parameters of machines in the energy generation industry, and the development of new technologies in the chemical, petrochemical, and processing industries, this group of materials will gain a more important role. The analysis of the current state of research on Fe-Ni alloys precipitation strengthened demonstrated that they are an interesting scientific problem, but also have utilitarian aspects, as their cost of manufacture is considerably lower in comparison to nickel and cobalt superalloys.

Alloys strengthened by precipitated ordered intermetallic phase γ' - $Ni_3(Al,Ti)$ with a regular structure fcc are the main group of creep-resistant Fe-Ni superalloys. The γ' phase in Fe-Ni alloys precipitates in the form of spheroid, highly dispersed particles coherent with the γ solid solution. In many reports [1–5], the γ' phase was considered to be the major phase strengthening iron-based alloys. In high-temperature applications, particles of the γ' phase strengthen Fe-Ni alloys up to 750°C. At higher temperatures, the γ' phase transforms into the η - Ni_3Ti phase, with a hexagonal structure and lamellar morphology. In Fe-Ni alloys, the strengthening effect caused by the precipitation of the η phase is less pronounced in comparison to strengthening by the γ' phase. Reduction in the nickel content in the matrix caused by the precipitation of γ' and η phases, at the simultaneous presence of molybdenum or wolfram as additives strengthening the solid solution, creates conditions for the precipitation of particles from TCP phases - σ, Laves, G, χ, and μ, decreasing the plasticity of alloy. At the same time, the precipitation of carbides (MC, $M_{23}C_6$, and M_6C) and borides or carboborides of different morphology can occur. The structural stability of Fe-Ni alloys improves as a result of the partial replacement of iron by nickel. This process allows for the increase in the concentration of components strengthening the solid solution γ and improves creep resistance, without the formation of undesirable TCP-phases.

Knowledge of the kinetics of the precipitation process of secondary phases in Fe-Ni alloys during heat treatment and/or service is necessary to forecast their functional properties and microstructural stability in operating conditions. Previous studies [3–6] and the results of the author's own research [7, 8] on Fe-Ni alloys were mainly focused on the analysis of the precipitation of intermetallic phases (γ', η, and G) and carbides ($M_{23}C_6$, M_6C, and MC) during heat treatment or service, and their effects on strength and plastic properties. However, there has been no comprehensive quantitative characterization prepared so far regarding changes in the diameter of γ' phase particles at elevated and high temperatures.

In the present study, research has been undertaken on the influence of prolonged aging on the course of precipitation and coagulation of γ' - $Ni_3(Al,Ti)$ phase in a high-temperature, creep-resisting Fe-Ni superalloy. The basic stereological parameters of the γ' phase particles have been determined and the growth step of the γ' phase has been analyzed on the basis of the LSW theory. It is assumed that the obtained test results will be used for forecasting the structural stability of the superalloy under the conditions of heat treatment and operation.

2. General Characteristic of Fe-Ni Superalloys

2.1. Basic groups of wrought Fe-Ni superalloys

Developmental work on creep-resistant wrought Fe-Ni alloys has been carried out in parallel with work on nickel superalloys. There are significant differences in physical, chemical, and mechanical properties between iron and nickel superalloys, caused mainly by differences in their chemical compositions. Nickel superalloys were created and developed through the chemical modification of nickel-chromium alloy (NiCr20), while Fe-Ni alloys were developed by modifying the chemical and phase composition of austenitic steel 18–8. Currently, many researchers [1, 2, 9, 10] claim that most creep-resistant Fe-Ni superalloys contain min. 36% Fe, max. 45% Ni and min. 12% Cr. However, the above contents of elements are not strictly followed. Recent research has led to the development of more than 20 types of iron alloys with complicated chemical compositions. Their original classification, based on the main mechanism of strengthening, chemical, and phase composition and development of Fe-Ni alloys is proposed in Table 1.

Analysis of the current state of research demonstrated that the group of creep-resistant wrought Fe-Ni superalloys includes a large number of materials diversified in terms of chemical composition, strengthening mechanism, obtained properties, and applications. The primary group of Fe-Ni superalloys includes materials strengthened by the precipitation of ordered intermetallic phases γ' - $Ni_3(Al,Ti)$ and/or γ'' - Ni_3Nb intended for work at temperatures from 540°C to 815°C. At temperature above 815°C, the precipitates of intermetallic phases γ' and γ'' become unstable and age into the equilibrium phases η - Ni_3Ti and δ - Ni_3Nb, which is associated with a significant decrease in their creep resistance. For this reason, at the upper range of working temperatures (900°C to 1100°C) Fe-Ni solid-solution strengthened alloys or ODS superalloys are mainly used. Both types of alloys can be hardened to achieve the parameters of precipitation-strengthened alloys. Values of the 1000-h stress-rupture for selected wrought Fe-Ni alloys are presented in Figure 1. The creep resistance of Fe-Ni superalloys mainly depends on the concentration of nickel and iron, and the content of elements that produce solid-solution or precipitation hardening of the alloy.

No.	Alloy	Nominal composition [wt. %]													
		C	Mn	Si	Cr	Ni	Fe	Co	Mo	W	Nb	Ti	Al	B	Other
		Group I. Fe-Ni alloys of solid-solution strengthened													
1.	19-9DL	0.30	1.1	0.6	19.0	9.0	66.8	–	1.25	1.25	0.4	0.3	–	–	–
2.	17-14CuMo	0.12	0.75	0.5	16.0	14.0	62.4	–	2.5	–	0.4	0.3	–	–	3.0Cu
3.	16-25-6	0.06	1.35	0.7	16.0	25.0	50.7	–	6.0	–	–	–	–	–	0.15N
4.	Incoloy 800	0.05	0.8	0.5	21.0	32.5	45.7	–	–	–	–	0.4	0.4	–	0.4Cu
5.	Incoloy 801	0.05	0.8	0.5	20.5	32.0	46.3	–	–	–	–	1.1	–	–	0.2Cu
6.	Incoloy 802	0.35	0.8	0.4	21.0	32.5	44.8	–	–	–	–	0.8	0.6	–	0.4Cu
7.	Carpenter 20Cb-3	0.07	0.8	0.4	20.0	34.0	42.4	–	2.5	–	1.0	–	–	–	3.5Cu
		Group II. Fe-Ni alloys of precipitation strengthened by γ' phase													
8.	W-545	0.08	1.5	0.4	13.5	26.0	55.8	–	1.5	–	–	2.85	0.2	0.080	–
9.	A-286	0.05	1.4	0.4	15.0	26.0	55.2	–	1.25	–	–	2.0	0.2	0.003	0.3V
10.	Discaloy	0.04	0.9	0.8	13.5	26.0	55.0	–	2.75	–	–	1.75	0.25	–	–
11.	Tinidur	0.04	1.0	0.75	14.5	26.0	54.0	–	1.25	–	–	2.15	0.2	0.003	0.03V
12.	V-57	0.08	0.35	0.75	14.8	27.0	48.6	–	1.25	–	–	3.0	0.25	0.010	0.5V
		Group III. Fe-Ni alloys with low thermal expansion													
13.	Incoloy 903	0.04			0.1	38.0	41.0	15.0	0.1	–	3.0	1.4	0.7	–	–
14.	Incoloy 907	0.01	0.3	0.15	0.1	38.0	42.0	13.0	–	–	4.7	1.5	0.03	–	–
15.	Incoloy 909	0.01		0.4	0.1	38.0	42.0	13.0	–	–	4.7	1.5	–	0.001	–
16.	Pyromet CTX-1	0.03			0.1	37.0	39.0	16.0	0.1	–	3.0	1.7	1.0	–	–
17.	Pyromet CTX-3	0.05	–	0.15	0.2	38.0	41.0	13.5	–	–	4.9	1.6	0.1	0.007	–
		Group IV. Fe-Ni and Ni-Fe alloys of precipitation strengthened by γ' and/or γ'' phases													
18.	CG27	0.05	0.1	0.1	13.0	38.0	38.4	–	5.7	–	0.7	2.5	1.6	0.010	–
19.	Incoloy 901	0.05	0.1	0.1	12.5	42.5	36.0	–	5.7	–	–	2.8	0.2	0.015	–
20.	Pyromet 860	0.05	0.05	0.05	12.6	43.0	30.0	4.0	6.0	–	–	3.0	1.25	0.010	–
21.	D-979	0.05	0.3	0.2	15.0	45.0	27.0	–	4.0	–	–	3.0	1.0	0.010	–
22.	Incoloy 706	0.03	0.2	0.2	16.0	41.5	40.0	–	–	–	2.9	1.8	0.2	–	–
23.	Inconel 718	0.04	0.2	0.2	19.0	52.5	18.5	–	3.0	–	5.1	0.9	0.5	–	–
		Group V. Fe-Ni and Ni-Fe alloys of precipitation strengthened by carbides													
24.	N-155	0.10	1.5	0.5	21.0	20.0	32.2	20.0	3.0	2.5	1.0	–	–	–	0.15N; ≤0.5Cu
25.	Haynes 556	0.10	1.5	0.4	22.0	20.0	29.0	20.0	3.0	2.5	0.1	–	0.3	–	0.5Ta; 0.02La; 0.002Zr

http://dx.doi.org/10.5772/61159

No.	Alloy	Nominal composition [wt. %]													
		C	Mn	Si	Cr	Ni	Fe	Co	Mo	W	Nb	Ti	Al	B	Other
		Group I. Fe-Ni alloys of solid-solution strengthened													
26.	Hasteloy X	0.10	0.5	0.5	22.0	47.0	18.5	1.5	9.0	0.6	–	–	–	–	–
		Group VI. Fe-Cr alloys of oxide dispersion strengthened (ODS)													
27.	Incoloy MA956	0.05			20.0		74.4					0.5	4.5		$0.5Y_2O_3$
28.	Incoloy MA957	0.05			14.0		84.4	0.3				1.0			$0.25Y_2O_3$
29.	PM2000 Alloy	0.05			20.0		73.4					0.5	5.5		$0.5Y_2O_3$

Table 1. Nominal chemical composition of selected wrought Fe-Ni, Ni-Fe, and Fe-Cr superalloys [11, 12].

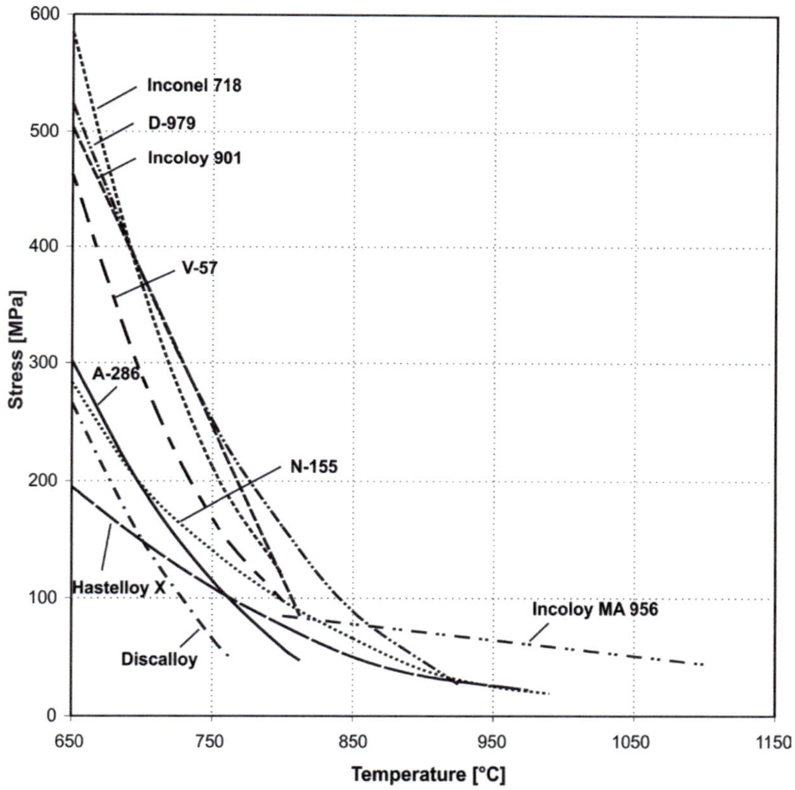

Figure 1. h stress-rupture curves of wrought superalloys of Fe-Ni, Ni-Fe and Fe-Cr type [2].

The increased content of nickel in Fe-Ni alloys is usually associated with good creep resistance, higher working temperatures, better thermal stability of microstructure, and higher price.

On the other hand, high iron content, despite reduced cost and improved machinability of the material, increases the melting point and decreases the resistance of the alloy to oxidation.

Creep-resistant Fe-Ni alloys, compared to nickel- and cobalt-based superalloys, have lower resistance to oxidation and gas corrosion, but have a high creep resistance at an intermediate range of temperatures, i.e. 540÷750°C.

Currently, novel types of ferritic ODS superalloys are strong competition for nickel and cobalt superalloys. They are popular due to their significantly lower price and good creep resistance (up to 1200°C). Fe-Ni superalloys, because of the similar phase composition, are frequently classified to the same group of materials as nickel superalloys [1, 2, 9]. Therefore, their further development is closely connected with the development of wrought creep-resistant nickel superalloys.

2.2. Elements interaction in Fe-Ni superalloys

High-temperature Fe-Ni superalloys are characterized by a complex chemical composition. There are as many as 15 significant elements, in the case of which control of their content enables solution strengthening and, at the same time, precipitation strengthening of the matrix and grain boundaries. Alloying elements can be divided into five principal groups, depending on their effect in the shaping of the microstructure and properties (Figure 2):

- matrix-strengthening elements: Fe, Ni, Co, Cr, Mo, W, and V;
- elements forming γ'- type intermetallic phases: $Ni_3(Al,Ti)$ and γ'' - Ni_3Nb – Al, Ti, Nb, and Ta;
- elements that enhance oxidation resistance: Cr, Al, and La;
- elements that effect grain boundaries: B, C, Zr, and Hf;
- harmful impurities: Si and Mn.

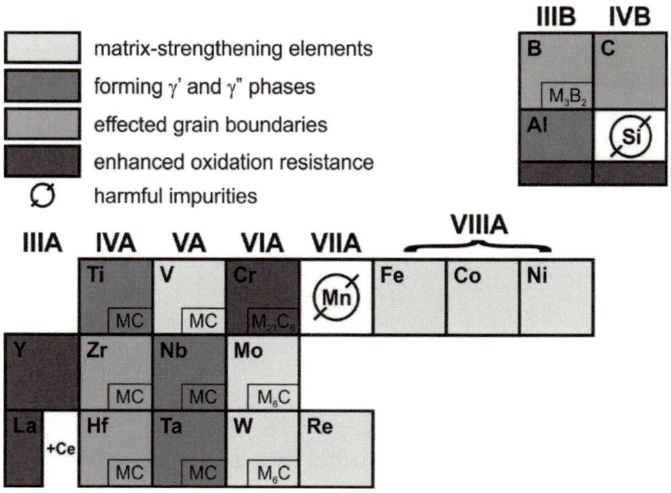

Figure 2. Interaction of alloying elements in iron- and nickel-base superalloys [1, 13].

Some elements, numbered among carbide-forming ones, form carbides and carbonitrides in the alloy, as well as borides and carboborides. The first group of alloying components, introduced mostly to strengthening the solid solution γ, includes most often [1, 2, 9, 10, 13]: Cr (12÷22%), Mo (up to 6%), W (up to 3%), and Co (up to 20%). The second group of elements introduced in order to form γ'- type intermetallic phases such as $Ni_3(Al,Ti)$, and/or γ''- type phases such as Ni_3Nb, and to induce precipitation strengthening includes Ti (up to 3%), Al (up to 1.2%), Nb (up to 5%), and sometimes Ta (up to 0.5%). The third group of alloy additions, whose purpose is to enhance oxidation resistance at high temperatures are: Cr (12÷22%) and aluminium (up to 5.5%). The fourth group, i.e., the elements added in order to improve the ductility of grain boundaries and creep resistance, includes micro-additions of B (up to 0.015%), Zr (up to 0.002%), and V (up to 0.5%) [1, 2, 13, 14]. Micro-additions of boron and zirconium segregate to the grain boundaries, reduce the energy of their mismatch, and influence the morphology of carbides and intermetallic phases. A vanadium addition improves resistance to notch effect at elevated temperatures and hot deformability of alloys. Deformability of alloys can be also increased by small contents, from 0.01% to 0.05%, of magnesium or rare-earth metals (REM). These elements also counteract brittleness caused by sulfur [1, 2, 6, 13, 21]. New generation iron superalloys contain elements from the lanthanide and actinide groups (e.g., La, Th, and Y), which improve alloys' resistance to high-temperature oxidation [2, 10, 15]. In the group of wrought superalloys, ODS alloys have been gaining importance over the latest period, as well as alloys produced by powder metallurgy methods (P/M) [2, 13, 15–19]. A detailed breakdown of the interaction of the main alloying elements in iron and nickel superalloys is presented in Table 2.

Effect	Iron-based superalloys	Nickel-based superalloys
Solid-solution strengtheners	Cr, Mo, W, Co	Co, Cr, Fe, Mo, W, Ta
Fcc matrix stabilizers	C, W, Ni	---
Carbide form:		
MC type	Ti	W, Ta, Ti, Mo, Nb
M_7C_3 type	---	Cr
M23C6 type	Cr	Cr, Mo, W
M6C type	Mo	Mo, W
Carbonitrides		
– M(C,N)	C, N	C, N
Forms γ' - $Ni_3(Al,Ti)$	Al, Ni, Ti	Al, Ti
Retards formation of hexagonal η - Ni_3Ti	Al, Zr	---
Raises solvus temperature of γ'	---	Co
Hardening precipitates and/or intermetallics	Al, Ti, Nb	Al, Ti, Nb, Ta
Forms γ'' - Ni_3Nb	Nb	Nb

Effect	Iron-based superalloys	Nickel-based superalloys
Oxidation resistance	Cr, Al	Cr, Al, Ta
Improves hot corrosion resistance	La, Y	La, Th
Sulfidation resistance	Cr	Cr
Increases rupture ductility	B[1)], Zr	B[1)], Zr
Causes grain-boundary segregation	---	Mg, B, C, Zr
Facilitates working	V	---

[1)] If present in large amount, borides are formed

Table 2. Role of elements in iron- and nickel-based superalloys [2, 13].

Silicon, phosphorus, sulfur, oxygen, and nitrogen are harmful in iron and nickel superalloys. First of all, they reduce the alloys' ductility during creep [1, 2, 13, 20, 21]. The content of these elements is limited and controlled by using relevant melting technologies. Also other trace elements, such as selenium, bismuth, lead, zinc, and arsenic should be kept at a low level (in the order of ppm). The content of harmful elements in iron and nickel superalloys is reduced by applying special vacuum metallurgy processes [13, 14, 22, 23]. Vacuum melting and casting prevent oxidation of the alloy, and allows better control over the principal, reactive elements. These processes also prevent an increase of the nitrogen content in the alloy during melting. As a result, the content of Al, Ti, and Nb is considerably increased through vacuum melting. In consequence, the relative volume of phase γ' and/or γ'' increases as well. However, the range of plastic working temperature of superalloys with a high content of Al, Ti, and Nb is narrow and close to the solidus line.

2.3. The basic phases in Fe-Ni superalloys and precipitation reactions

Creep-resisting Fe-Ni superalloys are basically Fe-Ni-Cr alloys with a number of added elements that form substitutional and interstitial solid solutions, and are introduced in order to ensure solution and precipitation strengthening of the matrix. Equilibrium phases in Fe-Ni alloys depend mostly on the content of three basic elements, i.e., Fe, Cr, and Ni (Figure 3). The analysis of the phase composition at a temperature of 1,050°C and 650°C shows that the Fe-Ni alloys with a content of ca. 13÷22% Cr, ca. 15÷42% Ni, and ca. 36÷60% Fe, have a structure of a disordered solid solution γ with a regular fcc structure.

In γ solid solutions, in the case of a majority of creep-resisting Fe-Ni superalloys, efforts are made to ensure an appropriate proportion of iron and nickel contents. The value of this proportion determines the thermal stability of the alloy microstructure, its operation temperature, and cost. An increased nickel content in Fe-Ni alloys usually allows increasing the concentration of elements strengthening the solid solution without the formation of undesirable TCP or other phases [1–3, 5]. Therefore, alloys with a high nickel content are characterized by good structural stability and creep resistance, which enables their use at high operating

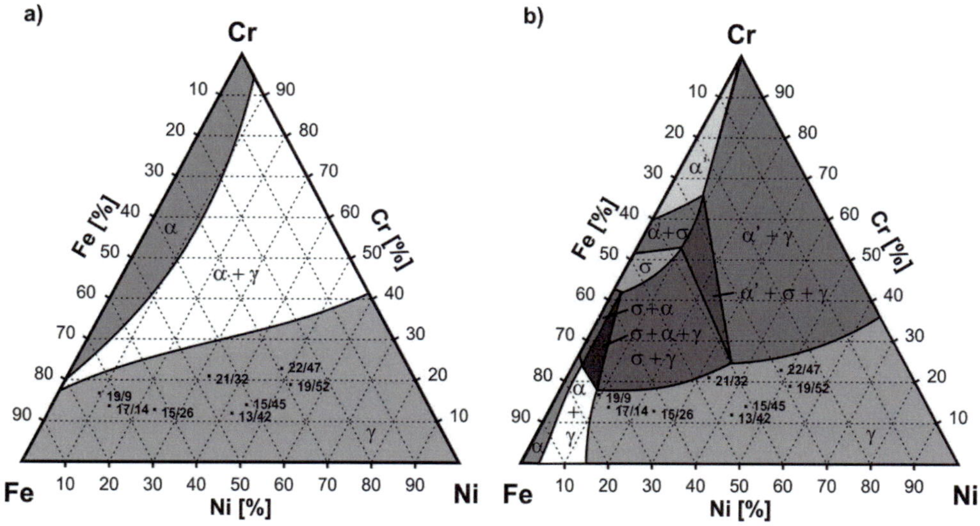

Figure 3. Cross-sections of isothermal phase diagram of Fe-Cr-Ni at a temperature [24]: a) 1,050°C; b) 650°C.

temperatures. The minimum nickel content that is necessary to form a stable γ solid solution in a majority of Fe-Ni alloys amounts to ca. 25% [1, 5, 14]. A cobalt additive or other additives stabilizing the γ phase allow reducing the required nickel level. At the same time, an increased iron content affects adversely on heat resistance. Iron oxides feature poor adhesion to the substrate. An increased iron content also enhances the susceptibility of Fe-Ni alloys to precipitation of the disadvantageous TCP phases, including in particular the σ phase.

The introduction of alloying elements to Fe-Ni alloys results in the formation in the γ solid solution of numerous precipitates of intermetallic phases, carbides, carbonitrides, borides, and carboborides, which do not always have a good effect on the properties of the alloys. Knowledge of the kinetics of the precipitation process of secondary phases in Fe-Ni alloys during heat treatment and/or service is necessary to forecast the changes in the characteristics of their mechanical, physical, and chemical properties. The issue of precipitation processes in creep-resisting Fe-Ni superalloys is addressed in a number of papers. The most important ones are those by Stickler [3], Dulis [4], Pickering [5], Sourmail [6], and the author's own studies [7,8,11]. Based on the research results obtained, it was found that a number of stable and metastable phase components, similar to those observed in nickel superalloys, are present in the microstructure of Fe-Ni alloys (Table 3). It should be pointed out that due to considerable differences in the chemical compositions of superalloys and their manufacturing technologies, the phases of the same type present in them may show significant differences in the chemical compositions and morphologies.

After heat treatment Fe-Ni superalloys consist of a matrix of a strengthened solid solution, γ, with dispersion particles of GCP intermetallic phases (mostly γ' and/or γ''), carbide precipitates (MC, M_6C, $M_{23}C_6$, M_7C_3) and, possibly, precipitates of nitrides, carbonitrides, borides, and carboborides [2,3,4,5,6,13].

Phases	Iron-based superalloys	Nickel-based superalloys
Matrix	γ – gamma, nonordered fcc, (Fe, Ni, Cr, ...).	γ – gamma, non ordered fcc, (Ni, Cr, Co, ...).
Geometrically close-packed (GCP)	γ' – gamma prime, ordered fcc, AB_3 type, $(Ni, ...)_3(Al, Ti)$; γ''– gamma double prime, ordered bct, AB_3 type, Ni_3Nb; η – eta, hcp, AB_3 type, Ni_3Ti; δ – delta, orthorombic, AB_3 type, Ni_3Nb; β – beta, bcc, AB type, NiAl.	γ' – gamma prime, ordered fcc, AB_3 type, $(Ni, Co, Fe, Cr, ...)_3(Al, Ti)$; γ''– gamma double prime, ordered bct, AB_3 type, Ni_3Nb; η – eta,hcp, AB_3 type, Ni_3Ti; δ – delta, orthorhombic, AB_3 type, Ni_3Nb.
Topologically close-packed (TCP)	σ – sigma, bct, AB type, Fe-Ni-Cr-Mo; Laves – hcp, A_2B type, Fe_2Mo, Fe_2W, $Fe_2(Ti, Nb)$; G – fcc, $A_{16}D_6C_7$ type, $Ni_{16}Ti_6Si_7$, $Ni_{16}Nb_6Si_7$; μ – mu, rhombohedral, A_6B_7 type, $(Cr, W)_6(Fe, Co)_7$; χ – chi, bcc, (Fe, Cr, Mo) or $M_{18}C$ type, $Fe_{36}Cr_{12}Mo_{10}$.	σ – sigma, bct, A_xB_y type, $(Fe, Mo)_x(Ni, Co)_y$; Laves – hcp, A_2B type, $(Fe, Cr, Mn, Si)_2(Mo, Ti, Nb)$; G – fcc, A_6B_{23} type, $Hf_6Ni_8Al_{15}$; μ – mu, rhombohedral, A_6B_7 type, $(Mo, W)_6(Ni, Co)_7$.
Carbides	MC – fcc, (Ti, Nb, V, Zr, Ta)C; M_7C_3 – trigonal, Cr_7C_3; $M_{23}C_6$ – fcc, $(Cr, Ni, Mo, Fe)_{23}C_6$; M_6C – diamond type fcc, $(Fe, Cr)_{21}Mo_3C$, Fe_3Nb_3C.	MC – fcc, (Ti, Mo, Nb, Ta, W)C; M_7C_3 – trigonal, Cr_7C_3; $M_{23}C_6$ – fcc, $(Cr, Mo, Co, W, Nb)_{23}C_6$; M_6C – diamond type fcc, $(Ni, Co)_3Mo_3C$, $(Ni, Co)_2W_4C$.
Borides and carboborides	$M_{23}(C,B)_6$ – fcc,	M_3B_2 – tetragonal, $(Mo, Ti, Cr, Ni, Co)_3B_2$; $M_{23}(C,B)_6$ – fcc.
Nitrides and carbonitrides	MX – fcc, Ti(C, N), Nb(C, N); MN_2 – hcp, Cr_2N.	MX – fcc, Ti(C, N); $M_{23}(C, N)_6$ – fcc.

http://dx.doi.org/10.5772/61159

Phases	Iron-based superalloys	Nickel-based superalloys
Others	M(C, N, P);	$Ni_x(Mo, Cr)_y(C, Si)$;
	$(M, P)_{23}C_6$;	$Ti_4C_2S_2$;
	Cr_3Ni_2Si;	ZrS_x.
	α' – alpha prim, bcc,	
	(Cr,Fe);	
	$Ti_4C_2S_2$.	

Table 3. Phases identified in iron- and nickel-base superalloys [2, 3, 6, 13].

Such a microstructure is stable in a narrow range of temperature and time parameters, and may be unsuitable for the operating conditions. The effect of such instability is the changes in the chemical composition, morphology, arrangement, and properties of the phases present in complex temperature, stress, and environment-related conditions. The undesirable phases may develop during heat treatment or thermo-mechanical exposure, depending on the reactions between the different phases that are affected by the segregation of alloying and trace elements, as well as the interaction with crystallographic defects. The sequence of the precipitation reactions taking place in iron and nickel superalloys is the effect of changes in the chemical composition of the matrix that occur during heat treatment, and the thermo-mechanical interactions that occur in service. Table 4 summarizes the most important precipitation reactions proceeding in iron- and nickel-based superalloys during heat treatment and during operation at elevated temperatures.

No.	Iron-based superalloys	No.	Nickel-based superalloys
A.	During initial heat treatment:	A.	During initial heat treatment:
1.	$\gamma \rightarrow$ GCP (γ', γ'', η, ...) + γ^*	1.	$\gamma \rightarrow \gamma' + \gamma^*$
2.	$\gamma \rightarrow M_xC$ (MC, $M_{23}C_6$, ...) + γ^*	2.	$\gamma \rightarrow M_xC$ (MC, $M_{23}C_6$, M_6C, ...) + γ^*
B.	During extended exposure to elevated temperatures:	B.	During extended exposure to elevated temperatures:
		1.	$\gamma^* \rightarrow$ TCP (σ, Lavesa, ...)
1.	$\gamma^* \rightarrow$ TCP (G, σ, χ, Lavesa, ...)	2.	γ^* + MC $\rightarrow \gamma' + \gamma^{**}$ ("/> 900°C)
2.	$\gamma' \rightarrow \eta$	3.	γ^* + MC $\rightarrow \gamma' + M_{23}C_6$
3.	$\gamma'' \rightarrow \delta$	4.	γ^* + MC $\rightarrow \gamma' + M_6C$
4.	$\gamma' \rightarrow \beta$	5.	$\gamma^* + M_6C \rightarrow M_{23}C_6 + \gamma^{**}$
5.	$M_{23}C_6 \rightarrow$ TCP + γ^{**} (C in solution)	6.	$\gamma' \rightarrow \eta$
		7.	MC + M $\rightarrow M_7C_3$ + M $\rightarrow M_{23}C_6$ + TCP(σ)

Where: γ^* – gamma phase depleted in certain alloying elements;
γ^{**} – gamma phase of slightly different composition

Table 4. The basic phase reactions proceeding in iron- and nickel-based superalloys [3, 5, 13].

When comparing creep-resisting iron and nickel superalloys, it can be noted that with an increasing iron content in Fe-Ni alloys, the following disadvantageous interactions appear [1, 2, 3, 13, 25]:

- the dissolution temperature of the main strengthening phases, γ' and γ'', decreases;
- susceptibility to overaging and precipitation of the η and δ phases increases;
- susceptibility to the formation of TCP – σ, Laves, G, χ, and μ phases also increases;
- the mismatch parameters of the γ/γ' and γ/γ'' lattice change because iron, compared to nickel, has a greater (by ca. 3%) atomic diameter.

It can be concluded that the final effect of these adverse interactions is the much lower operating temperature (815°C) of Fe-Ni alloys in creep-resisting applications compared to the maximum operating temperature (1,090°C) of wrought nickel superalloys.

3. Material and methodology

The examinations were performed on rolled bars, 16 mm in diameter, of an Fe-Ni superalloy of the A-286 type. The chemical composition of the material is given in Table 5.

Content of an element [wt. %]															
C	Si	Mn	P	S	Cr	Ni	Mo	V	W	Ti	Al	Co	B	N	Fe
0.05	0.56	1.25	0.026	0.016	14.3	24.5	1.35	0.42	0.10	1.88	0.16	0.08	0.007	0.0062	55.3

Table 5. Chemical composition of the investigated Fe-Ni superalloy.

Test specimens cut from the bars underwent solution heat treatment at 980°C for 2 h followed by water quenching. Next, they were aged at 715°C, 750°C and 780°C, with the holding time ranging between 0.5 h and 500 h (Figure 4). The aging temperature (in the range of 715–780°C) and time (up to 500 h) were selected based on the possibility of simulating the operation of the alloy for 100,000 h at temperatures of 630–690°C, according to the Larson-Miller parameter [26].

Examination of the structure was performed with a Jeol JEM-200 FX transmission electron microscope by means of the thin foil technique. A Philips PW-140 X-ray diffractometer was used for the phase analysis of the isolates that formed during anodic dissolution of the specimens in a 10% hydrochloric acid solution in ethanol, where the density of current amounted to 10–12 mA/cm^2.

Quantitative analysis of the secondary phase particles, γ' - $Ni_3(Al,Ti)$, was performed on thin foil images (TEM) in the bright field. Images of the microstructure (TEM) were transformed into binary images using the computer program Met-Ilo [27]. On their basis, the basic stereological parameters of the γ' phase were determined: the mean diameter of particles on the

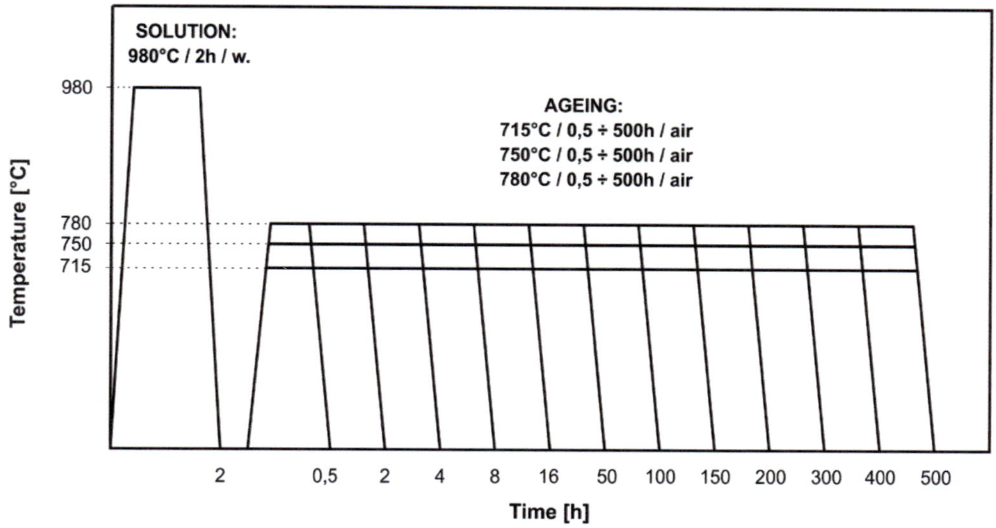

Figure 4. Scheme of heat treatment of the investigated Fe-Ni superalloy.

circles in the projected plane $\bar{d}$, the mean diameter of particles $\bar{D}$, the volume fraction of particles V_V, and the mean distance between particles, l_d.

The mean diameter of particles $\bar{d}$ on the circles in the projected plane of phase γ' was determined from the equation [28]:

$$\bar{d} = \sqrt{\frac{4A'}{\pi}} \tag{1}$$

where: A′ – particles surface on the circles (on image TEM) [nm^2].

The mean diameter $\bar{D}$ of the particles was evaluated with help of the relation given by Czyrska-Filemonowicz et al. [29]:

$$\bar{D} = \frac{\bar{d}t}{t - \bar{d} + \left(\pi A_A / L_A\right)} \tag{2}$$

.

where: t – the foil thickness [nm], A_A – the projection area, L_A – the perimeter density [1/nm].

The volume fraction V_V of γ' particles was calculated using the formula provided by Dubiel et al. [30]:

$$V_V = \frac{\pi}{6} \frac{\sum N_i d_i^3}{A\left(t + \bar{d}\right)} \tag{3}$$

where: N_i – number of particles with diameter d_i, A – total projection area [nm^2].

The mean distance between particles l_d was calculated by means of an equation given by Schröder and Arzt [31]:

$$l_d = \bar{d}\sqrt{\frac{\pi}{6V_V}\left(1+\frac{s^2}{\bar{d}^2}\right)} - \frac{\pi}{4}\bar{d} \tag{4}$$

where: s – standard deviation of the particle diameter distribution.

4. Results and discussion

4.1. Precipitation processes in the Fe-Ni superalloy during prolonged aging

After a 2-hour solution heat treatment at the temperature of 980°C and water quenching, the structure of the Fe-Ni superalloy resembles twinned austenite with a small amount of undissolved precipitates (ca. 0.3 wt. %) and an elevated dislocation density (Figures 5 and 6). Presence of titanium compounds such as carbide TiC, carbonitride $TiC_{0.3}N_{0.7}$, nitride $TiN_{0.3}$, carbosulfide $Ti_4C_2S_2$ and Laves phase Ni_2Si and boride MoB were found in the extracted precipitates (Figure 7).

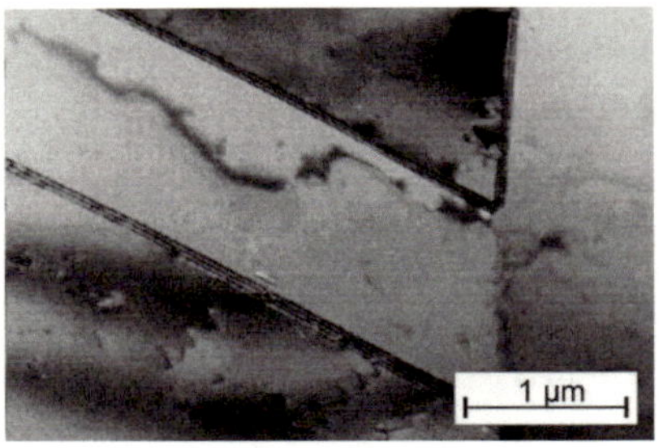

Figure 5. Alloy microstructure after solution heat treatment at 980°C/2 h/w. Twinned austenite with dislocations and undissolved carbosulfide $Ti_4C_2S_2$.

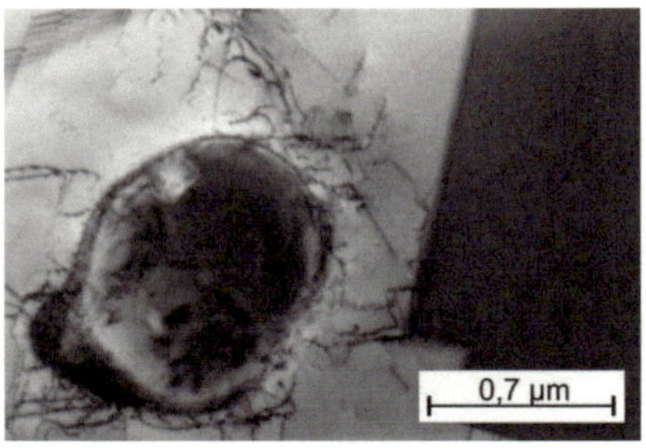

Figure 6. Alloy microstructure after solution heat treatment at 980°C/2 h/w. Twinned austenite with an increased dislocation density and undissolved carbide TiC.

http://dx.doi.org/10.5772/61159

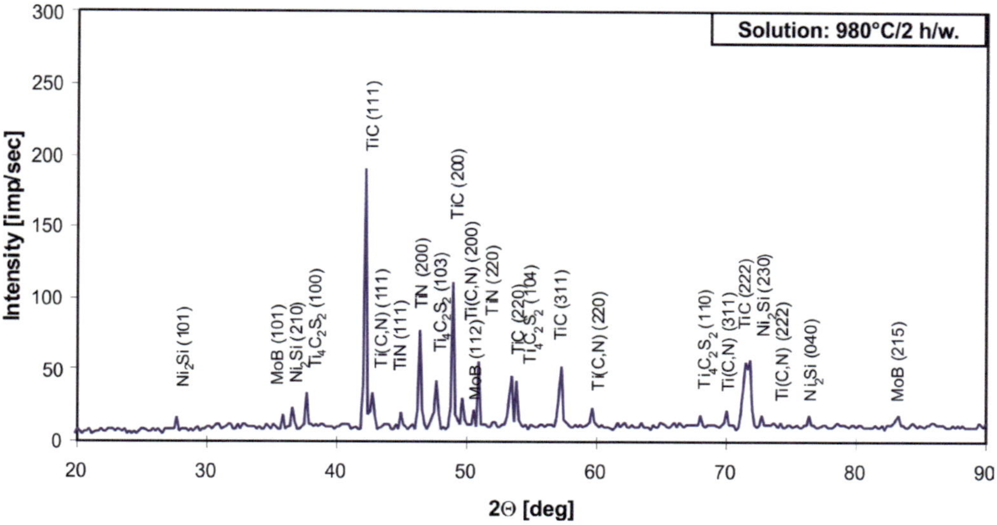

Figure 7. Isolate diffractogram of the alloy after solution heat treatment at 980°C/2 h/w.

One-stage aging applied after solution heat treatment at the above-specified temperatures of 715°C, 750°C, and 780ºC in time between 0.5 h and 500 h causes initiation of precipitation and coagulation, as well as overaging of carbides and intermetallic phases. After short-period (4-8 h) aging at 715°C, effects of the initial precipitation phases can be observed in the matrix, including clusters of the γ' -Ni_3 (Al, Ti) phase and coherent zones (Figure 8). As Pickering [5] and Wilson [32] claim, this phenomenon is connected with spinodal decomposition of supersaturated austenite. In this way, areas rich in Ti and Al were formed homogenously in the matrix. Fine lenticular lamellae of phase *G* ($Ni_{16}Ti_6Si_7$) in the form of chains, characteristic of discontinuous precipitation, appeared on grain boundaries (Figure 9). In the case of extending the aging time (within a range of 16 h – 500 h), the kinetics increased the homogeneous precipitation of the γ' phase particles in the matrix (Figure 10). In addition, presence of lenticular carbide particles $M_{23}C_6$ was detected in the region near the austenite boundary (Figure 11). The presence of carbide TiC, carbonitride $TiC_{0.3}N_{0.7}$, nitride $TiN_{0.3}$, carbosulfide $Ti_4C_2S_2$ and MoB and M_3B_2 borides were also detected in the alloy microstructure (Figure 12, Table 6).

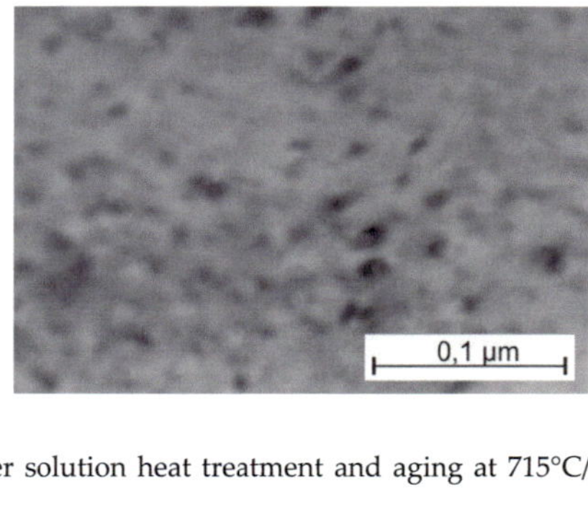

Figure 8. Alloy microstructure after solution heat treatment and aging at 715°C/4 h. Coherent zones and clusters of phase γ' in the austenite region.

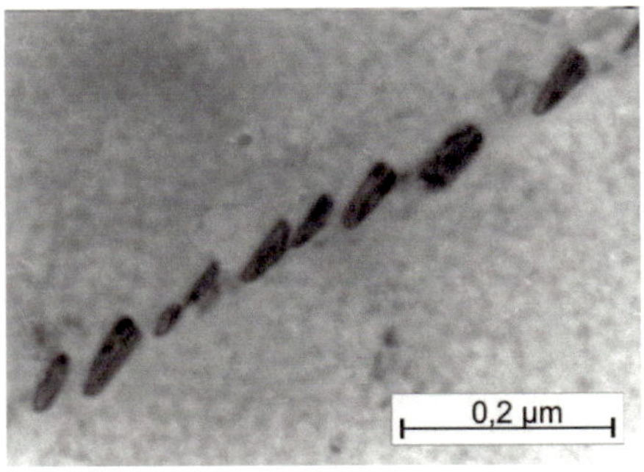

Figure 9. Alloy microstructure after solution heat treatment and aging at 715°C/8 h. Coherent zones of phase γ' in the matrix and lamellae of phase *G* on the grain boundary.

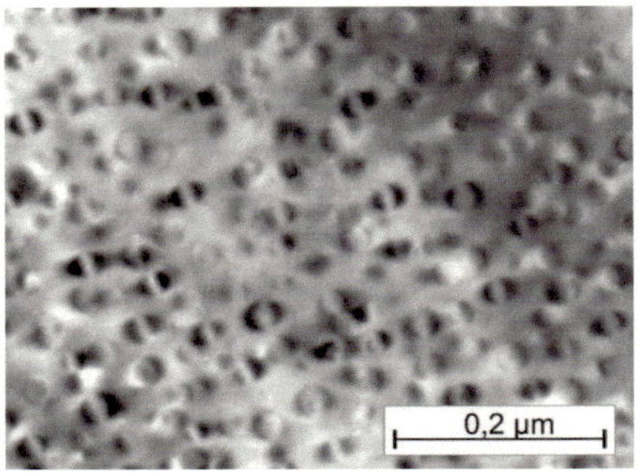

Figure 10. Alloy microstructure after solution heat treatment and aging at 715°C/150 h. Coherent spherical particles of phase γ' in the austenite.

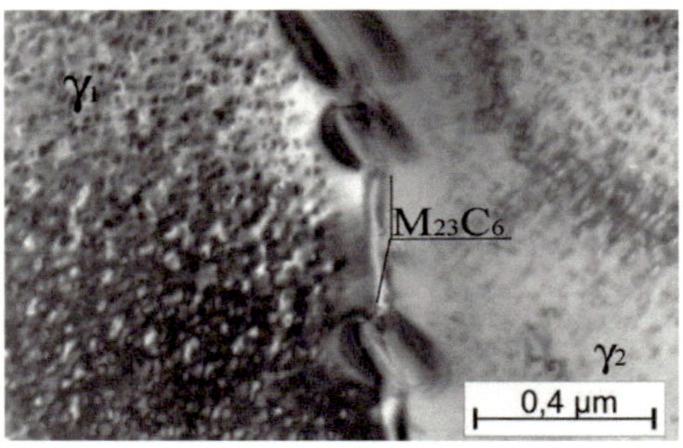

Figure 11. Alloy microstructure after solution heat treatment and aging at 715°C/300 h. Dispersive precipitates of phase γ' in the matrix and lenticular particles of $M_{23}C_6$ carbide on the grain boundary.

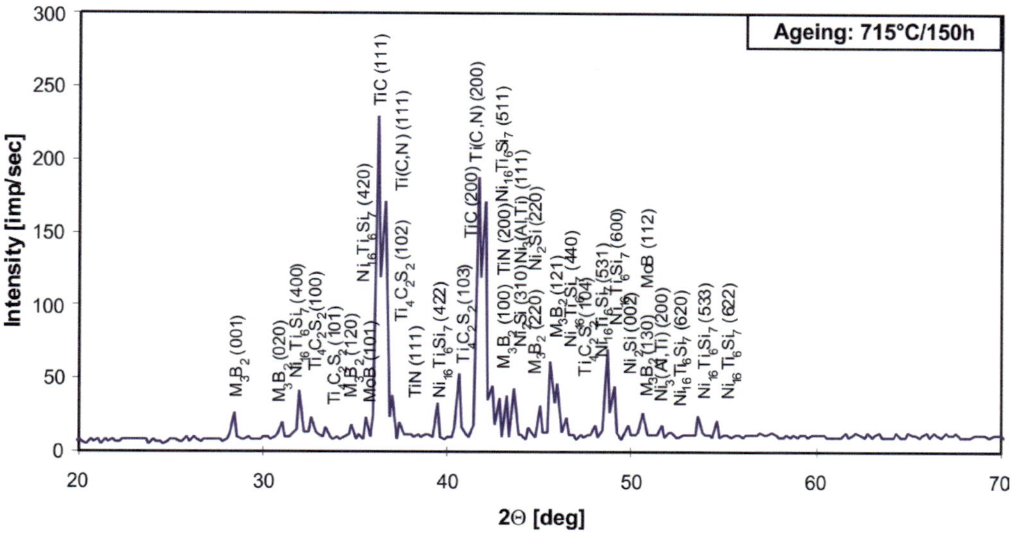

Figure 12. Isolate diffractogram of the alloy after solution heat treatment and aging at 715°C/150 h.

Alloy condition	Phase constituents
Solution heat treatment: 980°C/2 h/w	TiC; $TiC_{0,3}N_{0,7}$; $TiN_{0,3}$; $Ti_4C_2S_2$; Ni_2Si; MoB
Aging: 715°C/0.5 h	TiC; $TiC_{0,3}N_{0,7}$; $TiN_{0,3}$; $Ti_4C_2S_2$; Ni_2Si; MoB; $G - Ni_{16}Ti_6Si_7$
Aging: 715°C/2÷500 h	TiC; $TiC_{0,3}N_{0,7}$; $TiN_{0,3}$; $Ti_4C_2S_2$; Ni_2Si; MoB; G - $Ni_{16}Ti_6Si_7$; γ' - $Ni_3(Al,Ti)$; M_3B_2

Table 6. Phase composition of the Fe-Ni alloy isolates after solution heat treatment and aging at 715°C.

An increase in the aging temperature of the alloy to 750°C leads to intensification of the diffusive processes of coagulation, growth, and overaging of intermetallic phase and carbide particles. Where aging time is extended (within a range of 16 h – 50 h), lenticular and spheroidal particles of phase γ' precipitate, showing strong coherence with the matrix (Figure 13). The first signs of overaging were found in the structure of the alloy after 100 h – 300 h, where colonies of transcrystalline Widmanstätten lamellae were formed. The lamellae of phase η (Ni_3Ti) grew into the austenite grain in the <111> direction (Figure 14). The phase transition $\gamma' \rightarrow \eta$ was accompanied with dissolution of fine particles of the neighboring γ' phase. In addition, presence of carbide TiC, carbonitride $TiC_{0.3}N_{0.7}$, nitride $TiN_{0.3}$, carbosulfide $Ti_4C_2S_2$, and particles of the Laves phase (Ni_2Si), phase G ($Ni_{16}Ti_6Si_7$), phase β (NiTi), phase σ ($Cr_{0.46}Mo_{0.40}Si_{0.14}$), and boride MoB were also detected in the alloy microstructure (Figure 15, Table 7).

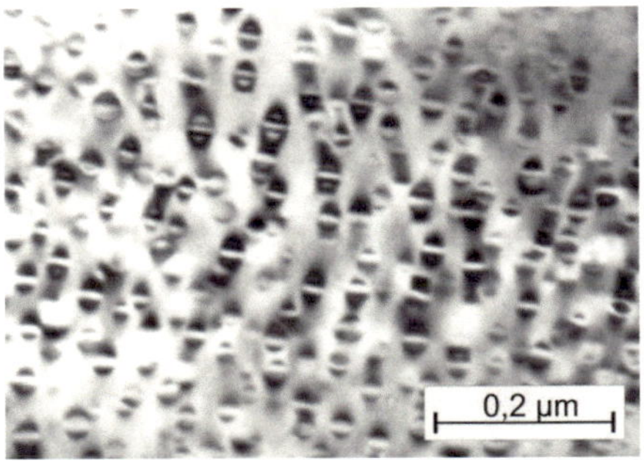

Figure 13. Alloy microstructure after solution heat treatment and aging at 750°C/50 h. Coherent spheroidal and lenticular particles of γ' in the matrix.

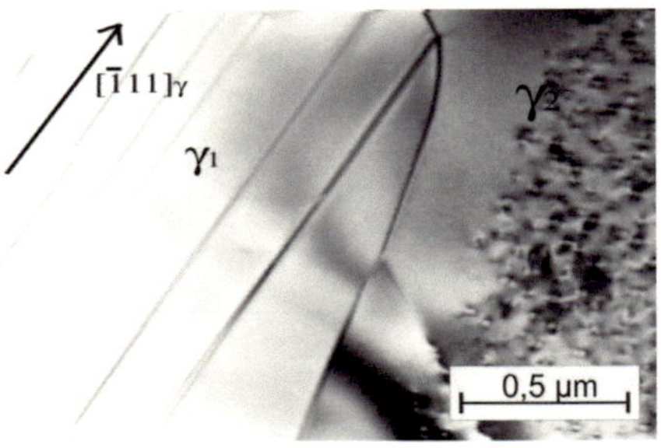

Figure 14. Alloy microstructure after solution heat treatment and aging at 750°C/100 h. Parallel lamellae of phase η in the matrix.

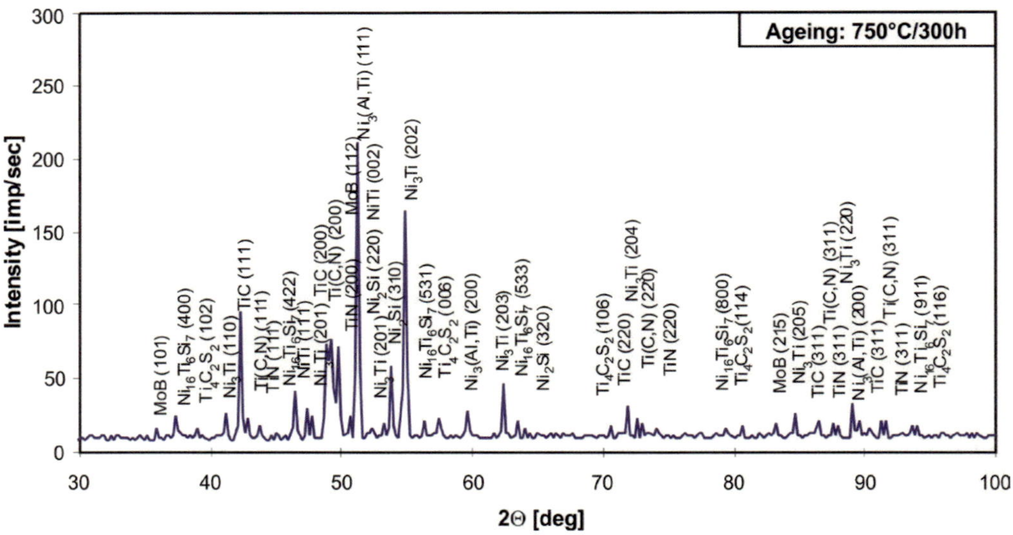

Figure 15. Isolate diffractogram of the alloy after solution heat treatment and aging at 750°C/300 h.

Alloy condition	Phase constituents
Solution heat treatment: 980°C/2 h/w	TiC; $TiC_{0,3}N_{0,7}$; $TiN_{0,3}$; $Ti_4C_2S_2$; Ni_2Si; MoB
Aging: 750°C/2÷50 h	TiC; $TiC_{0,3}N_{0,7}$; $TiN_{0,3}$; $Ti_4C_2S_2$; Ni_2Si; MoB; *G* - $Ni_{16}Ti_6Si_7$; γ' - $Ni_3(Al,Ti)$
Aging: 750°C/100÷150 h	TiC; $TiC_{0,3}N_{0,7}$; $TiN_{0,3}$; $Ti_4C_2S_2$; Ni_2Si; MoB *G* - $Ni_{16}Ti_6Si_7$; γ' - $Ni_3(Al,Ti)$; η - Ni_3Ti
Aging: 750°C/200÷300 h	TiC; $TiC_{0,3}N_{0,7}$; $TiN_{0,3}$; $Ti_4C_2S_2$; Ni_2Si; MoB *G* - $Ni_{16}Ti_6Si_7$; γ' - $Ni_3(Al,Ti)$; η - Ni_3Ti; β – NiTi
Aging: 750°C/400÷500 h	TiC; $TiC_{0,3}N_{0,7}$; $TiN_{0,3}$; $Ti_4C_2S_2$; Ni_2Si; MoB *G* - $Ni_{16}Ti_6Si_7$; γ' - $Ni_3(Al,Ti)$; η - Ni_3Ti; β - NiTi; σ - $Cr_{0,46}Mo_{0,40}Si_{0,14}$

Table 7. Phase composition of the Fe-Ni alloy isolates after solution heat treatment and aging at 750°C.

An increase in the aging temperature of the alloy to 780°C leads to intensification of the diffusive processes, such as coagulation and overaging of intermetallic phase and carbide particles. Extended aging time (from 50 h to 150 h) contributes to coagulation of phase γ' particles. Clear signs of non-dilatational strain were detected, which was connected with the occurrence of internal stacking faults (SF) (Figure 16), the latter being the nucleation sites of the η phase lamellae. The mechanism of the $\gamma' \rightarrow \eta$ transition was proposed by Merrick and Nicholson [33] who described it through the appearance of SF in the γ' phase whose lattice constants were slightly greater to those of the matrix. Furthermore, lamellar precipitates of the η phase were detected near the boundary of the cellular system (Figure 17). In addition, analysis of the phase composition of the isolated precipitates confirmed the presence of carbide TiC, carbonitride $TiC_{0.3}N_{0.7}$, nitride $TiN_{0.3}$, carbosulfide $Ti_4C_2S_2$, and particles of the Laves phase (Ni_2Si), phase *G* ($Ni_{16}Ti_6Si_7$), phase σ ($Cr_{0.46}Mo_{0.40}Si_{0.14}$), and boride MoB in the alloy microstructure (Figure 18, Table 8).

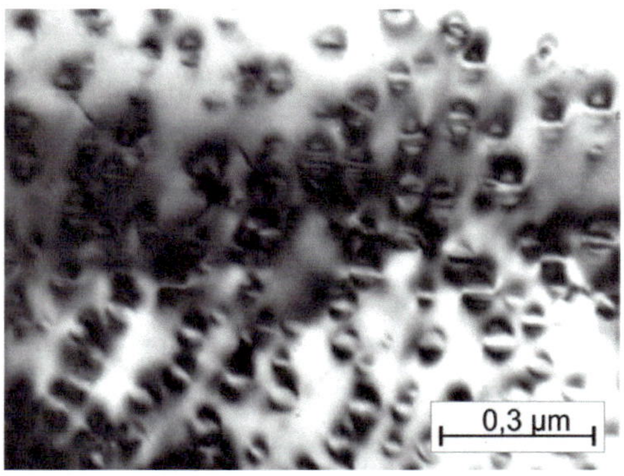

Figure 16. Alloy microstructure after solution heat treatment and aging at 780°C/150 h. Nucleation of transcrystalline lamellae of phase η by means of the SF mechanism in phase γ'.

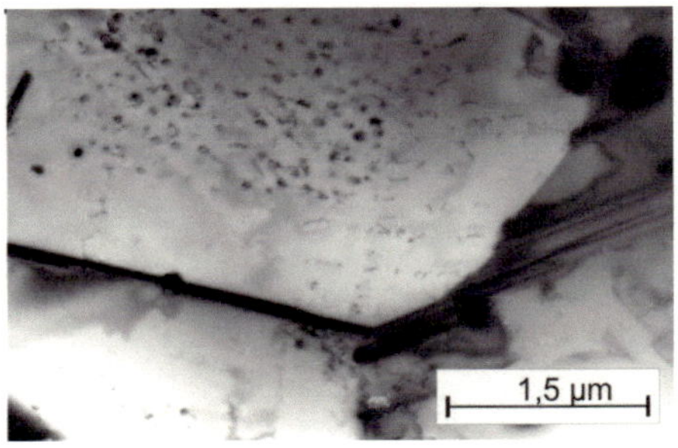

Figure 17. Alloy microstructure after solution heat treatment and aging at 780°C/500 h. Precipitates of lamellae η in the cellular system in the region near the boundary.

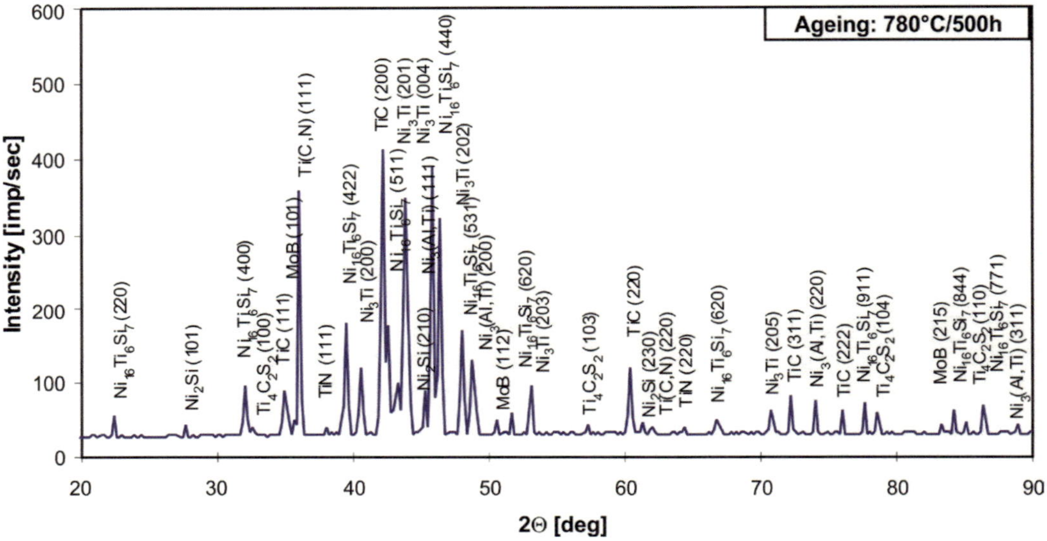

Figure 18. Isolate diffractogram of the alloy after solution heat treatment and aging at 780°C/500 h.

Alloy condition	Phase constituents
Solution heat treatment: 980°C/2 h/w	TiC; $TiC_{0,3}N_{0,7}$; $TiN_{0,3}$; $Ti_4C_2S_2$; Ni_2Si; MoB
Aging: 780°C/2 h	TiC; $TiC_{0,3}N_{0,7}$; $TiN_{0,3}$; $Ti_4C_2S_2$; Ni_2Si; MoB; G - $Ni_{16}Ti_6Si_7$; γ' - $Ni_3(Al,Ti)$
Aging: 780°C/4÷500 h	TiC; $TiC_{0,3}N_{0,7}$; $TiN_{0,3}$; $Ti_4C_2S_2$; Ni_2Si; MoB; G - $Ni_{16}Ti_6Si_7$; γ' - $Ni_3(Al,Ti)$; σ - $Cr_{0,46}Mo_{0,40}Si_{0,14}$; η - Ni_3Ti

Table 8. Phase composition of the Fe-Ni alloy isolates after solution heat treatment and aging at 780°C.

4.2. Growth kinetics of γ' phase in the Fe-Ni superalloy during prolonged aging

Observation of the microstructure by means of an electron microscope (TEM) using the thin foil technique showed that the main intermetallic phase that precipitates in the Fe-Ni alloy during aging was the γ'- $Ni_3(Al,Ti)$ phase. Particles of this phase is characterized by high dispersion, precipitated homogeneously in the matrix, and a spheroidal or lenticular shape. An increase in the temperature and prolongation of the aging time caused an increase in the kinetics of their growth, coagulation, and overaging.

Quantitative analysis of the γ' phase particles that precipitated in the alloy during aging was made for aging temperatures of 715°C, 750°C and 780°C with holding times in the range of 4–500 h. The results for the selected aging times, i.e., 4 h, 150 h, 300 h, and 500 h are shown in Table 9. As can be seen in the table, in the case of aging at 715°C with holding times of 4–500 h, the average diameter of particles $\overline{D}$ grew from 7.5 nm to 28.3 nm, while the volume fraction of the particles changed within a range of 6–11%. In the case of aging at 750°C with the same holding times, an increase was observed in the average diameter of the particles from 10.6 nm to 39.5 nm, accompanied by a change in the volume fraction from 7.1% to 10.1%.

Parameter		Time of aging [h] at 715°C				Time of aging [h] at 750°C				Time of aging [h] at 780°C		
4	150	300	500	4	150	300	500	4	150	300	500	
A_A [%]	10.2	14.3	14.8	16.4	11.5	15.8	16.6	15.6	9.5	13.5	13.1	11.5
$\bar{d}$ [nm]	7.4	15.7	19.6	24.6	9.2	24.4	27.0	32.1	12.7	41.2	45.0	51.2
$\bar{D}$ [nm]	7.5	17.0	22.5	28.3	10.6	28.8	32.1	39.5	13.1	45.7	52.2	58.4
V_V [%]	6.0	10.3	10.4	11.1	7.1	8.6	9.4	10.1	8.1	7.7	7.8	8.4
l_d [nm]	144	117	121	114	135	141	123	124	153	105	103	124

Table 9. Stereological parameters of the γ' particles in the tested Fe-Ni alloy.

When the aging temperature was the highest (780°C) with analogical holding times, the average particle diameter reached the highest values in the range from 13.1 nm to 58.4 nm, while the highest volume fraction ranged between 8.1% and 8.4%.

Examples of particle distribution histograms for the selected ageing times within the range of 4–500 h at 715–780°C are shown in Figures 19–22. As the aging time and temperature increase, a shift can be observed of the maximum towards classes with larger diameters. This corroborates that the mean diameter $\overline{D}$ of the γ' particles increases with a longer aging time and a growing temperature of aging.

Analysis of the obtained results of stereological examination of the γ'- $Ni_3(Al,Ti)$ phase particles was the basis for determining the kinetics of their growth, depending on the aging conditions of the Fe-Ni alloy (Figure 23). Based on the results from the obtained dependencies, the highest growth rate of the γ' phase particles was found at an aging temperature of 780°C, the medium rate at 750°C, and the lowest at 715°C.

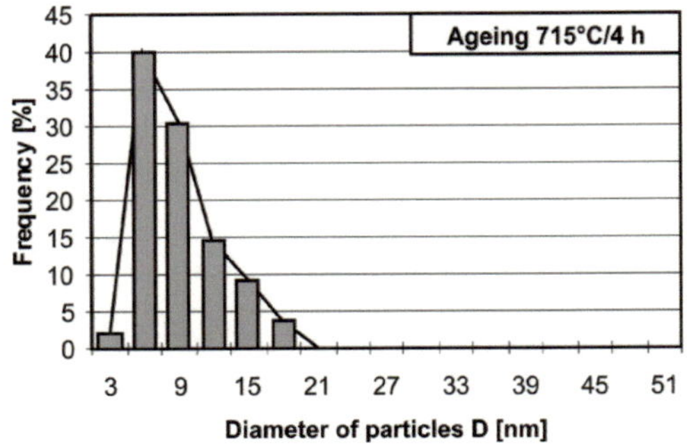

Figure 19. Distribution histogram of the diameter of γ' particles in the alloy after solution heat treatment and aging at 715°C/4 h.

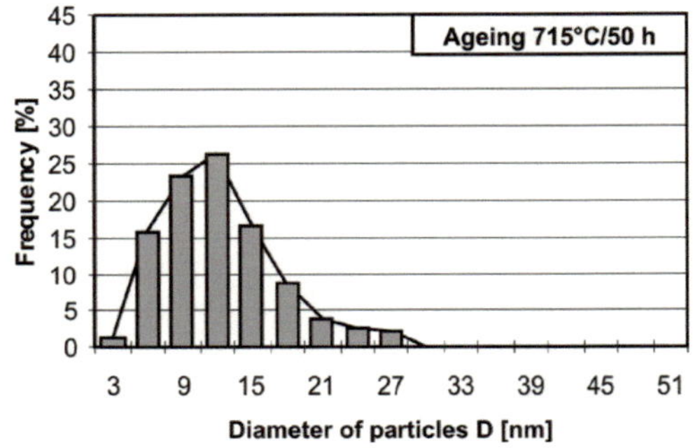

Figure 20. Distribution histogram of the diameter of γ' particles in the alloy after solution heat treatment and aging at 715°C/50 h.

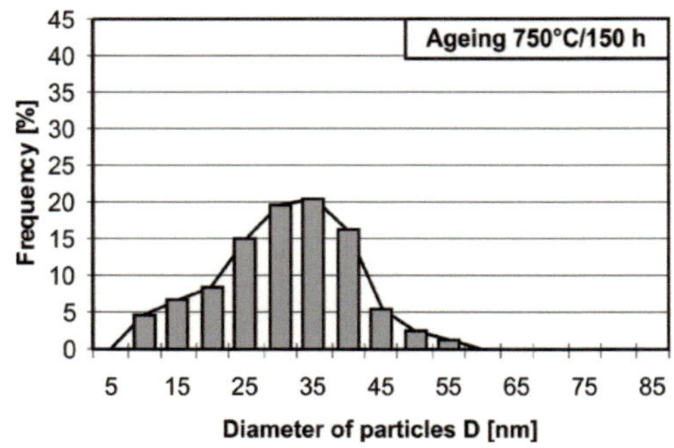

Figure 21. Distribution histogram of the diameter of γ' particles in the alloy after solution heat treatment and aging at 750°C/150 h.

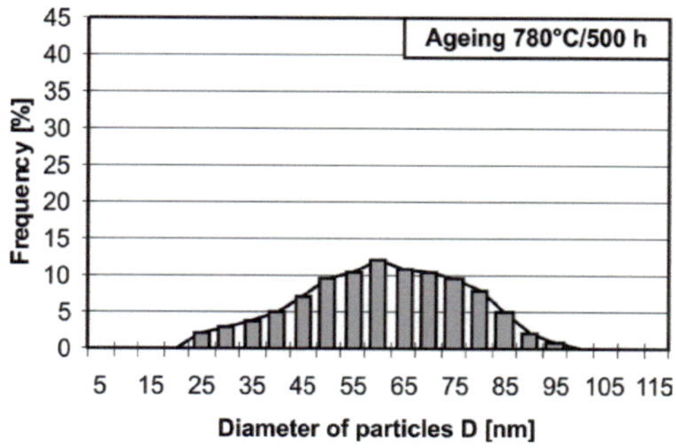

Figure 22. Distribution histogram of the diameter of γ' particles in the alloy after solution heat treatment and aging at 780°C/500 h.

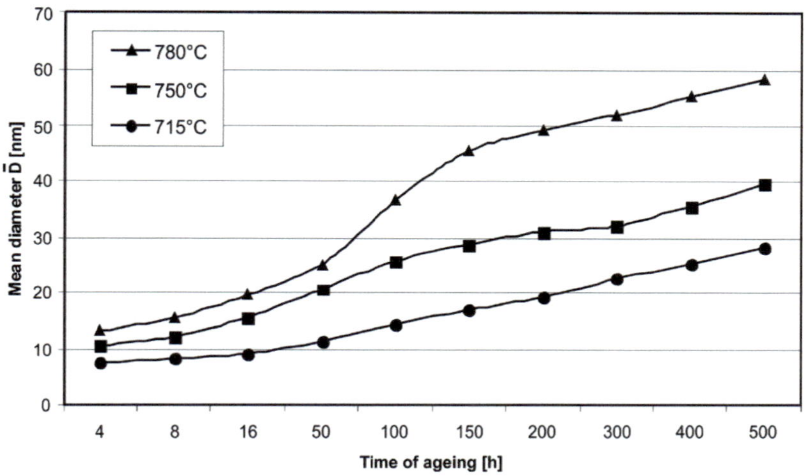

Figure 23. Dependence of the mean diameter of phase γ' precipitates in the Fe-Ni alloy on the temperature and aging time.

The analysis of the γ'- $Ni_3(Al,Ti)$ phase growth step was made on the basis of the LSW theory [34,35] for coagulation on volume diffusion control according to the relationship provided by Kusabiraki et al. [36,37]:

$$\bar{d}^3 - \bar{d}_0^3 = \frac{64\sigma D C_e V_m^2}{9RT} = K't \tag{5}$$

where: $\bar{d}$ and $\bar{d}_0$ – mean diameter of the precipitates at, respectively, time t and t = 0; σ – interfacial energy between the precipitates and the matrix; D – diffusion coefficient of the solute atom in the matrix; C_e – concentration of solute atoms in the matrix in equilibrium with a particle of an infinite size; V_m – molar volume of the precipitation phase; R – gas constant; T – absolute temperature; K' – growth rate constant.

The diffusion coefficient (D) can be generally expressed as:

$$D = D_0 \cdot exp\left(-E / RT\right) \tag{6}$$

where: D_0 – pre-exponential factor, E – activation energy of diffusion.

Hence, constant K' assumes the following form:

$$K' = \frac{64\sigma D_0 C_e V_m^2 exp\left(-E / RT\right)}{9RT} \tag{7}$$

On assumption that the values of σ, C_e, and V_m are nearly independent of temperature, the value of diffusion activation energy (E) can be obtained from the slope of Arrhenius plot being the linear relationship between ln(TK') and T^{-1}.

The dependence of the mean diameter $\bar{D}$ of the γ' phase particles on aging time ($t^{1/3}$) and temperature is presented in Figure 24. For the analyzed aging temperature of 715÷780°C, linear dependencies were obtained, which is consistent with the LSW theory for coagulation controlled through volume diffusion. Certain deviations from the LSW theory are possible, as the theory was developed for spherical precipitates with high dispersion and a low relative volume [34, 35]. The deviation from the straight line found at a temperature of 780°C and aging time of 16÷100 h may result from breaking the γ/γ' coherence, increasing the degree of lattice mismatch between the γ' particle and the γ matrix, and in consequence, the acceleration of coagulation of the γ' phase particles. Increasing the aging temperature leads to increasing the growth and coagulation of the γ' phase particles. Values of the mean diameter of the particles $\bar{D}$ for the aging time of 0.5 h and 2 h were determined through extrapolation (Figure 24).

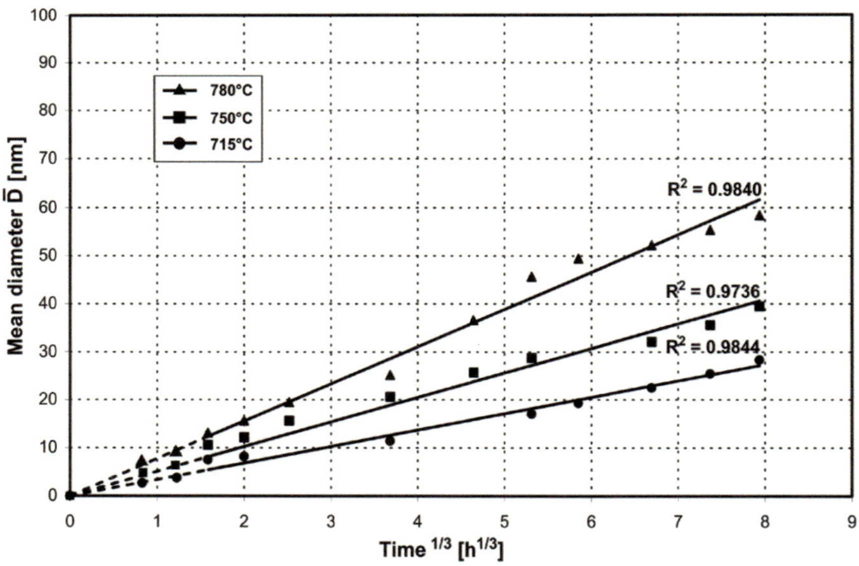

Figure 24. Dependence between the mean diameter of particles of phase γ' in the Fe-Ni alloy and the ageing time and temperature.

The activation energy of coagulation of the γ' -$Ni_3(Al, Ti)$ phase in the Fe-Ni alloy was determined from the slope of Arrhenius straight line (Figure 25) and its value (E = 297 kJ/mole) turned out to be close to that of the activation energy of phase γ' (245-298 kJ/mole), which was estimated for the chosen Fe-Ni and Ni-based superalloys [36, 37]. Moreover, the obtained value of phase γ' coagulation activation energy was close to that of the activation energy of diffusion in austenite γ of the solute elements, such as Al, Ti, Ni, Fe, Cr, and Mo, which contribute to the growth of phase γ'. In the case of the examined Fe-Ni alloy, the E value showed that it was the diffusion of the solution elements that controlled the growth of phase γ', which agrees with the LSW theory.

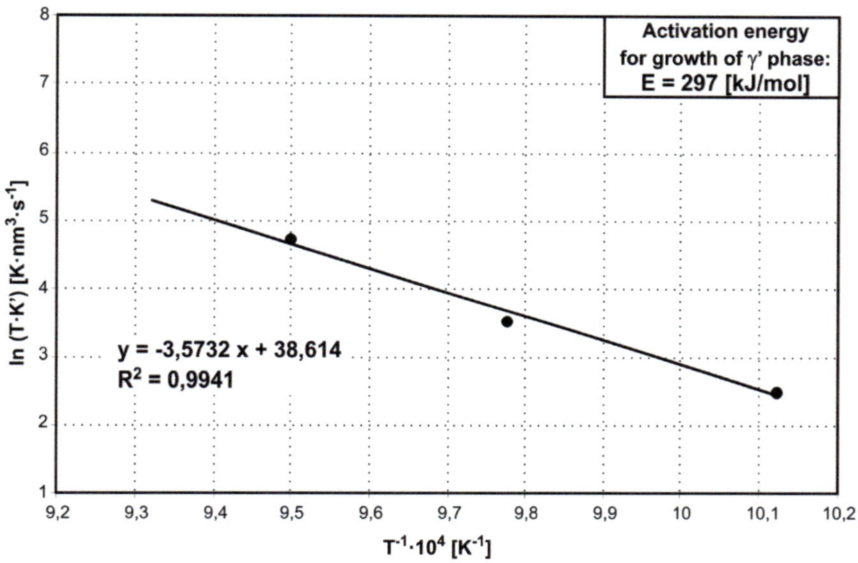

Figure 25. The plot for determination of the activation energy for coagulation process of phase γ' in the Fe-Ni alloy.

5. Summary

The analysis of the current status of the issue of wrought Fe-Ni superalloys showed that a modification of their basic chemical composition does not bring significant effects any longer. A significant improvement of technological and functional properties of Fe-Ni alloys can be obtained by introducing micro-additives that form precipitates of strengthening high-dispersion phases interacting with grain boundaries, and by applying modern manufacturing technologies that improve the quality of products. Further development of the material group of this type is closely connected with the development of wrought nickel superalloys.

The work analyzes the phase transitions and changes in the morphology of phase components of the microstructure in the selected Fe-Ni superalloy type A-286 in conditions simulating its heat treatment and operation processes. The research and analysis of its results showed a significant influence of solution heat treatment and extended aging at an elevated temperature on the course of the precipitation process and the microstructure of the creep-resisting Fe-Ni superalloy.

After a 2-hour solution heat treatment at 980°C/water, the structure of the Fe-Ni alloy is similar to that of twinned austenite with a small content (ca. 0.3 wt. %) of undissolved precipitates of titanium compounds, such as carbide TiC, carbonitride $TiC_{0.3}N_{0.7}$, nitride $TiN_{0.3}$, carbosulfide $Ti_4C_2S_2$ and Laves phase Ni_2Si and boride MoB.

The application of a single-stage aging after solution heat treatment at temperatures of 715°C, 750°C, and 780°C with holding times from 0.5 h to 500 h causes precipitation processes of γ' - $Ni_3(Al,Ti)$, η - Ni_3Ti, β - NiTi, G - $Ni_{16}Ti_6Si_7$ and σ - $Cr_{0.46}Mo_{0.40}Si_{0.14}$ intermetallic phases in the investigated alloy, as well as the carbide $M_{23}C_6$ and boride M_3B_2.

The intermetallic γ' - $Ni_3(Al, Ti)$ phase was the main phase that precipitated during the aging of the alloy. It is possible to distinguish three characteristic stages during the γ' phase precipitation, namely: coherent zones, coherent spheroidal particles (7–25 nm), and coagulated spheroidal particles (30–60 nm). On histograms of the γ' phase size distributions, a shift can be observed of the maximum towards classes with larger diameters as the aging time and temperature increased. The volume fraction of the γ' phase particles depended on the aging time and temperature, and changed within the range of ca. 6–11%.

The mean diameter of the γ' - $Ni_3(Al, Ti)$ phase precipitates was found to increase as a function of the cube root of the aging time, which is consistent with the LSW theory claiming that coagulation is controlled by volume diffusion. In the examined alloy, the value of the γ' phase coagulation activation energy amounted to E = 297 kJ/mole and was close to the value of the solute elements' diffusion activation energy in γ-*Fe*, whose elements contribute to the growth of phase γ'.

The phase transition $\gamma' \rightarrow \eta$, which determines the overageing stadium of the Fe-Ni alloy, takes place through the formation of stacking faults in the structure of the γ' phase particles. A growth of the η phase lamellae proceeds in austenite in the direction <110> γ and is accompanied by the dissolution of the neighboring γ' phase particles. The η phase lamellae form the microstructure of intercrystalline Widmanstätten lamellae or precipitates in a cellular system in the regions at austenite grain boundaries.

The results of the study show that the examined Fe-Ni superalloy type A-286 has a thermally stable microstructure in the range of operation temperatures of up to ca. 700°C and therefore, it should be considered for applications in such temperatures.

Author details

Kazimierz J. Ducki

Address all correspondence to: kazimierz.ducki@polsl.pl

Silesian University of Technology, Katowice, Poland

References

[1] Sims Ch.T., Stoloff N.S., Hagel W.C. Superalloys II. New York: John Wiley & Sons; 1987.

[2] Stoloff N.S. Wrought and Powder Metallurgy (P/M) superalloys. In: ASM Handbook, editors. Volume 1, Properties and Selection: Irons, Steels, and High Performance Alloys. Materials Park, OH: ASM International; 2005. p. 1478-1527.

[3] Stickler R. Phase Stability in Superalloys. Proc. of the Symposium: High-Temperature Materials in Gas Turbines, Baden. 1973;115-146.

[4] Dulis E.J. Age-hardening austenitic stainless steels. Proc. Conference: The Metallurgical Evolution of Stainless Steels, London. 1979;420-441.

[5] Pickering F.B. Some aspects of the precipitation of nickel-aluminium-titanium intermetallic compounds in ferrous materials. Proc. Conference: The Metallurgical Evolution of Stainless Steels, London. 1979;391-401.

[6] Sourmail T. Precipitation in creep resistant austenitic stainless steels. Materials Science and Technology. 2001;17:1-14.

[7] Ducki K.J., Hetmańczyk M., Kuc D. Analysis of precipitation process of the intermetallic phases in a high-temperature Fe-Ni austenitic alloy. Materials Chemistry and Physics. 2003;81:490-492.

[8] Ducki K.J. Analysis of the precipitation and growth processes in a high-temperature Fe-Ni alloy. Journal of Achievements in Materials and Manufacturing Engineering. 2008;31:226-232.

[9] Brooks J.W., Bridges P.J.. High Temperature Alloys for Gas Turbines and other Applications. Netherlands: D. Reidl Publish. Company; 1986.

[10] Rohrbach K.P. Trends in high-temperature alloys. Advanced Materials & Processes. 1995;10:37-40.

[11] Ducki K.J. Microstructural aspects of deformation, precipitation and strengthening processes in austenitic Fe-Ni superalloy (in Polish). Gliwice, Poland: Silesian University of Technology; 2010.

[12] Ducki K.J. Fatigue behaviour and creep resistance of Fe-Ni superalloy. Saarbrücken, Germany: LAP Lambert Academic Publishing; 2014.

[13] Schubert F. Mechanische Eigenschaften von Superlegierungen und ihren Verbunden (in German). VDI Berichte. 1987;600.4:85-136.

[14] Weβling W. Festigkeitseigenschaften von Stählen für eine langzeitige Beanspruchung bei hohen Temperaturen (in German). VDI Berichte. 1987;600.4:45-84.

[15] Capdevila C., Chen Y.L., Jones A.R., Bhadeshia K.D.H. Grain Boundary Mobility in Fe-Base Oxide Dispersion Strengthened PM2000 Alloy. ISIJ International. 2003;43:777-783.

[16] Wasilkowska A., Bartsch M., Messerschmidt U., Herzog R., Czyrska-Filemonowicz A. Creep mechanism of ferritic oxide dispersion strengthened alloys. Journal of Materials Processing Technology. 2003;133:218-224.

[17] McColvin G., Munasinghe D., O'Driscoll J., Jacobs M.. Fabrication of gas turbine combustion hardware in ODS ferritic materials. In: Proc. of the 7th Liege Conference: Materials for Advanced Power Engineering; Liege. 2002. p. 833-844.

[18] Pérez P. Influence of the alloy grain size on the oxidation behaviour of PM2000 alloy. Corrosion Science. 2002;44:1793-1808.

[19] Suryanarayana C. Mechanical alloying and milling. Progress in Materials Science. 2001;46:1-184.

[20] Choudhury I.A., El-Baradie M.A. Machinability of nickel-base super alloys: A general review. Journal of Materials Processing Technology. 1998;77:278-284.

[21] Kreitner P., Appelmaier K., editors. Einfluss von Spurenelementen auf die Warmverformbarkeit von Stählen und Nickelbasis-legierungien. Spurenelemente in Stählen ed. Düsseldorf: Verlag Stahleisen; 1985. 23-34.

[22] Sczerzenie F., Maurer G.E.. Developments in Disc Materials. Materials Science and Technology. 1987;3:733-742.

[23] Härkegård G., Guédou J.Y.. Disc Materials for Advanced Gas Turbines. In: Proc. of the 6th Liége Conference: Materials for Advanced Power Engineering; Liege. 1998. p. 913-931.

[24] Pickering F.B. Physical metallurgical development of stainless steels. In: Proc. of the Conference: Stainless Steels '84; 1984; Göteborg, Sweden. 1984. p. 1-28.

[25] Schubert F.. Temperature and Time Dependent Transformation: Application to Heat Treatment of High Temperature Alloys. In: Phase Stability in High Temperature Alloys. London, UK: Appied Science Publishers Ltd; 1981. p. 119-149.

[26] Penkalla H.J., Schubert F. Ni-base wrought alloy development for USC steam turbine rotor applications. Inżynieria Materiałowa. 2004;3(140):415-421.

[27] Szala J., editor. Computer program Quantitative Metallography. Katowice, Poland: Department of Materials Science, Silesian University of Technology; 1997.

[28] Ryś J., editor. Stereology of materials (in Polish). Kraków, Poland: Fotobit Design; 1995. 394 p.

[29] Czyrska-Filemonowicz A., Dubiel B., Wiencek K.. Determination of the oxide particle density in ODS alloys by means of transmission electron microscopy. Acta Stereologica. 1998;17:225-236.

[30] Dubiel B., Wosik J., Kruk A., Penkalla H.J., Czyrska-Filemonowicz A.. Quantitative TEM microstructural analysis of Ni-based superalloy Waspaloy. In: Proceedings of the Stereology and Image Analysis in Materials Science; Kraków, Poland. Fotobit Design; 2000. p. 135-140.

[31] Schröder J.H., Arzt E.. Electron-Microscopic Investigations of Dispersion-Strengthened Superalloys. Praktische Metallography. 1988;25:264-273.

[32] Wilson F.G. Age hardening of austenitic steels by γ'. In: Efect of second-phase particles on the mechanical properties of steels. London, UK: The Iron and Steel Instytute; 1971. p. 16-21.

[33] Merrick H.F., Nicholson R.B. Some Features of Precipitation in a Nickel-Chromium-Titanium Alloy. In: Proc. 5th International Congress for Electron Microscopy; Philadelphia. New York, USA: Academic Press; 1962. Paper K.8.

[34] Lifshitz I.M., Slyozow V.V. The kinetics of precipitation from supersaturated solid solution. Journal Phys. Chem. Solids. 1961;19:35-50.

[35] Wagner C. Theorie der Alterung von Niederschlägen durch Umlösen (Ostvald Reifung) (in German). Zeitschrift für Elektrochemie. 1961;65:581-591.

[36] Kusabiraki K., Takasawa Y., Ooka T. Precipitation and Growth of g' and h Phases in 53Fe-26Ni-15Cr Alloy. ISIJ International. 1995;35(5):542-547.

[37] Kusabiraki K., Zhang X., Ooka T. The Growth of γ' Precipitation in a 53Ni-20Co-15Cr Superalloy. ISIJ International. 1995;35:1115-1120.

Chapter 7

Assessment of Dental Alloys by Different Methods

Lavinia Ardelean, Lucien Reclaru,
Cristina Maria Bortun and Laura Cristina Rusu

Additional information is available at the end of the chapter

http://dx.doi.org/10.5772/61115

Abstract

Alloys are used in various areas of dentistry. The field of dental alloys is a very extensive one, encompassing both the materials themselves as well as the manufacturing methods, which are constantly developing. Our chapter focuses on corrosion and biocompatibility assessment, using various methods. At present there is no perfect dental alloy. Superalloys for dental use are not yet available, and only few studies concerning the new generation of superalloy candidates for medical applications have recently been developed, with promising results.

Keywords: dental alloys, corrosion, sintering, laser welding, citotoxicity

1. Introduction

Alloys are used in various areas of dentistry. Iron-based alloys are used for orthodontic wires and tools. Noble-metal alloys and base-metal alloys are used for manufacturing crowns, inlays, partial fixed dentures, and metallic frames of removable partial dentures. The field of dental alloys is a very extensive one, encompassing both the materials themselves as well as the manufacturing methods, which are constantly developing [1].

However, superalloys for dental use are not yet available, and only few studies concerning the new generation of superalloy candidates for medical applications have recently been developed. Three superalloy candidates [2] for medical application were assessed by means of corrosion behavior, cation release, and biological evaluation for a prolonged contact with skin, in comparison with two reference austenitic stainless steels 316L and 904L. The three superalloy candidates assessed were X1 CrNiMoMnW 24-22-6-3-2 N, NiCr21 MoNbFe 8-3-5 AlTi, and CoNiCr 35-20 Mo 10 BTi. Several electrochemical parameters were measured and determined (Eoc, Ecorr, icorr, ba, bc, Eb, Rp, Ecrev, and coulometric analysis) in order to assess corrosion behavior. The cation release evaluation and in vitro biological characterization have also been carried out. In terms of corrosion, the results reveal that the 904L steels presented the best behavior followed by the super austenitic steel X1 CrNiMoMnW 24-22-6-3-2 N. For the other two superalloys (NiCr- and CoNiCr-type alloys) tested in different conditions (annealed, work hardened, and work hardened + age hardened), their behavior to corrosion was found to be weak and close to the other reference stainless steel, 316L. As far as extraction is concerned, a mixture of cations in relatively high concentrations was noted, and, therefore, a cocktail effect was not excluded. The results obtained in the biological assays WST-1 and TNF-alpha were in correlation with the corrosion and extraction evaluation [2]. The encouraging results of this study may support the development of superalloys for medical applications.

2. Classification of dental alloys

Commonly used dental alloys may be classified as shown in Table 1

Noble alloys	Base metal alloys
With a high gold content	Ni-Cr alloys
With a low gold content	Co-Cr alloys
Based on Ag-Pd (with and without copper)	Titanium alloys
Palladium-based (with increased content of Pd, based on Pd-Cu and Pd-Ag)	

Table 1. Classification of dental alloys

At present, there is no perfect dental alloy. Alloys are mixtures of two or more metals. Metals, when melted, are usually inter-soluble; once the mixture has cooled, the result is a solid solution with higher values of hardness, resistance, and flexibility than the initial pure metals from which it was derived. The grain-type structure of the alloy predisposes to corrosion, which is quite an issue in the case of dental alloys. Other shortcomings involve biocompatibility and manufacturing technology, which, in case of casting, is quite laborious.

3. Corrosion assessment

Corrosion characterizes the chemical reactivity of metals and alloys, which results in a visible alteration of the material and affects the function of a metallic component or of the entire system, since the results of corrosion are metallic compounds. Such compounds are more stable than the metals concerned.

A major necessity that concerns any material of metallic nature used in the oral cavity is good resistance to corrosion.

Corrosion that occurs in the presence of atmospheric moisture or water is an electrochemical (galvanic) corrosion. The oral cavity provides an ideal environment for conducting electrochemical corrosion phenomena.

In the case of dental alloys, corrosion causes an unaesthetic change of the alloy surface. Metal restorations (crowns, inlays, amalgam fillings) are charging electrically in the oral environment, the degree of damage on the restoration depends on the metallic material (type of alloy), the presence on the surface of the restoration of soft or hard deposits, the connection with anatomical substratum, the existence of fractures in the veneering material, the age of the restoration, and the composition of saliva [3].

Metals and alloys are good electricity conductors, most of the corrosion processes involving the formation of an electrolytic cell, as the first stage of the process. Conditions for the formation of an electrolytic cell in the oral cavity include the presence of two or more metals with different electrode potentials and of an electrolyte (saliva and tissue fluids are good electrolytes). The two materials must be immersed in an electrolyte and must be in electrical contact for current to flow [1].

Bimetallic devices may present signs of corrosion after variable time [4]. Such degradations are quite frequent in the oral cavity, fortunately at a smaller scale in case of noble dental alloys [5], but may occur in various other situations (Figure 1). Such degradations correspond to the galvanic corrosion type.

Figure 1. Bimetallic wrist bracelet made of bimetallic rings gold/steel after corrosion in saline solution

Numerous factors influence the galvanic corrosion: internal factors (metal related) and external (electrolyte related).

Parameters related to the material (composition, structure, impurities, surface condition, etc.) [6]:

The nature of the metal: metals and alloys which when inserted into the oral cavity have a good stability are copper, silver, molybdenum, iridium, and palladium. Gold and platinum are very stable; copper and iron are relatively stable; chromium, zinc, and aluminum are very unstable.

The tendency of a metal to corrode is given by its electrode potential, as metals with increased negative potential are more likely to corrode, while metals with high positive values are much less reactive and are also called noble metals.

At first sight it seems incomprehensible why chromium, having one of the most negative potential values, is used as a component of many dental alloys. This apparent contradiction is explained by the fact that despite being electrochemically active, chromium reacts by forming a layer of oxide that protects the metal or alloy from further decomposition [7, 8].

The metal's structure depends on:

Grain size: if there is a precipitation of impurities between grains, the larger the grains, the more intense is corrosion along them.

Structural heterogeneity: the more homogeneous the distribution of metallic atoms in an alloy, the less present is the tendency to corrode. That is why most manufacturers submit the alloys to a homogenizing treatment using heat, in order to minimize the possibility of electrochemical corrosion.

The surface condition depends on the following: the polishing degree, surface properties, absence of pores or cracks in the primary protective layers. Deformations, and mechanical tensions in the metal contribute to decreasing the potential, as fatigue is more accentuated in corrosive environment [1].

Parameters related to the nature of the electrolyte (composition, concentration, temperature, oxygen content, pH, heterogeneity, etc.):

Chemical: pH- corrosion is poor in a neutral environment, generally high in acidic environment, and variable in basic environment; oxidants present in the solution determine, most frequently, an increase of the electrode potential [9].

Physical: temperature may have an important indirect effect; the movement of the solution may cause variations by changing concentration of the metal's specific ions, in the vicinity of the electrode surface, duration of exposure being a key factor.

Parameters related to topography (surface ratio of the two materials, geometric distribution, etc.) may also play a role in developing corrosion.

Corrosion effects

Chemical effects are manifested by loss of metallic luster and brown coloration of the surface.

The physical and mechanical effects, leading to the deterioration of mechanical properties, are the following:

Uniform corrosion decreases the metal thickness.

Localized corrosion in the form of plates or punctiform.

Intercrystalline corrosion: metal corrodes in depth.

Selective corrosion: only a component of the alloy is involved.

Biological effects of galvanic microcurrents translate as:

Subjective symptoms: metallic taste due to the release of ions, reflex salivation or xerostomia, burning sensation, headache due to the occurrence of galvanic current, and trigeminal neuralgia.

Objective symptoms: gingivitis and glossitis, hypertrophy and turgescence of taste papillae, erosions and ulcers of the oral mucosa, late leukoplasia, and changes of blood chemistry.

General manifestations: fatigue, dyspepsia, and headache.

An unwanted effect is an increased charge of metal ions, with negative effects, in particular regarding heavy metals like mercury and nickel [1].

Galvanic corrosion is rare in case of noble dental alloys. However, incorrect manufacturing technology or improper finishing may have unwanted results.

3.1. Corrosion of noble alloys due to incorrect manufacturing technology

A combined fixed prosthetic restoration of four elements consisting of a metal-ceramic part (23-24) and a cast part (25-26), bonded after firing the ceramics, presented a significant corrosion shortly after luting.

The two parts were made following strictly all the rules, using noble alloys with high content of gold (Table 2), bonded by soldering in the oven using a gold-enhanced content alloy.(Table 1). Restoration was then finished and luted.

Technique	Alloy	Manufacturer	Au	Pt	Pd	Ag	Cu	Zn	In	Cd
Metal-ceramic	Esteticor Royal	Cendres et Metaux	810	118	20	32	2	1	17	0
Conventional	Qualigold 4	Qualident	685	23	42	125	103	22	0	0
Soldering	Esteticor R No.2	Cendres et Metaux	785	0	20	25	X	X	0	X

Table 2. Composition of the alloys used for the fixed prosthetic restoration

One week after luting, we noticed the appearance of intense colored deposits, irregular in some superficial parts of the cast area. The restoration was removed from the mouth.

Surface analysis

The superficial analysis of the cast part (Figure 2) by the scanning electronic microscope (SEM) showed three different areas:

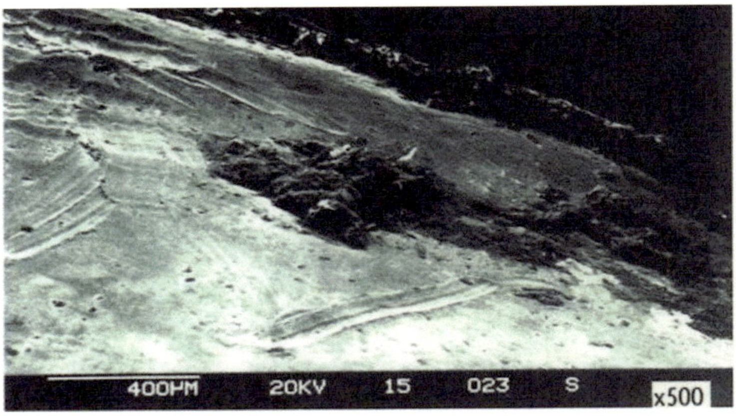

Figure 2. SEM observation (500 x magnification) of the corroded crown surface

Areas with no deposits and corrosion-free deposits

The spectrum obtained by energy-dispersive X-ray spectroscopy (EDX) reflects a normal nominal alloy composition (Figure 3).

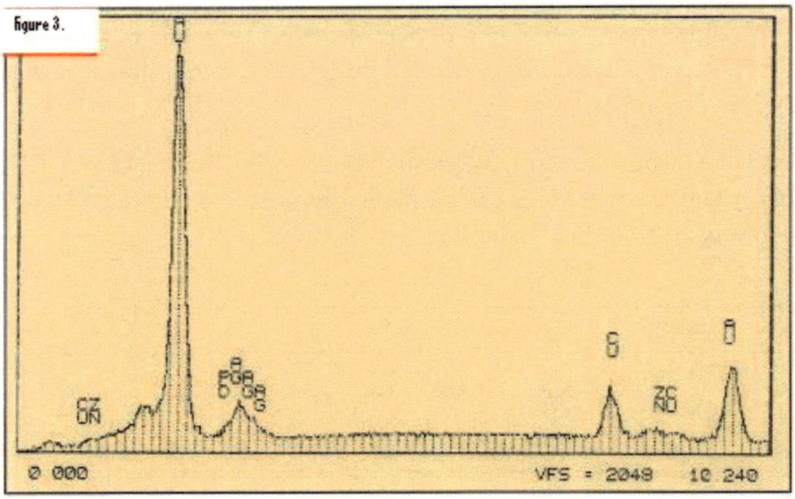

Figure 3. Energy-dispersive X-ray spectroscopy of the cast alloy

Black-bluish colored areas, without deposits

In these areas the laser ionization surface analysis revealed formation of copper and silver sulfides. Sulfur concentration decreases rapidly depending on the analysis depth.

Areas characterized by relatively thick whitish deposits

In these areas, the analysis shows the presence of sulfur, sodium and, potassium chlorides, as well as calcium phosphates (Figure 4). We also noticed a high concentration of copper (Figure 4) compared to that measured at the nominal composition (Figure 3).

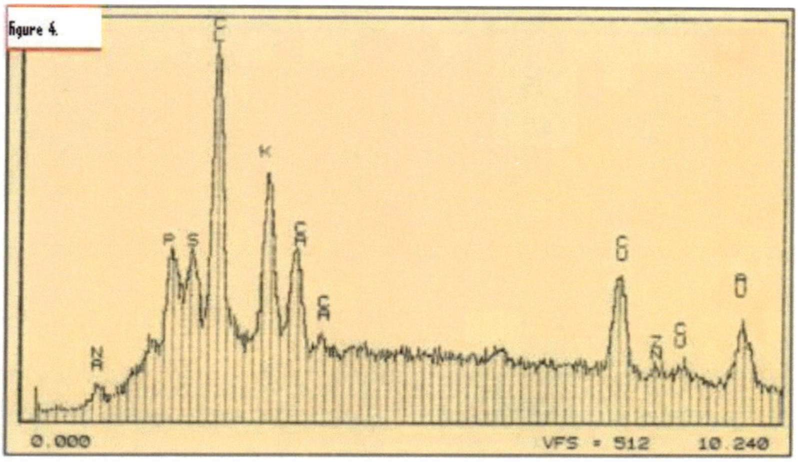

Figure 4. Energy-dispersive X-ray spectroscopy of the whitish deposit

Section analysis

Energy-dispersive X-ray spectroscopy shows that the segregations are nodules of copper and zinc oxide (Figure 5, 6).

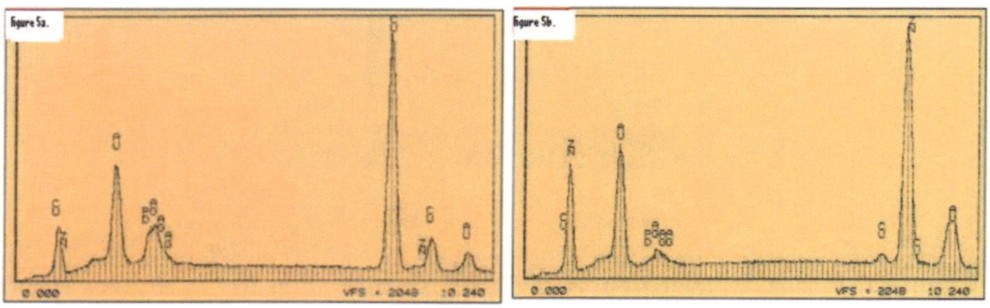

Figure 5. Energy-dispersive X-ray spectroscopy of a segregation consisted mainly of copper oxide (a) and of zinc oxide (b)

Optical microscopy of a metallographic section of the cast part does not reveal defects in the internal structure of the alloy. The negligible percentage of the internal porosity demonstrates a correct soldering [1]. On the contrary, the section observed by scanning electronic microscopy, with a 5000x magnification, shows an important segregation in the cast part, with a depth order of 10-20 microns (Figure 7).

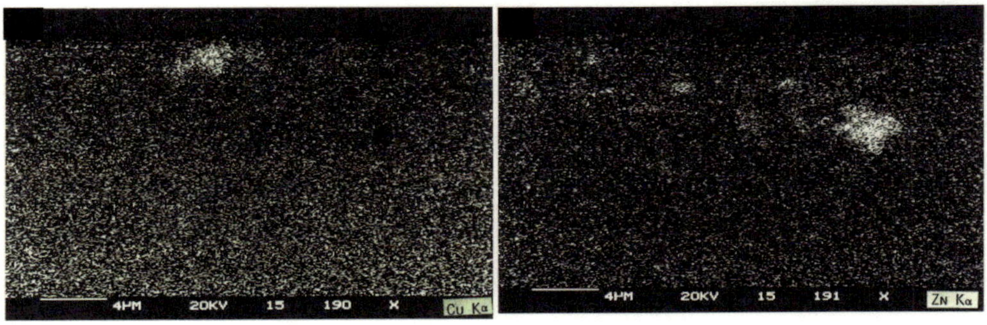

Figure 6. Picture of the copper density and of zinc density observed on the same area and same magnification as illustrated in Figure 7

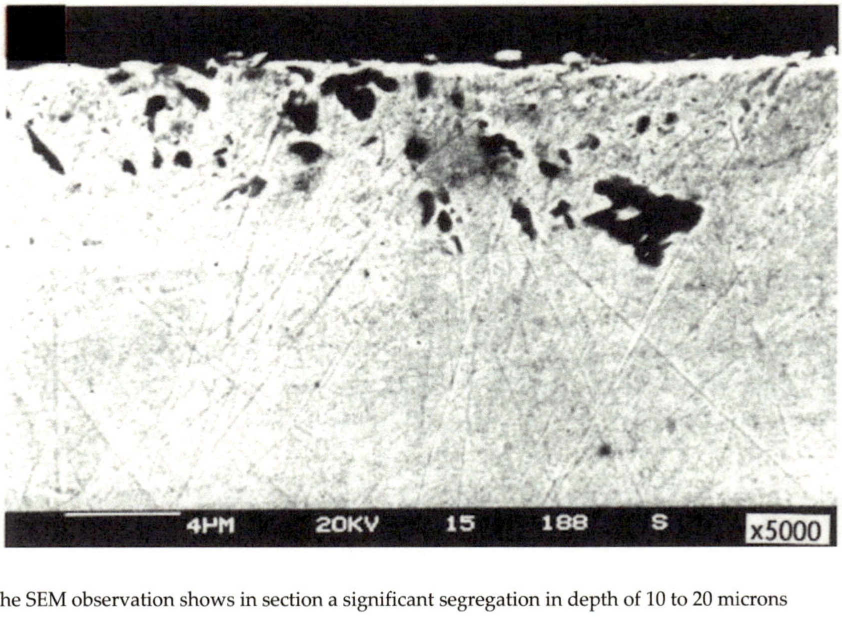

Figure 7. The SEM observation shows in section a significant segregation in depth of 10 to 20 microns

The nodules observed originate from the internal oxidation of the less noble alloy elements, under the effect of heat treatment represented by secondary bonding in the furnace [10]. We are dealing here with a case of localized corrosion [7] in which the less noble parts of the surface consist of copper and zinc oxide nodules.

Thus, a mixed potential (corrosion potential) has been created on the same surface due to the areas of different chemical composition [9]. In this case, the anodic parts are areas consisting of copper and zinc oxide nodules. We cannot take into account the galvanic corrosion between the metal-ceramic and the cast parts, because the localized corrosion described above could have also occurred in the absence of the metal-ceramic part.

A galvanic cell can still accompany a localized corrosion process. In this case, the main cause of corrosion is the appearance of a mixed potential on the surface of the cast part.

The same bridge rebuilt with the same materials, but with a correct finishing phase, including a deoxidation in an etching salt and planted to the same patient has not shown signs of corrosion [11].

The omission of certain technological phases can have severely adverse consequences to the quality and durability of prosthetic restorations. In this case, the consequence was the occurrence of a localized corrosion following luting in the mouth.

3.2. Surface condition influence on galvanic corrosion

In order to illustrate the importance of the surface condition on galvanic corrosion, we performed tests in sections consisting of two plates of different noble dental alloys, joined by a solder point (Figure 8).

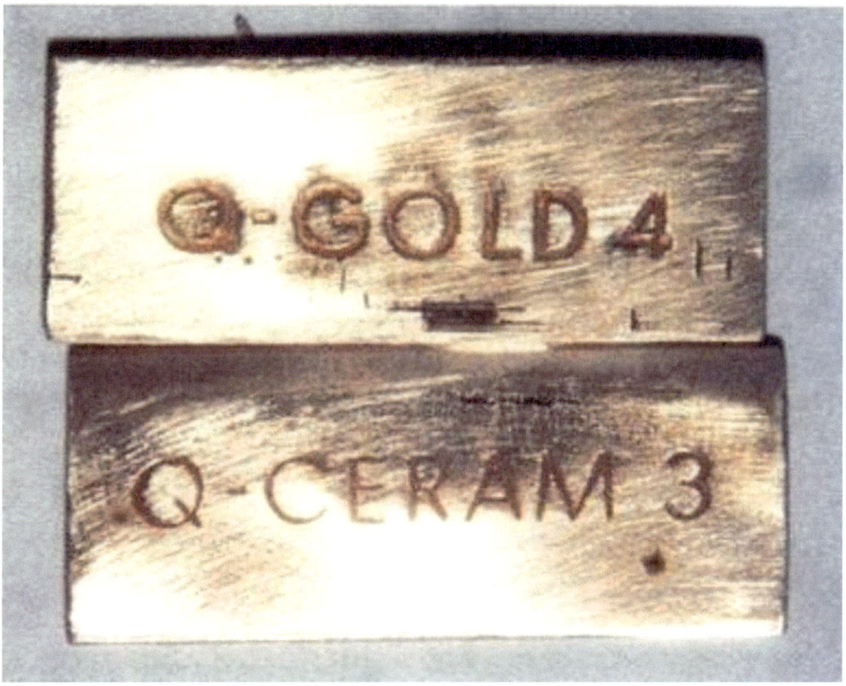

Figure 8. The corrosion test samples are two plates of noble alloy connected by soldering

The two chosen alloys (Table 3) are an alloy for the metal-ceramic technique (Qualiceram 3, Qualident) and an alloy for the conventional technique (Qualigold 4, Qualident).

The two alloys are characterized by very different electronic properties. The secondary soldering was done with a noble gold soldering alloy (Table 3).

Type	Alloy	Au	Pt	Pd	Ag	Cu	Zn	In
Metal-ceramic alloy	Qualiceram 3 (Qualident)	84,4	7,9	4,6	0	0,4	0	2,6
Conventional alloy	Qualigold 4 (Qualident)	68,5	2,3	4,2	12,5	10,3	2,2	0
Soldering alloy	Qualisold S2 (Qualident)	75	0	0,3	4,8	11,4	3	5,5

Table 3. Composition % weight of the couples tested for corrosion

Five samples were prepared by soldering in the oven at 860°C. The sections then underwent different surface treatments (Table 4): no treatment (1), cleaning by a slight polishing with a polishing paste (2 and 3), chemical etching in a hot etching salt (4) and abrasion with ceramic abrasive disks followed by polishing and chemical etching (5).

Sample	Couple	Surface treatment after soldering
1	Qualiceram 3/Qualigold 4	No treatment
2	Qualiceram 3/Qualigold 4	Cleaning by brush and polishing paste
3	Qualiceram 3/Qualigold 4	Cleaning with rubber and then with polishing paste
4	Qualiceram 3/Qualigold 4	Chemical etching in a hot etching salt
5	Qualiceram 3/Qualigold 4	Abrasion with ceramic abrasive disks followed by polishing and chemical etching

Table 4. Defining the surface treatments after soldering, before the corrosion testing

Human artificial saliva may vary to a considerable degree and is dependent on the age and sex of the patient, time of day, eating habits, medication and oral hygiene [12]. The corrosion test consisted in immersion of the samples in Fusayama artificial saliva (detailed composition in Table5) at 37°C for 30 days. We selected Fusayama artificial saliva because it was shown to produce results that were consistent with the clinical experience of dental alloys. The pH of such solution is about 5.6 [13].

The test results are illustrated in Figure 9.

Samples 1, 2, and 3. The samples subjected to only a slight polishing or no treatment have significant corrosion product deposit.

NaCl	0,40 g/l
KCl	0,40 g/l
$CaCl_2 \cdot 2H_2O$	0,79 g/l
$NaH_2PO_4 \cdot H_2O$	0,69 g/l
Urea	1,00 g/l
$Na_2S \cdot 9\ H_2O$	0,005 g/l
NH_2CONH_2	1.00 g/l
Distilled water	Up to 1000 ml

Table 5. Artificial saliva Fusayama type

The blue stains show the formation of copper-based compounds, deriving from the alloys for the conventional technique.

Sample 4. The sample subjected to chemical etching reacted very little. Only a slight gloss change is visible.

Sample 5. The sample subjected to surface abrasion, and then to a polishing followed by chemical etching showed no sign of corrosion or any other change.

We also observed that areas protected by a fusing agent during soldering were not corroded.

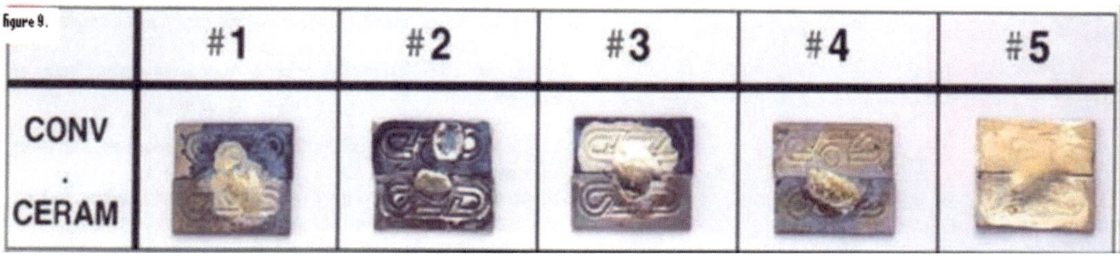

Figure 9. Illustration of galvanic corrosion observed in the 5 samples (1-5)

The corrosion observed on the artificially created surface shows that the passivity and non-passivity phenomena of the surfaces strongly influence the formation and functioning of a cell.

Soldering, which is linked to a thermic treatment, produces a more or less thick layer on the surface, layer consisting of various oxides. In case of alloys for the conventional technique, this layer is typically a few tenths of microns thick and contains highly corrosive oxides in saliva (as copper oxide, zinc, etc). Thus, the behavior of samples 1, 2, and 3, conditioned by nonremoval of the oxide layer, is under anodic control and leads to a significant corrosion despite the nobility of the two alloys [14].

Sample 4, for which this layer has been practically eliminated, reacted very little. Sample 5, for which the layer was completely removed, did not react at all and did not show any effect of corrosion.

For the noble dental alloy combinations, a surface treatment that removes the oxide layers properly is essential to prevent the consequences, sometimes catastrophic, of the galvanic corrosion incidence [14].

3.3. Corrosion due to improper melting conditions

Only a few months after luting, we noticed a severe color change and the loss of the metallic shine in case of four-element bridge (Figure 10). Its surface presented dark and shiny areas. The bridge (23-26) was manufactured using a low-gold-content conventional class E alloy (Au, Ag, Pd, Cu, Zn). We decided to remove it and have it analyzed for determining the causes that lead to its failure.

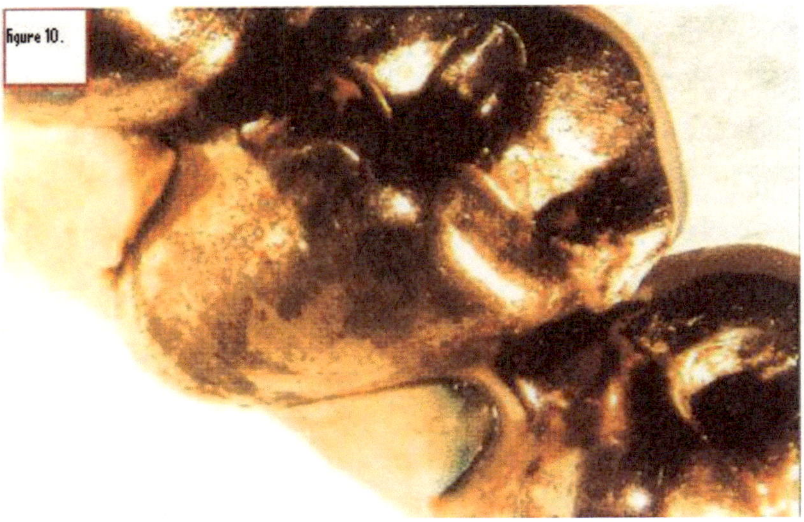

Figure 10. The deteriorated surface of the four-element bridge

For analytic purposes, we prepared two samples, one part of the bridge (24-25) was used to analyze the damaged surface using scanning electron microscopy (SEM) and energy-dispersive X-ray spectroscopy (EDX), without any surface alteration.

The second part of the bridge (26) was transversally sectioned in order to observe the superficial structure of the alloy.

The surface analysis performed with retrodiffused electrons (Figure 11) and energy-dispersive X-ray spectroscopy (EDX) (Figure 12) shows that the coloration is associated with a nonhomogeneous surface composition. We noticed areas without sulfur (area 1, Figure 12), areas rich in copper and sulfur (area 2, Figure 12), and areas rich in silver and sulfur (area 3, Figure 12).

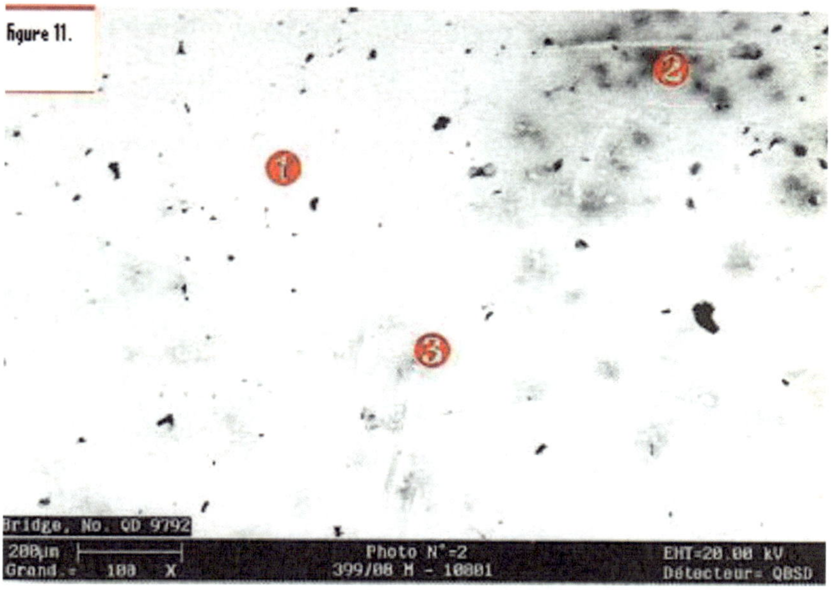

Figure 11. Retrodiffused-electron image (chemical contrast) of the 25-crown surface

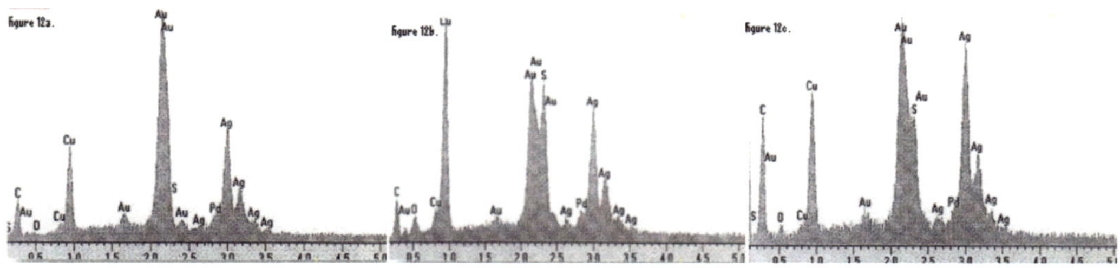

Figure 12. EDX spectrum of (a) area 1 (without sulfur), (b) area 2 (rich in copper and sulfur), and (c) area 3 (rich in silver and sulfur)

Retrodiffused electron observation and the location of the elements (Au, O, Cu, S and Ag) (Figure 13) indicate that the shine loss in area 2 was characterized by excess of sulfur and copper. The presence of oxygen was associated with copper localization.

SEM section analysis using secondary electrons (Figure 14) shows a very important millimetric internal porosity. The porosity area is localized on the external part, in some areas merging to the surface. SEM analysis with retrodiffused electrons (Figure 14) shows a dendritic nonhomogeneous structure, close to the surface.

The main alloy elements (Au, Ag, Cu) are positioned close to the surface, mainly in the proximity of the porosity areas, in a nonhomogeneous way. This nonhomogeneity is more important in the sub-superficial darkened areas than in the sub-superficial shiny areas. The

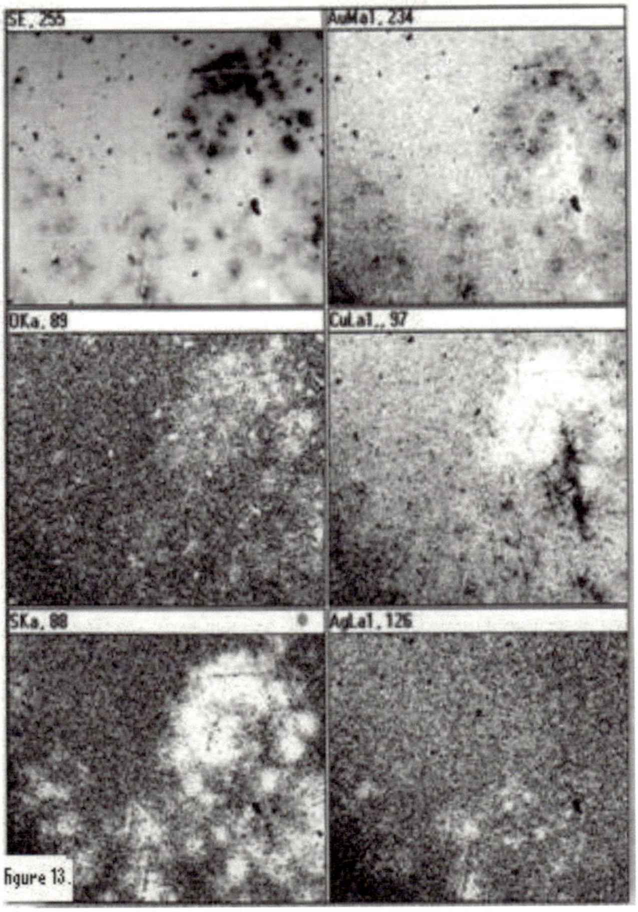

Figure 13. Retrodiffused electron analysis (chemical contrast) and allocation of the elements (Au, O, Cu, S, and Ag)

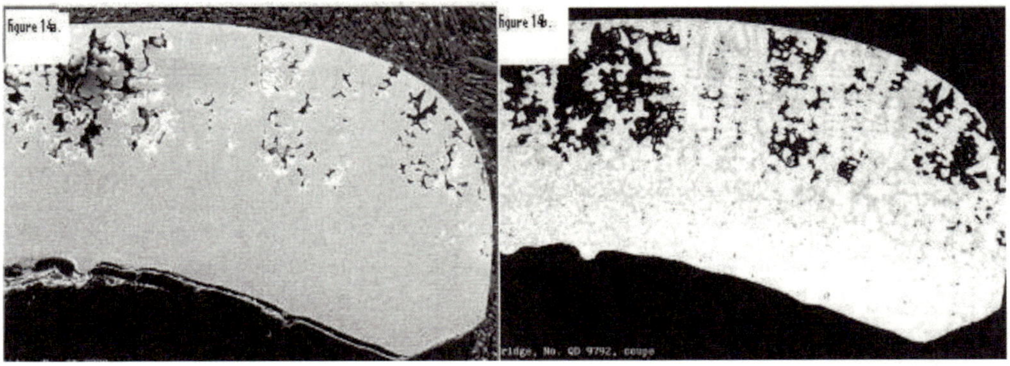

Figure 14. Section observation with (a) secondary electrons (topographic contrast) and with (b) retrodiffused electrons (chemical contrast)

http://dx.doi.org/10.5772/61115

porosities are colored in black and are characterized by the presence of tiny inclusions very rich in Cu (Figure 16).

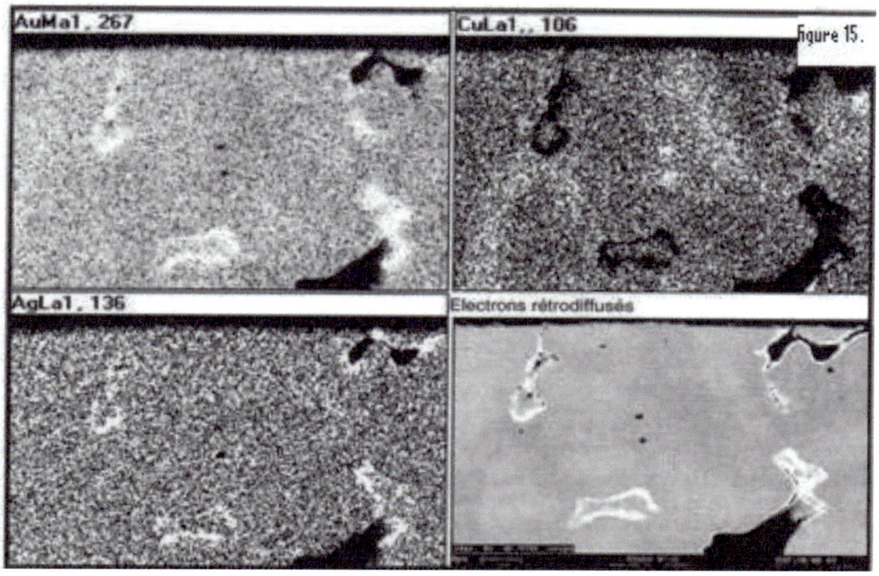

Figure 15. Observation of the sub-superficial layer of the dark area with retrodiffused electrons (chemical contrast) and element allocation (Au, Ag, and Cu)

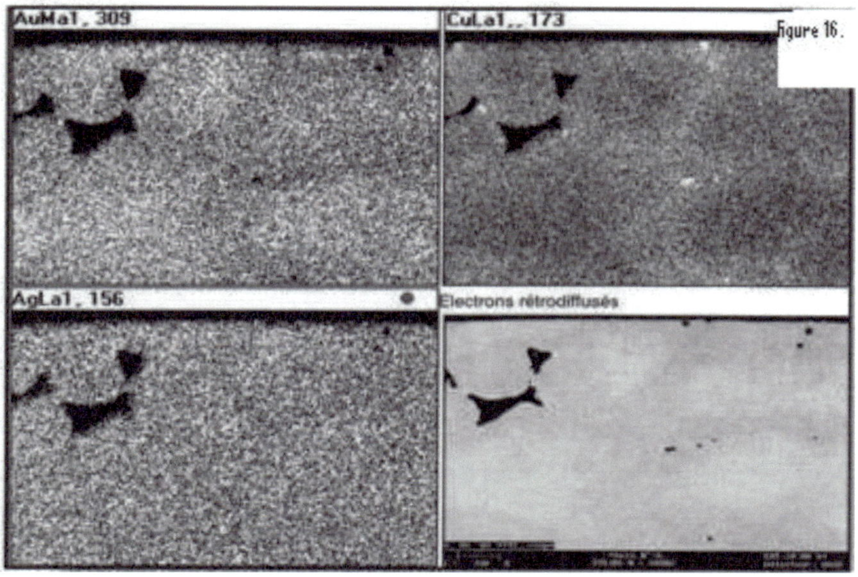

Figure 16. Observation of the sub-superficial layer of the shiny area with retrodiffused electrons (chemical contrast) and the element allocation (Au, Ag, and Cu).

Comparing the nominal chemical composition to that of the analyzed areas (areas I, II, III,and IV), we noticed differences (Table 6). As known, any change in the alloy's composition (decreasing or increasing with one or more elements) has negative consequences for the corrosion resistance [7]. The corrosion process will be accelerated by the unstable phases in the alloy structure. A galvanic cell usually appears between these unstable phases and the alloy matrix, with the appearance of subsequent selective corrosion [15, 7]. In our case, the shine loss represents its result. Composition analysis shows that sulfur and oxygen are present. Areas rich in copper or silver reacted with the sulfurous compounds in the saliva, which explains their presence in the darkened areas. In this case, a selective corrosion process is involved, with formation of insoluble chemical products (sulfides, copper, and silver) [16].

	Au*	Ag*	Cu*
I	560	243	197
II	574	237	189
III	611	354	35
IV	253	123	624
‰ weight			

Table 6. Relative chemical compositions in Au, Ag, and Cu measured in four areas: I and II internal composition of the matrix and external composition of the inclusions; III inclusions rich in Au and Ag (Figure 15); and IV inclusions rich in Cu (Figure 16).

The failure is due to the incorrect alloy usage. The alloy's normal melting conditions were not respected. This incorrect melting was the cause for the internal porosities and for the areas with very different superficial compositions. Furthermore, the surface was not properly polished. The surface, porous and chemically nonhomogeneous, as a result of incorrect melting and surface treatment, led to decreased corrosion resistance and subsequent degradation of the bridge, shortly after luting.

3.4. Corrosion resistance of cobalt-chromium alloys doped with precious metals

Cobalt-based alloys are often used to manufacture different types of devices implanted in the body by surgery. Their applications include hip prosthesis, knee plates, screws for osteosynthesis, and basic structures for heart valves [17]. In dentistry [17,18], cobalt-chromium alloys are mainly used for manufacturing removable partial dentures and metal ceramic fixed partial dentures. In both these cases fine framework constructions are involved. These alloys are characterized by excellent corrosion resistance and superior mechanical properties (e.g., high stiffness) [19]. The dental Co-Cr alloys have diversified over time, aiming at creating both new products and new technologies to process them [20].

A new generation of cobalt-chromium alloys enriched with precious metals (Au, Pt, Ru) are available, their goal being to improve corrosion resistance. In order to verify this hypothesis 4 different such alloys were tested [21,22].

http://dx.doi.org/10.5772/61115

The compositions of the commercial tested alloys and of a "classical" Co-Cr alloy are listed in Table 7.

	Chemical composition (wt %)				
Element	**Co-Cr**	1	2	3	4
Co	63.7	63.5	52.0	50.6	59.3
Cr	28.9	21.0	25.0	18.5	25.0
Mo	5.3		4.5	3.0	5.0
Ga		4.5	6.0		2.5
In		Trace	5.0		1.2
Au		2.0	2.0		2.0
Pt		Trace	2.0	15.0	
Ru				10.0	
Sn			1.0		
Mn	0.8	6.5	0.5	1.0	
Si			2.0	0.75	
W	0.1			0.5	4.0
Nb				0.5	
Al		2.5			
Ti				Trace	
Fe	0.4				

Table 7. Composition of the tested alloys (wt %).

Before electrochemical testing the alloys were analyzed micrographically, analysis of phases by energy-dispersive X-ray spectroscopy (EDX) was carried out, and hardness properties were also tested.

Metallographic structures of the tested alloys are shown in Figures 17-20.

The microstructures of alloys 1 and 4 showed round "inclusions" with a diameter up to 0.1 mm. The chemical analysis of these zones showed In (between 42 and 51 %), Pt (around 28 %), and Au (between 18 and 27 %).

The Vickers tests of such zones for alloy 4 gave a mean hardness value more than twice lower (147 HV) compared to the overall hardness value of the alloy (326 HV).

The hardness values are given in Table 8.

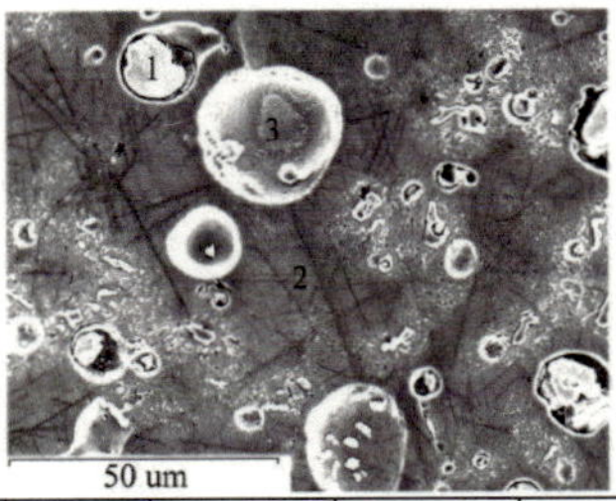

	Co	Cr	Mo	Si	Mn	Ga	In	Fe	Pt	Au
1	60.15	28.42	4.39	1.06	0.50	4.89	-	0.58	-	-
2	1.30	0.62	-	-	-	-	42.77	-	28.31	27.01

Figure 17. Alloy 1: Microstructure and phase composition (wt %)

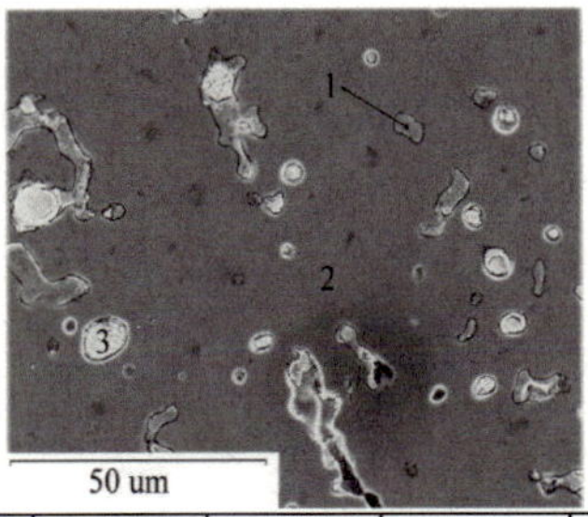

	Co	Cr	Mo	Si	Ti	W	Ga	In	Pt	Au
1	63.48	25.71	3.55	-	-	4.85	2.41	-	-	-
2	28.46	32.83	22.98	1.74	0.95	11.31	-	-	1.73	-
3	1.91	0.76	-	-	-	-	-	13.52	-	83.80

Figure 18. Alloy 2: Microstructure and phase composition (wt %)

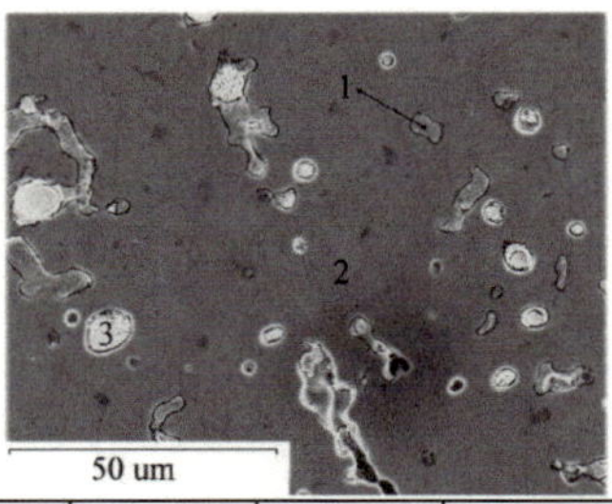

	Co	Cr	Mo	Si	Mn	W	Ti	Nb	Pt	Ru
1	48.96	17.48	3.44	1.02	0.70	1.25	-	-	16.44	10.72
2	42.75	19.51	11.76	4.36	0.88	1.56	-	-	8.25	10.93

Figure 19. Alloy 3: Microstructure and phase composition (wt %)

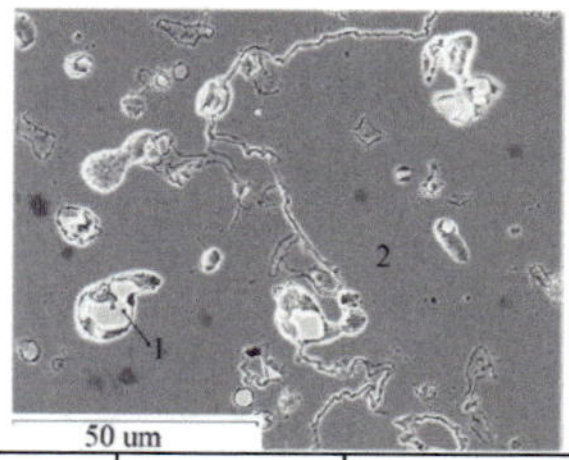

	Co	Cr	Mo	Si	Ga	In	Pt	Au
1	59.61	28.04	5.37	0.63	6.35	-	-	-
2	1.44	0.76	-	-	-	51.17	28.87	17.77

Figure 20. Alloy 4: Microstructure and phase composition (wt %)

Alloys	1 overall	2 overall	3 overall	4 overall	4 zone 2
$HV_{0.2}$	333	435	338	326	147

Table 8. Vickers hardness of the tested alloys (n=5)

Electrochemical measurements were conducted in artificial saliva of the Fusayama type (Table 4) using the rotating electrode technique. The cathodic and anodic potentiodynamic polarization curves were measured from 1000 mV to + 1250 mV vs. saturated calomel electrode (SCE).

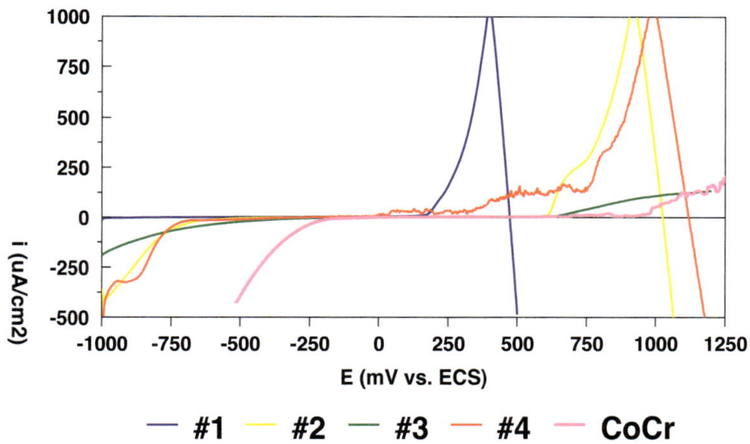

Figure 21. Potentiodynamic polarization curves in linear system of alloys #1 to # 4 in comparison with a Co-Cr alloy

The potentiodynamic curves (Figure 21) show important differences in the behavior of the 4 studied alloys (1 - 4), compared to the Co-Cr alloy. The worst behavior was shown by the alloys containing only gold (1 and 4). Au is not miscible to Co and Cr, confirming existing results [23].

Alloys 1 and 4 showed a very complex microstructure compared to the other studied alloys. The round "inclusions" with a diameter up to 0.1 mm are in part nonmiscible phases with a very low corrosion resistance. Regarding the corrosion behavior, the conventional Co-Cr alloy is the best, followed by alloys 2 and 3 (addition of, respectively, 4 % and 25 % precious metals). The worst behavior was noticed in cases of alloys 1 and 4 (with only addition of 2 % of Au). The presence of precious metals can deteriorate the corrosion behavior of Co-Cr alloys in a significant way. Gold doping, in particular, produces heterogeneous microstructures that are vulnerable to corrosive attack [21].

Scientifically speaking, Co-Cr dental alloys enriched with precious metals do not make sense. Due to their poor behavior, we conclude that using dental Co-Cr alloys enriched with precious metals, as an alternative to conventional Co-Cr alloys, is not justified [24].

4. Biocompatibility assessment

Conceived for a biological environment, dental alloys should essentially be integrated without developing adverse effects and maintaining their function without degrading within an acceptable time limit [25]. Dental casting alloys are widely used in applications that place them into contact with oral tissues for many years. With the development of new dental alloys over the past 15 years, many questions remain about their biologic safety [26].

A biocompatible material may be defined as: inert, nontoxic, non-mutagenic, non-recognizing, nonirritating, and nonallergenic [25].

The single most relevant property of a casting alloy when it comes to its biologic safety is its corrosion. The potential systemic and local toxicity, allergy, and carcinogenicity result from releasing elements into the mouth during the corrosion process. Little evidence supports concerns of casting alloys causing systemic toxicity. Local toxic effects (adjacent to the alloy) are not well documented, but it represents a higher risk, because local tissues are exposed to greater concentrations of released metal ions. Several elements such as nickel and cobalt have relatively high allergic potential, but the true risk of using alloys containing these elements is not certain. Prudence dictates that alloys containing these elements should be avoided as much as possible. Several elements in casting alloys are known mutagens, and a few, such as beryllium and cadmium, are known carcinogens in different chemical forms. Carcinogenic effects due to dental casting alloys usage have not been yet demonstrated [26].

The negative effects that the materials could have on the tissues or on the body at cellular level are precisely induced by the presence of certain components released as degradation products, especially metal cations in solution [27]. In demonstrating that the surface of the materials is degrading in contact with the biological environment due to corrosion, it still has not been established which constitutive elements of the respective materials are more apt to dissolve or in what quantity. The degradation of metallic dentures in a biological environment is accompanied by the release of cations such as Cr, Co, Ni, and Ti. Therefore, we are faced with a cumulative effect: allergic, irritating, mutagenic, and toxic [25]. The release of cations like nickel may induce allergies. To prevent contact dermatitis, the European Directive 94/27/EC prohibits selling products that come into prolonged contact with the skin, if they release more than 0.5 ìg/cm^2/week of nickel. Some dental alloys as copper-aluminum bronzes, which appeared in the 1980's as a substitute for conventional gold-rich alloys and were used for a certain time, release about 100 times this amount [28].

The number of published studies addressing the mutagenicity of dental materials is low. So far, no published clinical reports document the carcinogenic effect of certain dental materials. The long exposure time that is necessary for a malignant tumor to show is a very aggravating factor for the clinical assessment of potential carcinogenic properties of dental materials [29].

A mutagenic substance is a substance capable of inducing mutations, in other words capable of inducing the occurrence or loss of a hereditary characteristic. For a chemical substance to become mutagenic, it should generally induce a primary DNA damage. Mutations can affect a single gene (genic mutations) or the number of chromosomes in a cell (chromosomal mutations). Several researches show that the majority of mutagens are potentially cancerous, and the majority of carcinogens have mutagenic properties [25]. According to the International Cancer Research Centre, metals are classified in two major families: metals having definite (Ni derivates, Cr+6, Cd and it's derivates, Be and its derivates) or potentially cancerous properties in humans and metals that have no cancerous potential. Metals not classified as cancerous but relevant as mutagens are: Sn^{+2}, Cu^{+2}, and Fe^{+2}. Metals not relevant as mutagens include Zn. Metals with none or limited references as mutagens are Cu^{+1}, Sn $^{+4}$, Au, Pt, Ag, Pd, In, and Ga [25].

In Europe, the main threat to health is nickel allergy. In the last years the dental alloys market has undergone dramatic changes for reasons of economy and biocompatibility. In certain countries, nickel-based, cheaper alloys have increasingly been subjected to more and more regulations or even banned. In the other countries, less expensive alloys are still the most used. Despite the restrictions imposed by the EU, the use of Ni-Cr alloys is on the increase [4]. In these current circumstances, there are some questions to be faced regarding the safety risk of nickel contained in dental alloys. In recent years, palladium and, to a lesser extent, nickel were frequently viewed as harmful when used in dental alloys [30,31]. All evidences currently available indicate that dental alloys for prosthodontic restorations are not carcinogenic [29].

Our studies are up to date, concerning cytotoxicity of dental alloys and their components involved in in vitro and in vivo tests.

4.1. Ni-Cr alloy testing for cation release

In [4] eight Ni-Cr dental alloys were evaluated for corrosion resistance (generalized, crevice, and pitting) and the quantities of cations released, in particular nickel and Cr+6 in relation with their microstructure. Cytotoxicity tests and an evaluation specific to TNF-alpha were made. The results showed that their corrosion behavior is weak and that nickel release is high. The quantities of nickel released are higher than the EU limits, in cases when contact with the skin or piercing is involved. The biological tests do not show any cytotoxic effect on HeLa and L929 cells nor any change in TNF-alpha expression in monocytic cells. The alloys do not show any proinflammatory response in endothelial cells as demonstrated by the absence of ICAM-1 induction. Therefore no direct relationship between the in vitro biological evaluation tests and the physicochemical characterization of these alloys may be proved [4].

4.2. Ni-Cr and Co-Cr alloy assessment on cell cultures

At sufficiently high concentrations, metal ions alter cellular metabolism or lead to cell death [29].

Tests for Ni-Cr and Co-Cr dental alloys, the most commonly used in practice, were carried out [32]. The test cell cultures of pure cell line dermal fibroblasts HDFa and of those obtained from skin biopsies were used, for both dental alloys and their eluates. The results were compared with control samples. Seven days after inoculation, the relative similarity between the Ni–Cr alloy and the Co–Cr alloy was observed. The cells did not detach from the plate and grew to the edge of the material. (Figures 22, 23) In the case of the eluates, there were no fragments detached, the cells having a relatively high confluence. The cytotoxic effects of the two alloys seem similar, even if there are speculations in the literature according to which Ni–Cr alloys have a higher effect [33]. All the samples we used during the study maintained their initial characteristics. The cytotoxicity of the alloys tested had minimal in vitro effects on fibroblasts from cell culture.

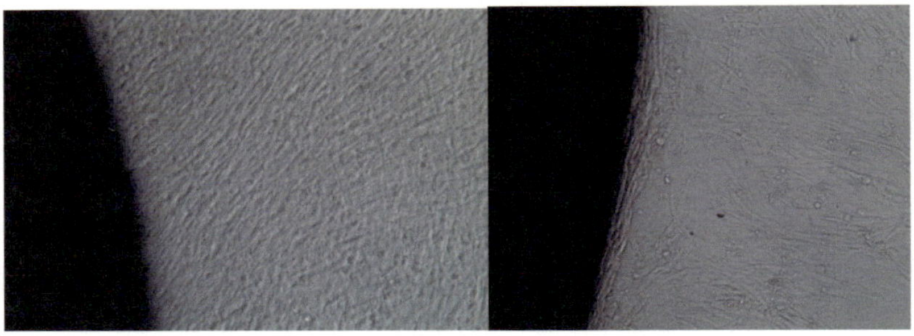

Figure 22. Pure cell line of dermal fibroblasts in direct contact with Ni-Cr and Co-Cr alloys, after 7 days (10 x)

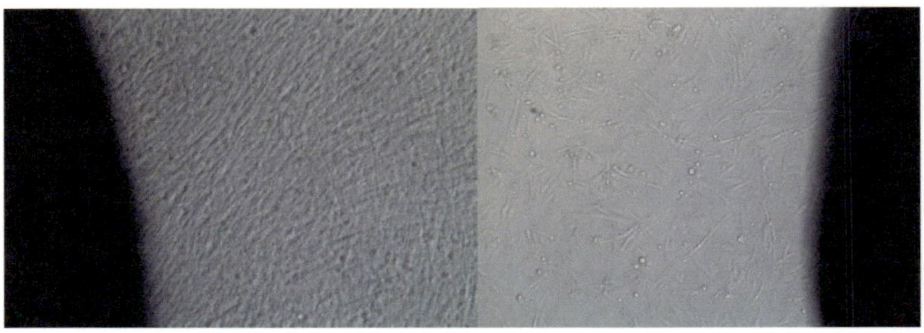

Figure 23. Primary culture of skin fibroblasts in direct contact with Ni-Cr and Co-Cr alloys, after 7 days (10 x)

4.3. In vivo tests

For in vivo testing we used the patch test (Figure 24). The patch-testing led us to the following results: sensitivity to gold, 5 %; palladium, 7 %; nickel, 30 %; and cobalt, 2 % [32]. It is not recommended to patch test an alloy before applying it ("prophetic examination") because the patch itself may sensitize the patient. Furthermore, a negative patch test is no guarantee of current or future absence of hypersensitivity [29].

In order to furthermore test the in vivo tolerance to metals we also used the MORA electronic device as an alternative to patch testing [34]. We considered, for testing with MORA, 55 different dental alloys, as well as the chemical elements which are components of these alloys in 100 individuals with good oral health condition. The results obtained in percentages are shown in Table 9 [35].

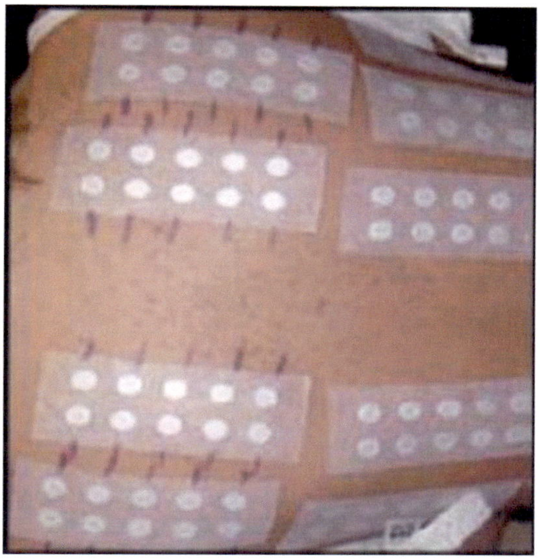

Figure 24. Patch test

Element	Au	Ag	Pt	Pd	Ti	Cr	Co	Ni	Cu	Ru	Zr
Well tolerated	10%	30%	10%	0%	20%	10%	0%	20%	10%	0%	40%
Intolerated	10%	20%	20%	40%	10%	30%	20%	30%	10%	10%	20%

Table 9. Results in percentages

4.4. Risk assessment of Pd genotoxicity

Regarding risk assessment of the genotoxicity of metals and metal alloys, mutagens, oncogenes, and toxic materials for reproduction, our current work is limited to the aspects of the toxicity from the point of view of the mutagenic potential of the palladium in $PdCl_3$ form [34].

The essays of gene mutation were conducted on cell cultures of mouse lymphoma L5178Y according to the protocol Test No. 476 OECD: In vitro Mammalian Cell Gene Mutation. The test allows the detection of chemically induced gene mutations. In the cell lines, the most commonly used genetic endpoints measure mutation at thymidine kinase (TK) and hypoxanthine-guanine phosphoribosyltransferase (HPRT) and a transgene of xanthine-guanine phosphoribosyltransferase (XPRT). The test is conducted on mouse lymphoma L5178Y. Cell L5178Y is treated with $PdCl_3$ until it reaches the limit of the toxicity. After several subcultures, trifluorothymidine (TFT) is added to the cultures and the cloning of mutant cells is carried out on a multiwall cell culture. At the same time there is a cloning of a sample of cells in the absence of TFT to determine how the cellular component survives the treatment. The number of mutant cells compared to the number of surviving cells allows the calculation of the frequency of transfer and is compared to untreated control cells. The results (Table 10) show that in the absence of metabolic activation no significant mutagenic effect is observed [36].

Quantity of $PdCl_2$ in ug/l dose	With metabolic activity		Without metabolic activity	
	% survival	Incidence of mutation**	% survival	Incidence of mutation
0	100	97	100	155
5	189	139	112	172
10	159	136	93	185
20	155	137	76	121
45	151	151	98	170
Positive witness cells*	97	1487	55	166

*10 ug/l methylmethanesulfonate without metabolic activation; 2 ug/l cyclophosphamide with metabolic activation

**Calculated for 10^6 cells

Table 10. Test results

5. Conclusions

The biocompatibility of dental casting alloys is a critical issue because these alloys are in long-term intimate contact with oral tissues [36].

Alloys used in dental restoration must have an appropriate corrosion resistance in order to avoid the release of cytotoxic or sensitizing elements into the biological milieu [28]. Of great importance, besides the alloy itself, are the manufacturing conditions as well as the environmental ones. The corrosion of dental alloys may be significantly increased by improper processing (e.g., formation of pits, crevices, or gold coating). Nickel alloys are especially susceptible to lower pH [29].

Corrosion of dental alloys is a necessary but not sufficient condition for adverse tissue reactions. Good corrosion resistance should be considered important criteria for alloy selection.

Electrochemical corrosion of alloys involves the ionization of elements that are released into the environment, e.g., saliva [37]. Initially uncharged elements loose electrons and become positively charged ions as they are released into solution. Corrosion influences other properties of an alloy, such as esthetics and strength. When biocompatibility is concerned, corrosion indicates that some of the alloy components may affect the surrounding tissues. The released elements may or may not cause problems [29]. Elemental release and corrosion occurs regardless of type or composition of the alloy, but the amount may vary a lot.

One of the main factors that influence element release from an alloy is its composition. Some elements, including copper, zinc, and nickel, have higher tendencies to be released than elements such as gold and palladium [29]. Another factor influencing element lability is the phase structure of the alloy. The presence of multiple phases in an alloy shows a greater risk of element release because of the potential electrochemical corrosion among the phases [38]. Surface characteristics of an alloy, (roughness and the presence of oxides) also influence element release. Surface roughness tends to increase elemental release because it concludes in exposing more atoms to the external environment. The oral environment also influences corrosion. Reduced pH significantly increases the corrosion of some alloys, particularly nickel-based ones. Corrosion is also particularly high in crevices, gaps, and pits and in the area of the gingival sulcus ("pitting corrosion" or "crevice corrosion") [4].

Furthermore, the surface composition can be significantly different from the composition of the bulk of the alloy [39]. The surface composition may have a direct influence on which elements are released [40,41].

Non precious dental alloys (Co-Cr) compared to conventional gold alloys reveal generally an inferior corrosion resistance (pitting, crevice corrosion).

Attempts to obtain alloys with better properties by doping Co-Cr alloys with precious metals were not successful. Some cheaper alloys as copper aluminum bronze, developed to substitute gold dental alloys, proved to be a great failure.

Studies concerning superalloys as candidates for medical applications are encouraging and this may be the future in dental alloys.

The cytotoxicity of the alloys is complex and is still not fully understood. The biocompatibility of dental materials is a critical concern with the development of new products [42].

Dental alloys may, depending on their composition, damage cells in culture. It has been documented that some alloys are cytotoxic over a longer period of time in vitro [38] or change human gingival cells in vitro [41]. Cellular damage may be correlated to elemental release from the alloys [43,44,38]. In accordance with present studies, multiple-phase alloys, which generally have higher corrosion rates, are more cytotoxic than single-phase alloys [29].

Correlations between the release of metal ions and cytotoxicity are very complex. Dental alloys that do not cause cell damage also release metal ions into the cell culture medium. Obviously, ion concentration and exposure time are not sufficient in these cases to cause cell damage. The release of metal ions is not sufficient to cause cell damage in every case. Results from in vitro tests have limited value in predicting a patient's reaction when exposed to certain alloys [29]. Most in vitro tests use only short-term exposures, in contrast with long in vivo exposure. For in vitro tests single-cell types that are often specifically altered to grow outside the body are used, and it is very uncertain if these cells react the same way as body cells. In vitro tests do not generally cover interactions between various cell types, which is a frequent feature of biological reactions in vivo. Obviously, these tests can only evaluate the general biological characteristics of materials [29].

Information on the carcinogenic activity of elements in dental alloys is incomplete or unavailable. There is little or no evidence from the dental literature that indicates that dental alloys are carcinogenic [29]. On the other hand, there is literature that documents the mutagenic potential of metal ions. Mutagenicity can be measured in bacterial systems or in mammalian cells. The reliability of these in vitro systems in predicting in vivo mutagenesis or carcinogenesis is currently limited at best [29].

Overall, there is no evidence that dental alloys cause or contribute to neoplasia in the body. As with toxic and allergic reactions, alloys must release elements for mutagenesis to occur. The form of the metal is critical to its mutagenic activity. Cr^{3+} is not a mutagen, but Cr^{6+} is. The molecular form of the metal is also important. Nickel ions are weak mutagens, but nickel subsulfide (Ni_2S_3) shows a high mutagenic potential [43]. Therefore, a metal is not mutagenic or carcinogenic per se because its mutagenicity depends on the specific form and oxidative state of the metallic element in question [29]. More recent in vitro studies with palladium and gallium chloride indicated a weak mutagenic potency of these ions [45].

It is obvious that metallic ions may act as mutagens or carcinogens in certain forms or using specific routes of exposure. Literature data were primarily determined through long-term epidemiological studies and are subject to limitations associated with these types of studies. For example, it is not correct to presume that a correlation between metal ion exposure and carcinogenesis proves a cause-and-effect relationship. Further evidence is needed, this being an active area of research, particularly regarding nickel, arsenic, cadmium, and other environmentally important metals [29].

To minimize biological risks, dental surgeons should select alloys that have the lowest corrosion potential. This goal can be achieved by using high-noble or noble alloys with single-phase microstructures. Selection of an alloy should be made on a case-by-case basis using corrosion and biologic data from dental manufacturers [29]. Success or failure depends on a variety of factors, some of them depending on the producer, some on the proper manufacture and some close related to the patient itself.

http://dx.doi.org/10.5772/61115

Author details

Lavinia Ardelean[1*], Lucien Reclaru[2], Cristina Maria Bortun[3] and Laura Cristina Rusu[1]

*Address all correspondence to: lavinia_ardelean@umft.ro

1 Department of Technology of Dental Materials and Devices in Dental Medicine, "Victor Babes" University of Medicine and Pharmacy, Timisoara, Romania

2 Department R&D, PX Holding SA, La Chaux-de-Fonds, Switzerland

3 Department of Dentures Technology, "Victor Babes" University of Medicine and Pharmacy, Timisoara, Romania

References

[1] Ardelean L, Rusu LC. Materiale, instrumente si aparate in laboratorul de tehnica dentara. Timisoara: Eurostampa; 2013

[2] Reclaru L, Ziegenhagen R, Unger RE, Eschler PY, Constantin F. New generation super alloy candidates for medical applications: Corrosion behaviour, cation release and biological evaluation. Materials Science and Engineering 2014;45(12): 411-420

[3] Laurent F, Grosgogeat B, Reclaru L, Dalard F, Lissac M. Comparison of corrosion behaviour in presence of oral bacteria. Biomaterials 2001;22(16): 2273-2282.

[4] Reclaru L, Unger RE, Kirkpatrick CJ, Susz C, Eschler PY, Zuercher MH, Antoniac I, Luthy H. Ni-Cr based dental alloys; Ni release, corrosion and biological evaluation. Materials Science and Engineering 2012;32(6): 1452-1460.

[5] Galo R, Ribeiro R.F, Rodrigues RC, Rocha LA, de Mattos MGC. Effects of chemical composition on the corrosion of dental alloys. Brazilian Dental Journal 2012;23(2): 141-148.

[6] Ardelean L, Reclaru L, Susz C, Rusu LC. Scanning electron microscopy and energy-dispersive X-ray spectroscopy in determining the metallic-shine-loss causes of a low gold content conventional alloy. Revista de Chimie 2012;63(2): 232-234.

[7] Susz C, Reclaru L, Ardelean L, Ghiban B, Rusu L. Aliaje Dentare. Timisoara: Victor Babes; 2010

[8] Huang HH. Surface characterization of passive film on NiCr-based dental casting alloys. Biomaterials 2003;24(9): 1575-1582.

[9] Bayramoglu G, Alemdaroglu T, Kedici S, Aksüt AA. The effect of pH on the corrosion of dental metal alloys. Journal of Oral Rehabilitation 2000;27(7): 563-575.

[10] Pop DM, Rominu M, Topala FI, Sinescu C, Dodenciu D, Rominu RO, Ardelean L, Rusu LC, Andoni M, Petrescu EL, Negrutiu ML. Laser weldings versus electrical weldings in dental technology. A corrosion approach study, Revista de Chimie 2011;62(12): 1203-1205.

[11] Reclaru L, Ardelean L, Rusu LC. Scanning electron microscopy and energy-dispersive X-ray spectroscopy in determining corrosion causes of a fixed prosthetic restoration. Revista de Chimie 2013;64(5): 545-547.

[12] Mareci D, Nemtoi G, Aelenei N, Bocanu C. The electrochemical behaviour of various non-precious Ni and Co based alloys in artificial saliva. European Cells & Materials 2005;10(2): 1-7.

[13] Mareci D, Chelariu R, Iacoban S, Munteanu C, Bolat G, Sutiman D. The estimation of localized corrosion behaviour of Ni-based dental alloys using electrochemical techniques. Journal of Materials Engineering and Performance 2012,21(7): 1431-1439.

[14] Reclaru L, Ardelean L, Rusu LC. Surface condition influence on galvanic corrosion. Medicine in evolution 2013;19(1): 152-157.

[15] Schmalz G, Garhammer P. Biological interactions of dental cast alloys with oral tissue. Dental Materials 2002;18(5): 396-406

[16] Ardelean L, Reclaru L, Susz C, Rusu LC. Scanning electron microscopy and energy-dispersive X-ray spectroscopy in determining the metallic-shine-loss causes of a low gold content conventional alloy. Revista de Chimie 2012;63(2): 232-234.

[17] Ghiban B, Bortun C. Aliaje dentare de cobalt. Bucuresti: Printech; 2009.

[18] Reclaru L, Eschler PY, Reto L, Blatter A. Electrochemical corrosion and metal ion release from Co-Cr-Mo prosthesis with titanium plasma spray coating. Biomaterials 2005;26(23): 4747-4756.

[19] Mareci D, Nemtoi G, Aelenei N, Bocanu C. The electrochemical behaviour of various non-precious Ni and Co based alloys in artificial saliva. European Cells & Materials 2005;8(10): 1-7.

[20] Gupta KP. The Co-Cr-Mo system, Journal of Phase Equilibria and Diffusion 2005;26(1): 87-92.

[21] Reclaru L, Luthy H, Eschler PY, Blatter A, Susz C. Corrosion behaviour of cobalt-chromium dental alloys doped with precious metals. Biomaterials 2005;26(21): 4358-4365.

[22] Ardelean L, Reclaru L, Bortun C, Ghiban B, Rusu L. New aspects concerning Co-Cr alloys, Metalurgia International 2010;15(9): 31-35.

[23] Kappert HF, Schuster M. Edelmetallhaltige Unedellergierungen. Dental Labor 2000;43(3): 352 -354.

[24] Reclaru L, Ardelean L, Rusu LC. Cobalt-Chromium dental alloys doped with gold or platinum. Medicine in evolution 2010;16(1): 69-72.

[25] Reclaru L, Ardelean L, Rusu L. Toxic materials, allergens and mutagens and their impact on the dental field. Medicine in evolution 2008;14(3): 98-102.

[26] Wataha JC. Biocompatibility of dental casting alloys: a review. Journal of Prosthetic Dentistry 2000;83(2): 223–234.

[27] Al-Hiyasat AS, Bashabsheh OM, Darmani H. Elements released from dental casting alloys and their cytotoxic effects. International Journal of Prosthodontics 2002;15(5): 473–478.

[28] Eschler PY, Lüthy H, Reclaru L, Blatter A, Loeffel O, Süsz C, Boesch J. Copper-aluminium bronze – a substitute material for Gold dental alloys? European Cells and Materials 2003;5(suppl. 1): 49-50.

[29] Schmalz G, Arenholt-Bindslev D. Biocompatibility of dental materials. Berlin/Heidelberg: Springer-Verlag; 2009.

[30] Berzins DW, Kawashima I, Graves R, Sarkar NK. Electrochemical characteristics of high-Pd alloys in relation to Pd-allergy. Dental Materials. 2000;16(4): 266–273.

[31] Syverud M, Dahl JE, Herø H, Morisbak E. Corrosion and biocompatibility testing of palladium alloy castings. Dental Materials. 2001;17(1): 7–13.

[32] Rusu LC, Bortun CM, Tanasie G, Podariu AC, Baderca F, Solovan C, Ardelean L. The cytotoxicity of dental alloys studied on cell culture. Romanian Journal of Morphology and Embriology 2014;55(1): 3-6.

[33] Rusu L, Reclaru L, Ardelean L, Podariu AC. Common contact allergens in dental materials, Medicine in evolution 2010;16(1): 15-18.

[34] Ardelean L, Rusu L, Lupas D, Petcov B. Electronic homeopathy in testing compatibility to dental alloys. European Cells&Materials 2008;16(suppl.5) 19.

[35] Ardelean L, Bratu DC, Rusu LC. MORA Device as an Alternative in Testing Compatibility to Dental Materials. Revista de Chimie 2012;63(7): 730-731.

[36] Reclaru L, Ardelean L, Govor I,Tigmeanu CV, Rusu LC. In vitro Toxicity Testing of Palladium Using Trifluorothymidine. Revista de Chimie 2014;65(10): 1149-1153.

[37] Craig RG, Powers JM. Restorative Dental Materials, 11th edn. St. Louis: Harcourt Health Sciences; 2002.

[38] Wataha JC. Principles of biocompatibility for dental practitioners. Journal of Prosthetic Dentistry 2001;86(2): 203–209.

[39] Messer RLW, Lucas.C. Cytotoxicity of nickel-chromium alloys: bulk alloys compared to multiple ion salt solutions. Dental Materials 2000;16(3): 207–212.

[40] Chen CC, Chen RGS, Huang ST. Effect of chemical composition on the corrosion behaviour of Ni-Cr-Mo dental casting alloys. Journal of Biomedical Materials Research. 2002;60(3): 458–465.

[41] Benatti OFM, Miranda WG, Muench A. In vitro and in vivo corrosion evaluation of nickel-chromium- and copper-aluminum-based alloys. Journal of Prosthetic Dentistry. 2000;84(3): 360–363.

[42] Can G, Akpınar G, Aydin A. The Release of Elements from Dental Casting Alloy into Cell-Culture Medium and Artificial Saliva. European Journal of Dentistry 2007;1(2): 86–90.

[43] Messer RLW, Doeller J., Kraus W, Lucas LC. An investigation of fibroblast mitochondria enzyme activity and respiration in response to metallic ions released from dental alloys. Journal of Biomedical Materials Research 2000;50(4): 598–604.

[44] Schmalz G, Garhammer P. Biological interactions of dental cast alloys with oral tissues. Dental Materials 2002;18(5): 396–406.

[45] Al-Hiyasat AS., Darmani H, Bashabsheh OM. Cytotoxicity of dental casting alloys after conditioning in distilled water. International Journal of Prosthodontics 2003;16(6): 597–601.

Chapter 8

Phase Equilibrium Evolution in Single-Crystal Ni-Based Superalloys

Lembit Kommel

Additional information is available at the end of the chapter

http://dx.doi.org/10.5772/61102

Abstract

The phase equilibrium evolution resulting from the interdiffusion of atoms in single crystals of nickel-based superalloys was studied with the aid of microstructural, chemical composition, and micromechanical property investigations. The experimental observation methods—optical microscopy, scanning electron microscopy, transmission electron microscopy, energy-dispersive spectroscopy, microchemical analyses, X-ray diffraction, hard cyclic viscoplastic deformation, and nanoindentation—were combined to obtain new insights into the phases' chemical composition and micromechanical properties' characterization that depend on strain-stress levels which are induced by tension-compression cycling in viscoplastic conditions at room temperature. The test samples with differences in the strain-stress parameters were received on the tension-compression stepped sample with four different cross-section areas. The strains with four levels of intensivity were added by using strain amplitudes of 0%–0.05%, 0%–0.2%, 0%–0.5%, and 0%–1% for 30 cycles, respectively. Microstructural investigations show that dendrite length decreased significantly in samples with minimal cross-section and accordingly at maximal strain-stress amplitudes. The main dendrites of the (001) direction were separated by $(\gamma + \gamma')$-eutectic pools. The length of newly formed dendrites depends on cumulative strain-stress amplitudes. The chemical composition and micromechanical properties of phases were changed as a result of the atoms' interdiffusion between different phases. These changes were influenced on the phases' equilibrium evolution of the single-crystal superalloy during testing.

Keywords: phase equilibrium, evolution, nickel-based superalloys, atom interdiffusion

1. Introduction

Cast single-crystalline (SC) Ni-based superalloys are promising high-strength refractory materials for manufacturing turbine blades (Figure 1) and vanes of turbojet engines as well as power gas turbine blades and polycrystalline turbine discs.

Figure 1. Air-cooled and thermal barrier-coated high-temperature turbine blade for turbojet engine Al31-F.

It is well-known that these materials possess extraordinary strength [1-3], high fracture toughness at fretting failure [4], good resilience under thermomechanical fatigue [5-8], low-cycle fatigue (LCF) [9], and high-cycle fatigue (HCF) [10, 11] as well as high oxidation resistance [12, 13] in gas environments at high temperatures during long-term exposures. The excellent creep-rupture lifetime properties of superalloys are presented in a large number of investigations [14-17]. These excellent exploitation properties of the superalloys depend on the chemical composition [1, 18, 19], solidification parameters [20-22], and casting conditions [23] of the manufacturing process. The dendrite microstructures of SC Ni-based superalloys differ based on their withdrawal, solidification, and cooling rates. The withdrawal rate and temperature gradient have effects on the dendritic microstructure evolution in directionally solidified SC of superalloys. The primary dendrite arm spacing for some commercial superalloys depends on the temperature gradient and withdrawal velocity, and their spacing varied from 50 to 550 μm for a number of commercial superalloys [20-22]. The withdrawal rate and melt overheating temperature, as well as cooling rate, also influence the primary dendrite length [22]. Therefore, the tensile and creep properties of these materials are not simply a function of the withdrawal rate. Usually, the withdrawal rates varied from 50 to 350 mm/h [20-22]. The primary and secondary dendrite arm spacing decreases with increasing withdrawal rate [21, 22]. The fine dendritic structure can lead to increased mechanical properties and higher solution and liquidus temperatures [21]. The cooling rate increase influences both the solidification behavior and the resultant structural and increased chemical microheterogeneity of the as-cast structure of superalloys [23]. As a result, the creep and fatigue strengths were increased by chemical microheterogeneity increases. Therefore, the number of solidification defects, like silvers and freckles in castings, can possibly be decreased by the addition of C, B, and N [18]. Therefore, for superalloys the selection of elements is an importance. The major constituent of commonly cast superalloys is nickel (Ni) and the main alloying elements are Cr, Co, Mo, Nb, Ta, Ti, Fe, V, W, Ru, Re, Ir, Hf, B, C, and Al [1, 20]. These elements have BBC, HCP, and FCC crystal structures (Figure 2). The listed elements with atomic numbers have the highest melting temperatures (Figure 3) and the lowest normalized activation energies for diffusion (Figure 4). To provide the best exploitation properties (lowered creep, high resistance to thermomechanical fatigue, and so on) for turbine blades, the elements can have high activation energies for high-temperature creep [24, 25] and activation energy for self-diffusion [25, 26], as shown in Figure 5. Diffusion in SC Ni-based superalloys under viscoplastic deformation at room temperature is studied in [27, 28]. The elements (W, Ta, Nb, and Re) with the needed higher activation energies for self-diffusion control mass transport via diffusion on the scale of the microstructure of the superalloy. In contrast, Ti has the lowest activation energy (Figure 5) for diffusion, high-temperature creep, and is not a suitable alloying element for casting SC superalloys [28]. The Al forms high-temperature intermetallic compounds in γ'-phase with Ni as well with other alloying elements (Ti, Ta, and Nb). Metal carbides (MC) [29] are formed on the base of C, with Nb and Ta or Ti and Ta.

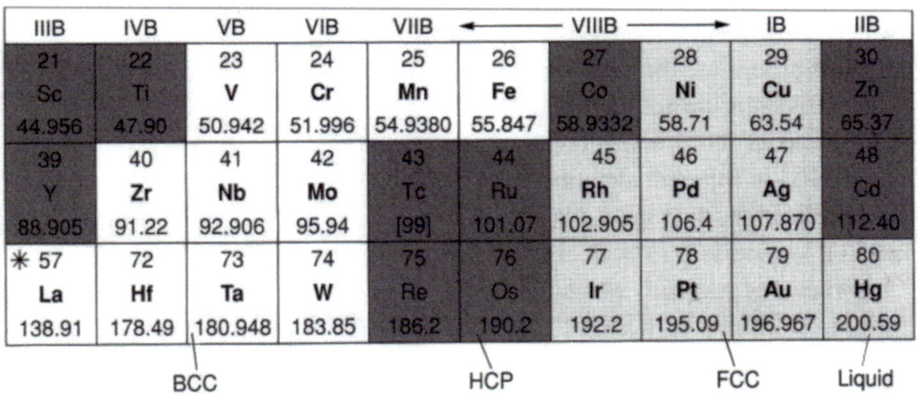

Figure 2. Correlation of the crystal structures of the transition metals (position in the periodic table) used for SC superalloy casting. Adapted from reference [1].

It is shown that MC is beneficial for creep property improvement, but transformed to M_6C, provides micropore formation, crack initiation, and led to final fracture. Thus, the choice of temperature regime for solution/homogenization heat treatment is the key to achieving the best properties for SC cast superalloys [30]. The historical development of superalloys (during 75 years) from their emergence in the 1940s is important [1]. The turbine entry temperature (TET) was improved from 700°C (wrought, uncooled) to 1850°C (SC, cooled, thermal barrier coated) at present (Figure 1). During this period, the high-temperature capability increased as the production technology was improved from wrought to conventional casting, to directional

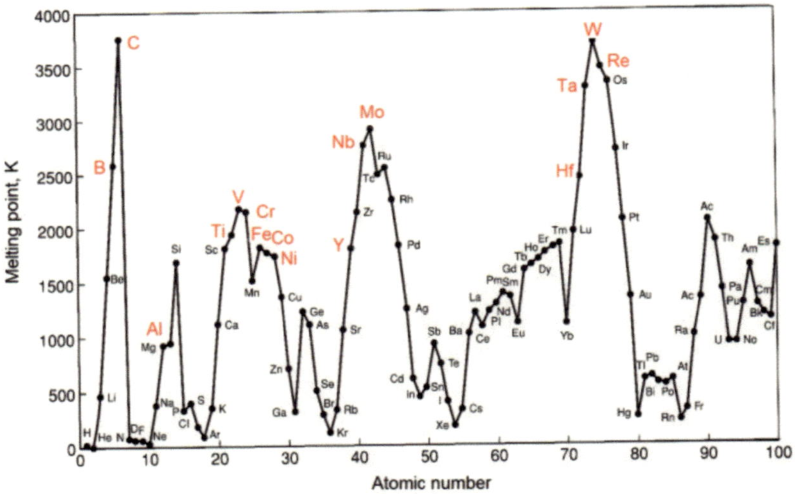

Figure 3. Alloying elements (with increased font, red-colored, mainly used for SC Ni-based superalloy casting, Fe was used for polycrystalline superalloys) melting temperatures for corresponding atomic number. Data taken from reference [1].

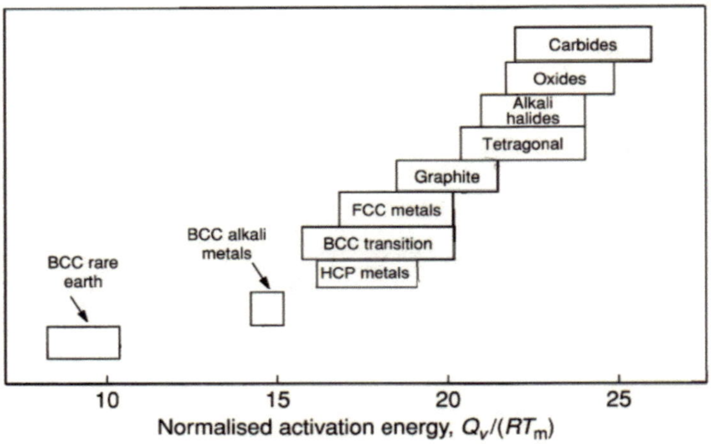

Figure 4. Normalized activation energies of diffusion for the various crystal classes. Adapted from reference [1].

solidification and then to single-crystal superalloy casting in directional solidification installation under high vacuum [20-23, 28-30].

2. Review of conventional high-temperature test methods

During exploitation, the gas turbines experience different temperature cycles and the turbine-entered temperature (TET) varies from ±60°C at ground idle during relative short time [1] up

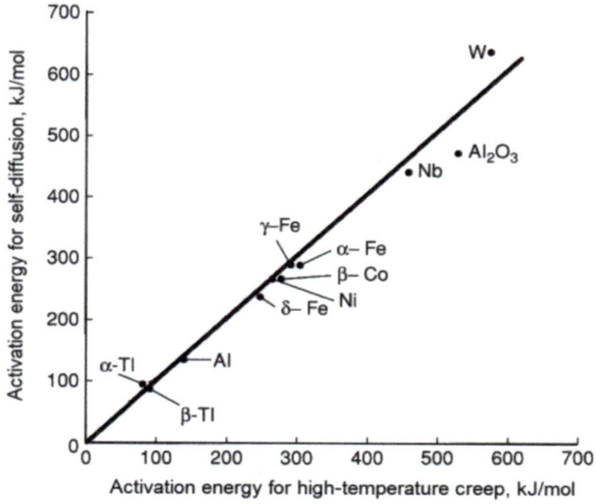

Figure 5. Correlation between the high-temperature creep activation energy and the activation energy for self-diffusion for several elements. Adapted from reference [1].

to 1850ºC during the take-off and climb to cruising altitude during a typical flight cycle. According to the turbine blade and disc materials, with single crystalline and polycrystalline microstructures, respectively, mainly are studied at high temperatures. The high-temperature creep [31-33] and metallurgical failure [34] of turbojet turbine rotor blade materials under mechanical tension, vibration, buffeting, and thermal tension-compression loadings at centrifugal forces during rotation are some of the most important problems in aircraft engines. Gas turbine rotor blades generally experience thermomechanical fatigue [6, 7] as well stresses at different locations in the turbine blade body under loadings in gaseous surroundings at high temperatures. Under low cycle fatigue (LCF) testing conditions at strain control regimes, the fatigue performance depends on the loading axis direction to crystal orientation (Figure 6). The yield stress at tension is maximal for the (001) direction of crystal orientation. The yield stress in tension depends on the temperature and exceeds that of compression. The strain softening of materials increase as temperatures increase. Determining all these influences during the viability testing of a blade material is complicated. Therefore, the test method that is currently used to determine the mechanical properties of Ni-based superalloys is creep testing at elevated temperatures and during very long (100-1000 h) exposure times [14-17, 31-33]. The high-temperature creep testing of superalloys is typically performed at 800°C-1100°C under tension load of 600-200 MPa, respectively. Usually, increased temperatures used lowered tension stresses. The microstructural aspects of high-temperature deformation of mono- or single crystalline nickel-based superalloys are well studied by Mughrabi [31]. The resulting creep dislocations help to form a raft structure (Figure 7), which is elastically unstable. The rafts forming direction depends on the stresses in PDA or SDA of dendrites and the orientations of rafts are "N" or "P" type (Figure 8). Thus, the rafts' orientation also depends on various γ' volume fractions.

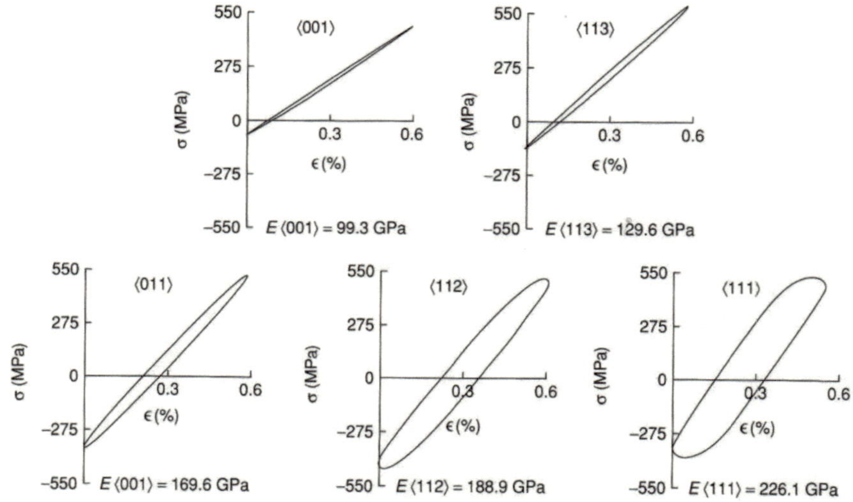

Figure 6. Effect of crystallographic orientation on fatigue hysteresis loops for an experimental SC superalloy at 980°C and with $\Delta\epsilon_{total} = 0.6$, $R = 0$. Data taken from reference [1].

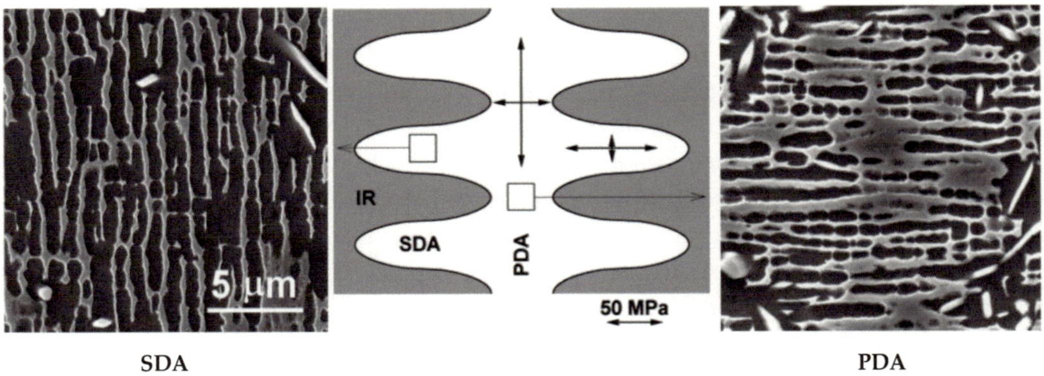

Figure 7. The γ/γ'-microstructure in the dendrite arms of CMSX-4 rafted during annealing at 1100°C under denritic stresses. Dendritic stresses estimated at 1100°C by FE modelling are shown by arrows. Adapted from reference [35].

The mechanical deformation of SC Ni-based superalloys has been theoretically studied by Huang et al. [36] and Yashiro et al. [37] via discrete dislocation dynamics (DDD) modeling. A transversely isotropic viscoelasticity model for a directionally solidified polycrystalline Ni-based superalloy has been developed by Shenoy et al. [38] to characterize the function of temperature and the stress-strain response has been studied at temperatures ranging from 427°C to 1038°C. These models indicate that mechanical deformation occurs because of interactions between dislocations and the internal microstructure. The channel width and precipitate size both affect the flow stress and cyclic hardening or softening of SC Ni-based superalloys. The mechanical deformation in the viscoelastic-plastic field of SC Ni-based

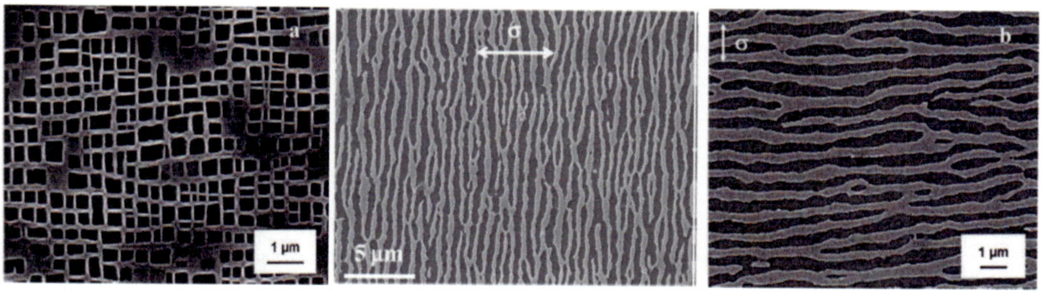

Figure 8. Microstructure of γ/γ′-phase in initial state (a) and at creep deformation formed different raft types. Adapted from reference [15].

superalloys under tension-compression cyclic loading [39] shows that the phases' chemical composition was changed. Instrumented indentation testing at elevated temperatures [40] and nanoindentation testing [41-43] at room temperature have been used to study the mechanical properties of these materials. The phases' micromechanical properties of SC Ni-based superalloys were studied, and the results show that Young's modulus, evaluated by the Oliver–Pharr method, depends on the testing temperature [40], and it was lowered by increases in temperature. The micromechanical properties depend on chemical composition and collected strain [41-43]. It is well known that the turbine-entered temperature (TET) varies from ±60°C at ground idle for relatively short times. Some new methods have been developed for SC NI-based superalloys testing [44, 45] and the different gradually shaped specimens were used. Therefore, in this chapter, we studied the tension-compression stress amplitudes in the field of viscoelastic as well viscoplastic deformation during low number and low frequency of cycles at room temperature. By this method, the cyclic stresses influenced the interdiffusion of elements and atoms between different phases. This method of SC Ni-based superalloy testing is called hard cyclic viscoplastic (HCV) deformation [28, 39, 44, 46]. We estimated the microstructure evolution, phase equilibria shifts, chemical composition phase changes, and changes in the micromechanical properties of different phases in SC Ni-based superalloys during HCV deformation at room temperature.

3. Materials and experimental procedures

3.1. Casting of SC Ni-based superalloy

The test material that was subjected to HCV deformation was the commercially available SC Ni-based superalloy ZS32-vi (Russian standard), which had the following composition (in at. %) of main elements: 12.1% Al, 5.3% Cr, 9.4% Co, 0.8% Nb, 0.9% Ta, 0.7% Mo, 2.5% W, and 0.7% Re, with the remainder being Ni. No Ti was present in this superalloy. The casts of the SC Ni-based superalloy samples were produced by NPO "A. Lylka-Saturn" via an electromagnetic induction melting technique in a directional solidification furnace under high vacuum. The withdrawing rate was ~3.4 mm/min, which yielded solid casts with dendrites of up to ~4.5 mm in length. The as-received SC cast rods were 15 mm in diameter and ~160 mm in length. We speculate that all of the cast test parts of samples possessed identical microstructures and properties before HCV deformation testing; as they were produced using identical technological parameters of casting [see 20-23].

3.2. Initial microstructure of SC cast Ni-based superalloy

The as-cast optical microstructure is shown in the longitudinal (001) direction in Figure 9 and in cross-section in Figure 10a, whereas the γ/γ' microstructure of a cast SC Ni-based superalloy ZS32-vi sample is presented in Figure 10b. The microstructure mine constituent's parts or details are shown in Figure 9 by arrows. As shown, the primary dendrites' length is about 4.5 mm and dendrite arm spacing is about 400-500 μm. The eutectic pools are situated near the primary dendrite axis and in the interdendritic region. Therefore, metal carbides with high concentrations of Nb/Ta content situated near eutectic pools (Figure 12 a) and metal carbides with lower Nb/Ta content in interdendritic region (Figure 12 b) between primary and secondary dendrite axis with γ/γ' microstructure. The phases under consideration in this study are presented in Figure 11. These phases are listed in Figure 11b: (a) primary dendrite axis (PDA) in (001) direction; (b) interdendritic region (IDR) with C-Nb/Ta-arrows (Figure 10); (c) secondary dendrite axis (SDA); (d) eutectic pools (EUT) with partly single γ'-phase in interdendritic region; and (e) Nb/Ta-rich metal carbides (MC) near EUT.

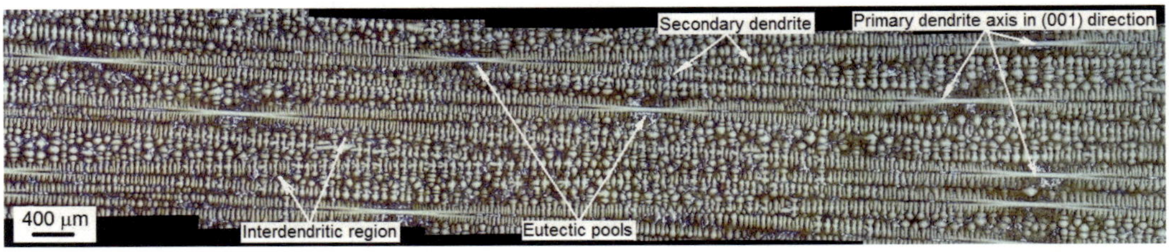

Figure 9. Microstructure of as-cast SC Ni-based superalloy ZS32-vi in longitudinal (001) direction. The microstructure elements are shown by arrows. Adapted from reference [46].

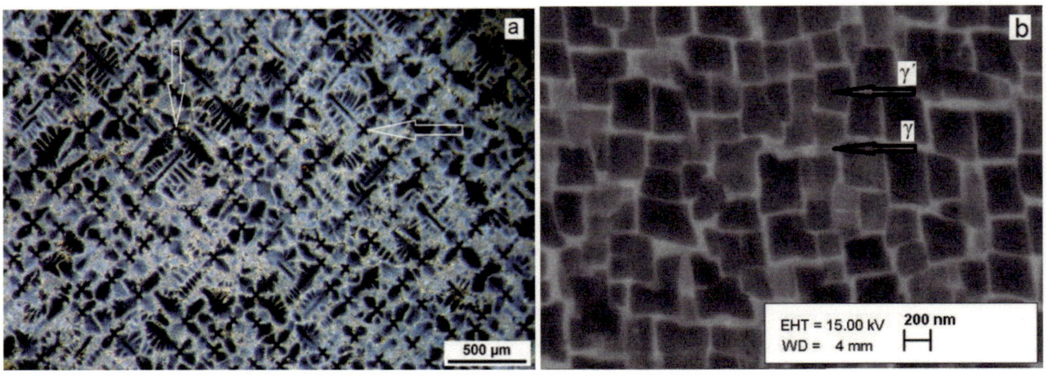

Figure 10. Microstructure of as-cast SC Ni-based superalloy ZS32-vi in cross-section of rod (a) and γ/γ'-phase (b). The primary dendrite quadrature axis and γ and γ' phases are shown by arrows.

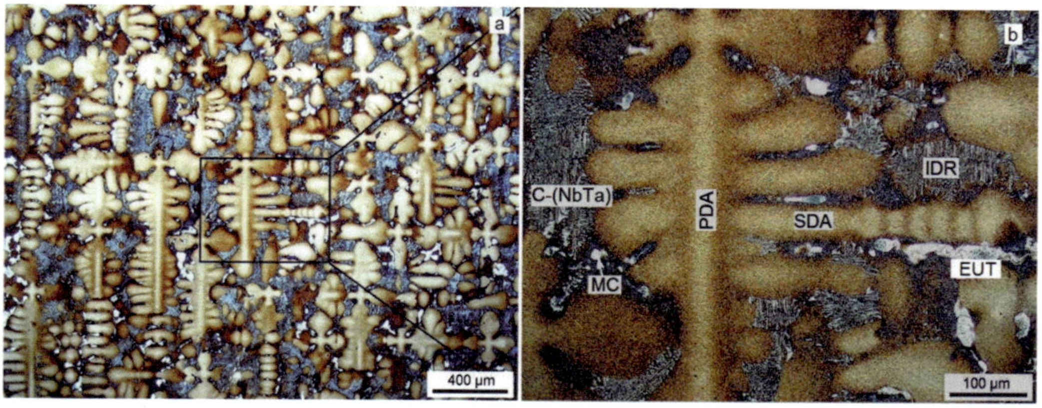

Figure 11. Microstructure of SC in the (001) direction (a) and magnification of dendrite microstructure (b) with the main tested phases which are under consideration. PDA, primary dendrite axis in (001) direction; IDR, interdendritic region with C-Nb/Ta-arrows; SDA, secondary dendrite axis; EUT, eutectic pools in interdendritic region; MC, Nb/Ta-rich metal carbides. Adapted from references [39, 46].

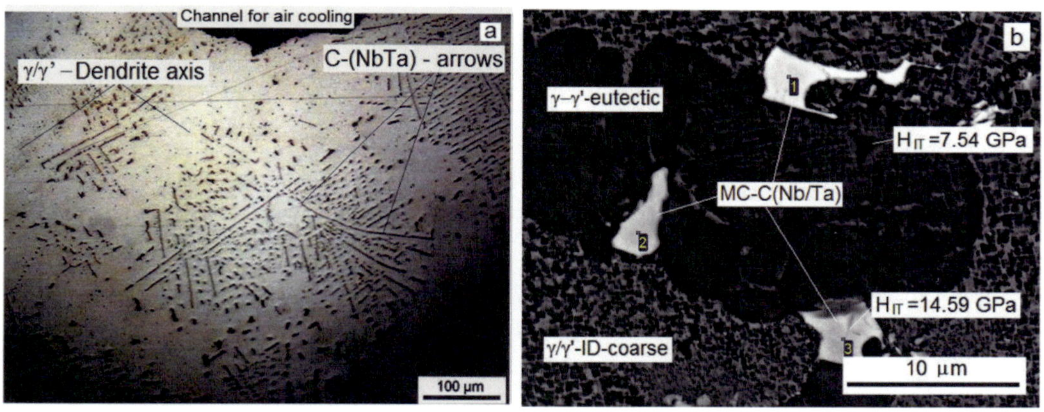

Figure 12. Interdendritic region in turbine blade cross-section with air-cooling channel (top) and C-(Nb/Ta)-metal carbide compound (arrows) for reinforcement of the interdendritic region (a) and MC near the eutectic pool in the interdendritic region (b).

4. Proposed test methods and results

4.1. Hard cyclic viscoplastic deformation

HCV deformation was done using an Instron 8516 materials testing installation at room temperature in the Materials Testing Laboratory at the Department of Material Engineering of Tallinn University of Technology, Estonia. In the present chapter, we studied the behavior of single crystalline Ni-based superalloy ZS32-vi for turbine blade application in aircraft turbojet engine AL-31-F. The SC material was tested at room temperature. We estimated the shift in the phase equilibria of an SC Ni-based superalloy during HCV deformation. The method we used was based on the Bauschinger effect. The objective of the present work was to study the role played by the deformation-assisted interdiffusion of atoms among different phases in shifting the phase equilibrium, which was characterized based on the evolving chemical composition and micromechanical properties of each investigated phases (see Figure 11 b). In my opinion, such an investigation can provide realistic information concerning the shifts in the phase equilibria of SC Ni-based superalloys under HCV deformation in start-up regime at ground idle temperature (±60°C) of turbine blades of turbojet engines. The specimens for HCV deformation were cut with stepped cross-sectional areas from the SC casts with dimensions of 16 mm in diameter and ~160 mm in length. The cross-sectional areas of the tension-compression specimens for HCV deformation (Figure 13) were 200 mm^2, 154 mm^2, 77 mm^2, and 38.5 mm^2 for test samples S1, S2, S3, and S4, respectively. The test parts were 12 mm long with a radius of 3 mm. The extensometer that was used to measure the true strain had a base length of 10 mm and was mounted on the test part (S4) with the smallest cross-section (38.5 mm^2). The stress values for the other cross-sections were calculated and found to decrease proportionally with the increase in the cross-section (Figure 14).

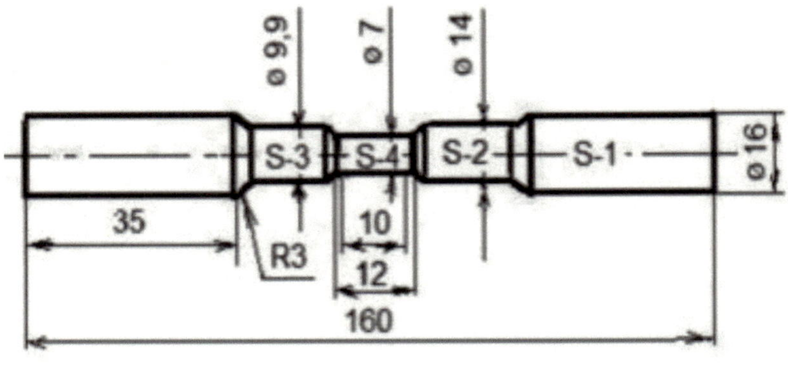

Figure 13. Scheme of the sample for HCV deformation. Designations in mm. Adapted from references [44, 46].

The cross-sections for samples S4, S3, and S2 increased twice as much per step relative to the next section. The HCV deformation tests were conducted using tension/compression-strain amplitudes of 0%–0.05%, 0%–0.2%, 0%–0.5%, and 0%–1% for 30 cycles at each strain amplitude. After such straining, the samples were fractured at tension strain of 1.5%. The cycling frequency was 0.5 Hz and remained constant for all strain amplitudes. The strain rates increased with increasing strain amplitudes. These strain amplitude values were experimentally chosen. The specimen length for HCV deformation was identical before and after cycling. Because the test sample (see Figure 13) has gradually shaped cross-sections, the stress and strain amplitudes decrease proportionally to increases in the cross-sectional areas. The collected maximal stress amplitude curves in sample S4 of testing for all strain amplitudes in Figure 15 are presented. As shown in the graphs at strain amplitudes of 0%-0.05%, 0%-0.2%, and 0%-5%, the

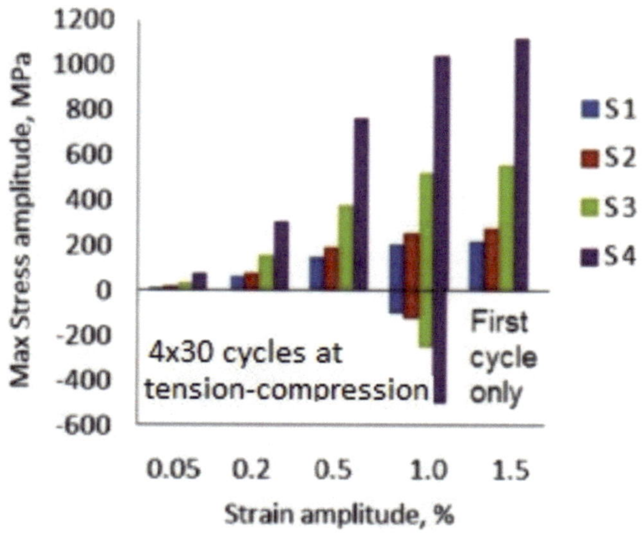

Figure 14. Maximal stress amplitude values (calculated) at tension and at compression for corresponding strain amplitudes in samples S1, S2, S3, and S4, respectively. Adapted from reference [3]

maximal stress values decrease slightly but the material shows fully viscoelastic behavior. The hysteresis loops were not formed as previously seen during conventional test method, which is presented in Figure 6. By increasing the strain amplitude, the stress amplitude increased proportionally. By increasing the strain amplitude to 0%-0.5%, the stress at tension increased up to 770 MPa at the first cycle and then decreased to 745 MPa at the end of cycling. The stress amplitudes decreased slightly under constant strain. This result indicates that the material does not harden due to an increase in the dislocation density. This standpoint (dislocation density increase) was presented by a large number of works listed in the references [36-38]. A discrete dislocation dynamic model [36, 37] was used to characterize the mechanical deformation and softening of the Ni-based superalloys SC. This model shows that the precipitates sheared by super dislocations were responsible for softening the SC Ni-based superalloy. Such softening behavior is presented in Figure 15b, d, and f. At maximal strain amplitude of 0%-1%, the tension stress at first cycle was maximal and then decreased during 2-3 cycles. Afterward, the stress amplitude maximal values do not decrease during the following cycles. The time-stress curves at strain amplitude of 0%-1% for the first, second, and 30 cycles are shown in Figure 16, and the summarized strain-stress curves are shown in Figure 17.

Experimental works [39, 44] demonstrate an identical behavior for SC Ni-based superalloys ZS32-vi during HCV deformation. During the first cycle of the present study (Figures 17 and 18), sample S4 was plastically elongated during the tension cycle starting at a strain of 0.65% to a strain of 1.03%. After this plastic elongation, sample S4 was compressed to zero (to the initial length) using a ~500 MPa stress. The curves for 0-I, 0-II, and 0-III overlap, and the material demonstrates a fully viscoelastic behavior. The plastic elongation (PL) of the material during viscoelastic cycling occurs from strains from 0.65% to 1.03%. For the first three test series (0-I,

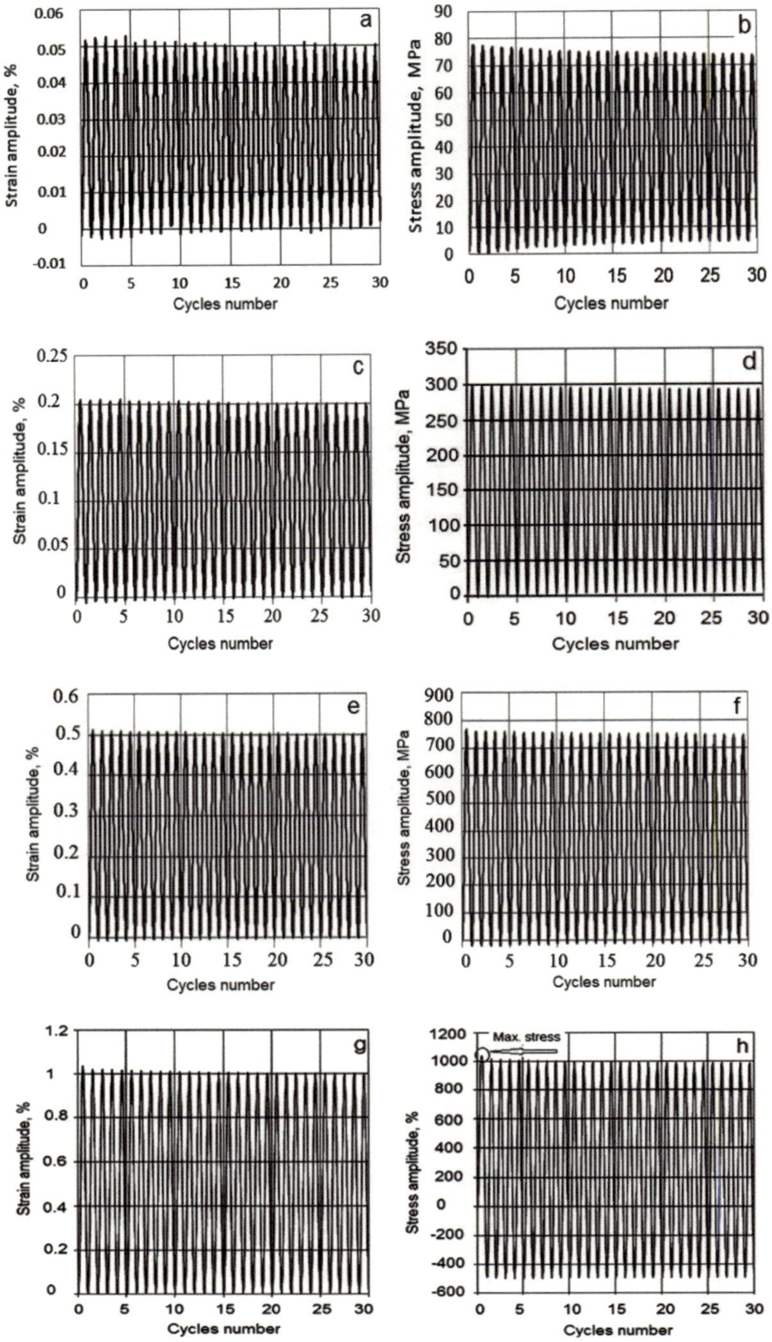

Figure 15. The strain amplitudes at 0.05%, 0.2%, 0.5%, and 1% (a, c, e, and g, respectively), and corresponding to the stress amplitudes (b, d, f, and h, respectively) in the SC of Ni-based superalloy ZS32-vi.

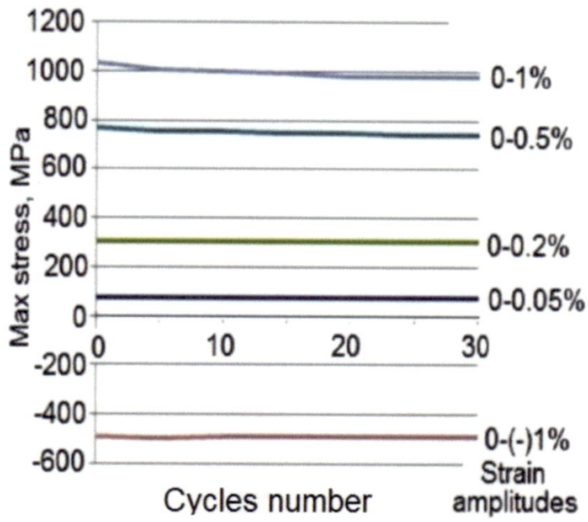

Figure 16. Maximal stress values evolution of 30 cycles measured in sample S4 for tension-compression strain amplitudes of 0%-0.05%, 0%-0.2%, 0%-0.5%, and 0%-1%, respectively. Adapted from reference [46].

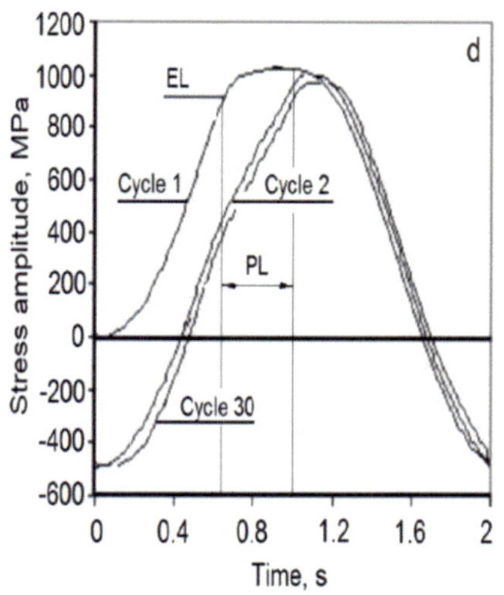

Figure 17. Time-dependent stress amplitudes shift at first, second, and 30 cycles for strain amplitude of 0%-1%, respectively.

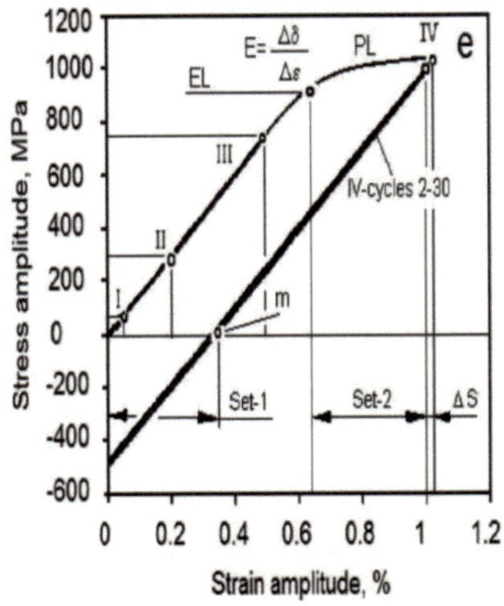

Figure 18. The graph summarizes the strain-stress curves of the SC Ni-based superalloy for the 38.5 mm² cross-section during the HCVD testing. The definitions are as follows: 0-I, stress-strain curve at a strain amplitude of 0%-0.05% for 30 cycles; 0-II, at 0%-0.2%; 0-III, at 0%-0.5%; 0-IV, the portion of the tension curve during the first tension cycle at a strain of 1%; and IV, cycles 2-30–stress-strain curves at a strain amplitude of 0%-1%; ΔS, ~0.03% strain is the system error; Set-1 ≅ Set-2, part of the plastic deformation during the first compression cycle in the fourth series; PL, proportional limit; EL, elastic limit; and "m", starting point of the material compression. Adapted from references [28, 39, 44].

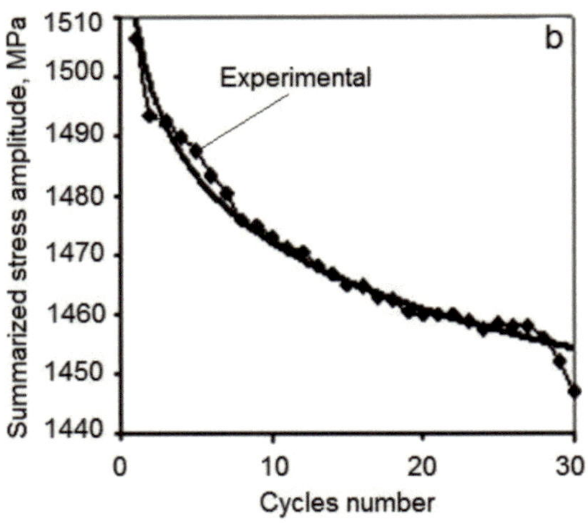

Figure 19. The absolute (summarized) reduction of the tension-compression stress amplitude for strain amplitudes from 0% to 1% for sample S4 during HCV deformation. Adapted from reference [44]

0-II, and 0-III), all of the stress-strain curves overlap and exhibit a linear dependence; therefore, the material is showing a fully elastic behavior and Young's modulus does not change. Increasing the tension strain of the first cycle from the start of the plastic deformation, 0.65%, elongates the sample (in the center, which is a cross-section of 38.5 mm) up to a strain of 1.03%. The extra 0.03% (over 1%) is shown in the graph as ΔS in Figure 18. The curve decreased from point "m" to the stress axis once the sample was compressed using ~500 MPa during the first cycle. The absolute (summarized) tension-compression amplitude reduction is shown in Figure 19. At the first two cycles and at the end of three cycles, the summarized stress amplitude decreases sharply, which indicates the softening behavior of SC materials at these cycles.

4.2. Nanoindentation: Micromechanical properties of phases

The micromechanical properties of the phases were characterized using a nanoindentation device from the NanoTest NTX testing center (Micro Materials Ltd.). This installation is presented in Figure 20. A diamond Berkovich tip with a three-sided pyramid apex angle of 142.3° and a radius of 100 nm was used for these measurements. This installation was equipped with an optical microscope to search for nanometer measurements of indented points on the diamond-ground surfaces of the samples. The nanoindentation tests were conducted on diametric sections of the samples. The following phases were investigated via scanning electron microscopy in combination with the energy-dispersive spectrometry system and nanoindentation: fine γ/γ', coarse γ/γ', eutectic γ-γ', and metal carbide (MC) phases. The MC phases contained Nb/Ta intermetallic compounds. The nanoindentation tests were conducted on diametric sections of the samples. The samples were mechanically ground and etched to indicate the dendrite direction.

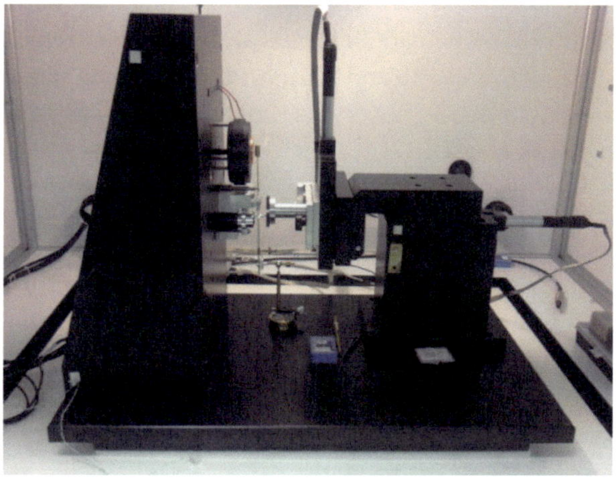

Figure 20. Nanoindentation setup NanoTest NTX testing center (Micro Materials Ltd.) used for phase micromechanical property measurements in this investigation.

An optical image of the as-cast microstructure is presented in Figure 9. The microstructures with corresponding indent points of phases in the testing regions of the samples are presented in Figure 21a, b, c, and d. In one study [3], 100 indents were made in all samples in a row with a 20-μm step over a distance of 2 mm using a load of 15 mN. The rows of nanoindentation are perpendicular to the (001) direction of the cast. The micromechanical properties of the phases were studied for all 400 test points (100 points per sample) from the SEM pictures of microstructures across all four samples. The maximal- and plastic depths of the indents were measured with high-precision installation (see Figure 20). The micromechanical properties were calculated automatically by computer according to the Oliver-Pharr method and presented in tables. According to table dates, the corresponding graphs of micromechanical properties of phases were built and are presented in Figures 22, 23, and 24, respectively. Changes in the microstructure occur under the viscoplastic strain conditions (shown in Figure 28). The micromechanical properties and their ratios from both before and after the HCV deformation study are shown [3] to demonstrate how the micromechanical properties of the phases were changed by the interdiffusion of various atoms (studied below). The nanoindentation test results show (Figure 22a) that the maximal and plastic indentation depths of the phases decreased with increases of the collected strain in samples, which depends on stress amplitude increases (see Figure 14). The mean of the micromechanical properties for all of the phases across all samples were calculated and are presented in Figures 22, 23, and 24. The micromechanical properties of the investigated phases were calculated according to the received/measured maximal and plastic indentation depths under a 15 mN load. Our previous work [3] showed that, for nanoindentation, the results depend on the load used. Therefore, the results of nano- and microindentation measurements (see in [3], Figure 9a and b) show approximately identical distributions of hardness values for all analyzed phases. The dependence of the nanoindentation test results on the applied load was previously studied by Neumeier et al. [43], Schöberl et al. [47] and Kommel et al. [3]. The mean (M) nanohardness (Figure 22b) shows that it increased stepwise with increasing cumulative strain (decrease of indentation depth; Figure 22a). The nanohardness and elastic modulus also increased for all of the samples and phases (Figures 23b and 24a). Additionally, the plastic and elastic work of the indentation (Figure 23b), elastic recovery parameter (Figure 24a), and contact compliance parameter (Figure 24b) all decreased.

The micromechanical properties and elemental ratios of the samples and phases were initially studied in [3]. The microhardness of the coarse γ/γ'-phase increased sharply relative to the fine γ/γ'-phase microhardness. The microhardness ratio decreased from 1.3 to 1.09 as the hardness of the coarse γ/γ'-microstructure increased. This increased hardness was predicted by the increase in the γ-γ'-eutectic pools surfaces at higher strain (shown below). For example, increasing the nanohardness decreased the contact compliance parameters but significantly increased the plastic and elastic portions of the indentation work. The elastic recovery parameter (Figure 24a) decreased during the HCV deformation similar to the elevated temperatures measured during microindentation in [39]. The results from this nanoindentation (Figure 23a) indicate that the modulus of elasticity (E_r) increased because the cumulative sample strain increased. The mean E_r increased from 200 GPa in as-cast SC to ~340 GPa and mean nanohardness from 12.8 GPa to 16 GPa after HCV deformation. The micromechanical properties and their ratios before and after the HCV deformation study are presented to

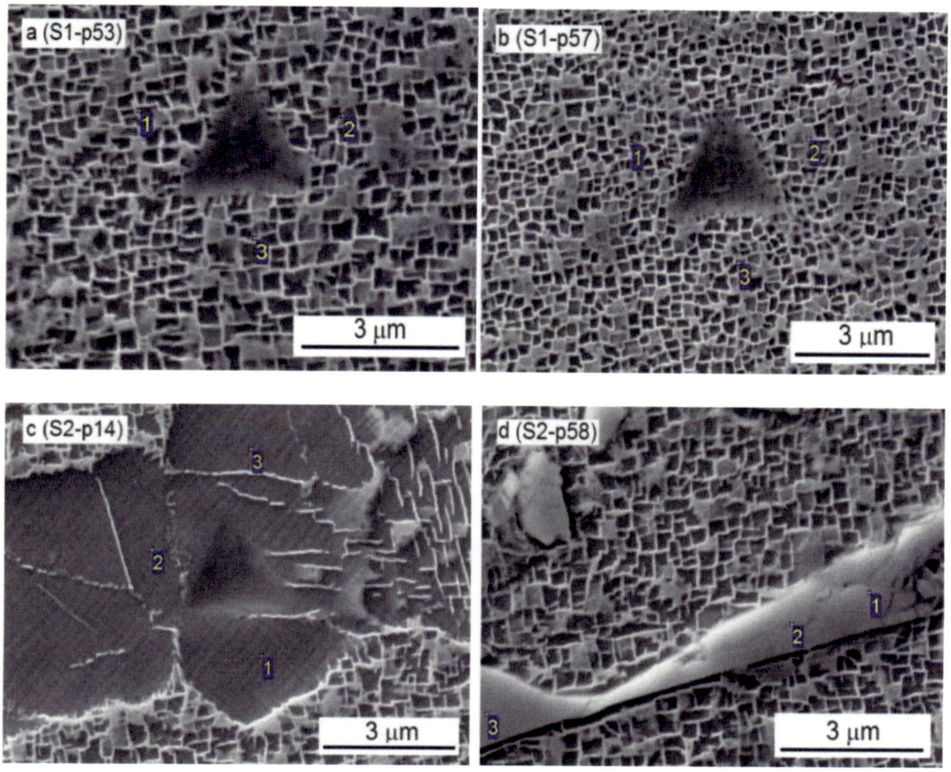

Figure 21. Nanoindentation triangular test points in the γ/γ′- coarse (a), γ/γ′- fine (b), γ-γ′- eutectic pool (c), and C(Nb/Ta)-MC (d) phases. The chemical elemental compositions of the three measured points by SEM EDS in these phases are shown near the indents. Adapted from reference [46].

demonstrate the change in the micromechanical properties of the phases caused by the interdiffusion of various atoms.

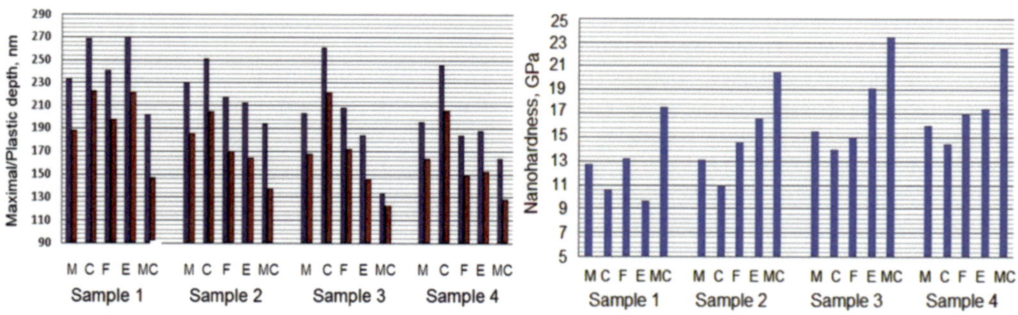

Figure 22. In addition to the maximal and plastic indentation depths, the "M", (mean value for 100 indents) of the test points was measured under a 15 mN load (a) and calculated nanohardness (b). Designations: M, mean values across 100 tests (according to the tabulated data); C, coarse γ/γ′-phase; F, fine γ/γ′-phase; E, γ-γ′ eutectic phase; and MC, Nb/Ta-MC phase.

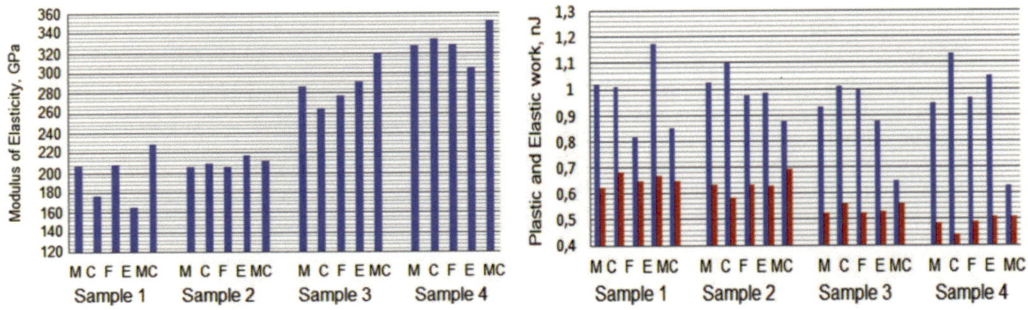

Figure 23. The phase dependence of the elastic modulus (a) and both the plastic and elastic work (b) for samples from collected strains. Designations such as those in Figure 22 are shown.

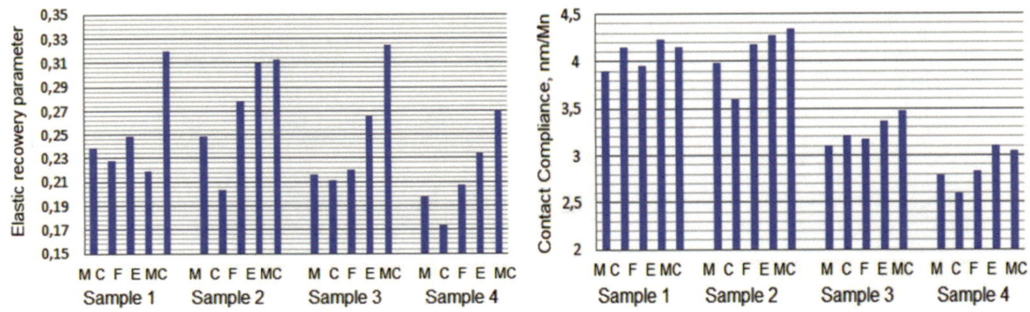

Figure 24. The phase dependence of the elastic recovery (a) and contact compliance (b) parameters for samples under cumulative strain. Designations such as those in Figure 22 are shown.

This work strongly indicates that such changes in the micromechanical properties of the phases depend on the change in the chemical concentration of the phase after interdiffusion. The micromechanical properties and elemental ratios of the samples and phases initially evolved are studied. The effect atomic interdiffusion had on the evolution of the micromechanical phase properties has been defined as a ratio of these properties [25, 28]. The microhardness of the coarse γ/γ'-phase increased sharply relative to the fine γ/γ'-phase microhardness. The microhardness ratio decreased from 1.3 to 1.09 as the hardness of the coarse γ/γ'-microstructure increased. This increased material hardness was predicted by increases in the γ-γ'-eutectic phase pool's surfaces (Figure 26) at higher strain.

4.3. Microstructural analysis of phases

Depending on the processing parameters (temperature gradient, withdrawal velocity, etc.) by directional solidification [20-23], the solidification parameters influence the length of the dendrites and the spacing of the microstructure. The experimental results of measured dendrite arm spacing [1, 47] are presented in a graph (Figure 25). The dendrite spacing of some commercial superalloys lies from 50 to 600 μm (Figure 25). In this study, SC superalloy ZS32-

vi with a dendrite spacing of 350-400 μm and dendrite length of up to 4 mm, respectively, was used (see Figures 10-13). The width of a γ-phase channels was about 100-200 nm in the PDA region and about 300-400 nm in the ID region (see Figure 21a and b), respectively. During HCV deformation (at room temperature) the atoms interdiffusion between different phases was induced and the dendrite measures decreased (Figure 26 and 27).

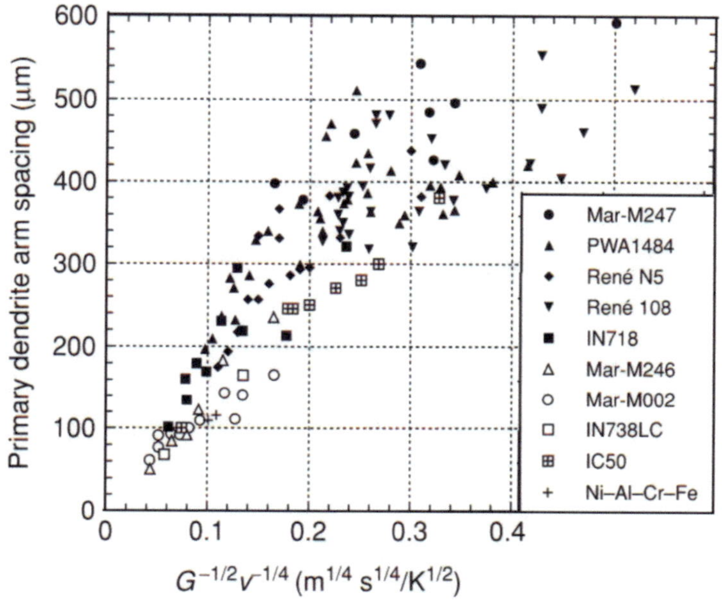

Figure 25. Some commercial superalloys' primary dendrite arm spacing with the combination $G^{-1/2}v^{-1/4}$, where G is the temperature gradient and v is the withdrawal velocity. Adapted from references [1, 47].

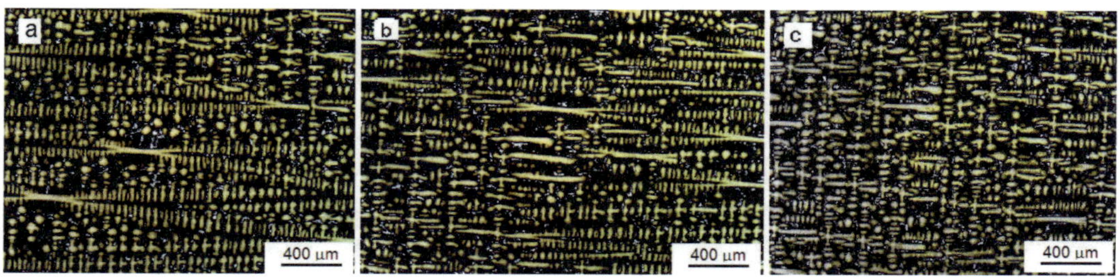

Figure 26. Optical micrographs of the SC Ni-base superalloy ZS32-vi dendrites measure stepwise decreases in samples S2, S3, and S4 during HCV deformation. The corresponding stress-strain amplitudes are presented in Figure 14. Adapted from reference [46]

The number and area of the γ-γ′-phase eutectic pools relative to the microstructure increased stepwise: S1, 1.05%; S2, 1.14%; S3, 1.86%; and S4, 2.06%. This development of the γ-γ′-phase eutectic pools is presented in Figure 27. The formation and distribution of these pools were

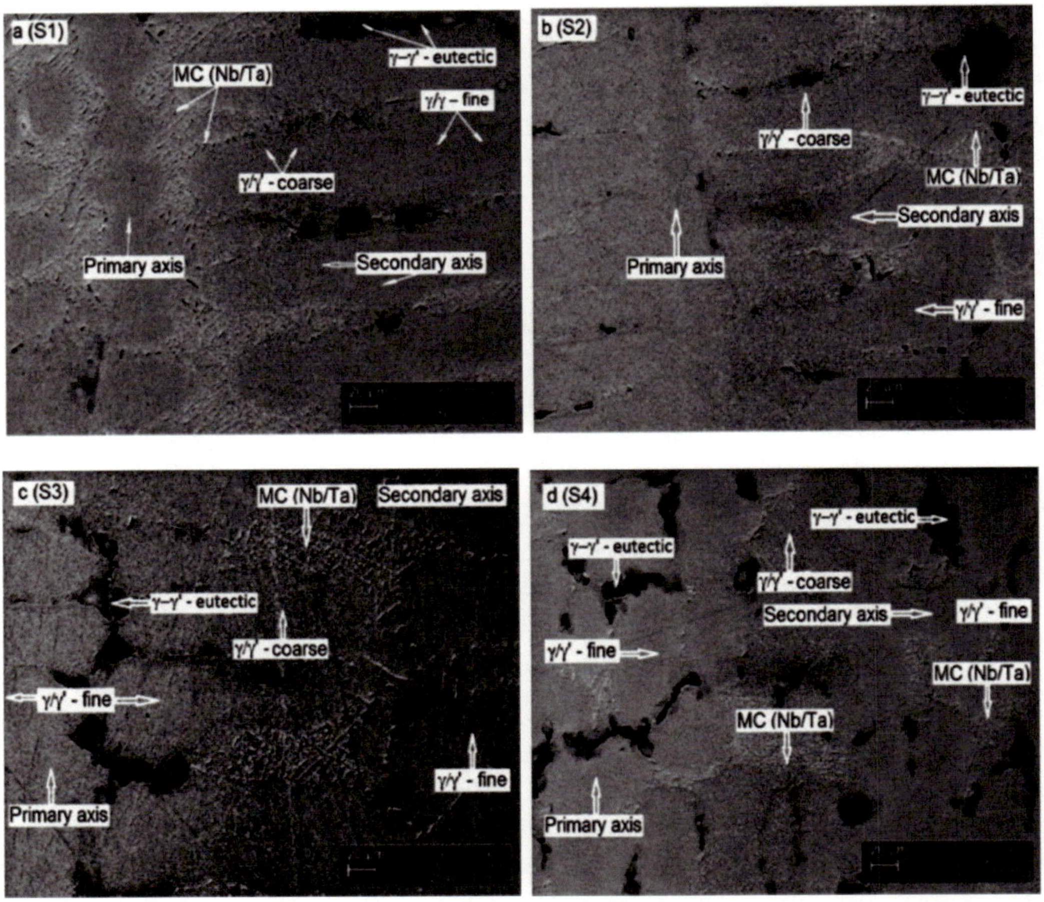

Figure 27. Eutectic pools evolution on primary dendrite area with γ/γ'- fine microstructure of samples S1, S2, S3, and S4, respectively, as a result of atom interdiffusion during HCV deformation. Adapted from reference [46 (a and d)].

investigated on the 3 mm^2 surfaces of all samples. Starting from sample S2, the new γ-γ'-phase eutectic pools took form with increasing strain amplitudes. As shown in the figure, the primary axes of samples S3 and S4 were crossed with new γ-γ'-phase eutectic pools (Figure 27c and d), which formed the finer microstructure of the primary dendrites (Figure 26c). Such micro-segregation-induced inhomogeneity in the γ-γ'-phase eutectic pools has been found to occur in Ni-based SC superalloys at high temperatures [26]. There are two approaches to studying the diffusion and melting processes in SC superalloys: the absolute reaction-rate theory and the dynamic theory of diffusion [27]. These theories indicate that the diffusive motion of atoms primarily depends on the amplitudes of oriented vibrations and the surroundings. The diffusion activation energy varies with the size, atomic weight, and charge of the diffusing species. Hence, it is different for different elements (see Figures 2-5). In this study, this activation energy was provided by HCV deformation at room temperature and was expected to vary with the strain amplitude.

4.4. Elemental concentration analysis

The changes in the elemental concentrations of the coarse γ/γ′- phase (a), γ/γ′-fine phase (b), γ-γ′-eutectic phase (c), and Nb/Ta-rich MC phase (d) are plotted in Figure 28. The Al concentration was higher in the coarse γ/γ′-phase (Figure 28a) than in the fine γ/γ′-phase (Figure 28b). The increase in the Al content of the coarse γ/γ′-phase can be attributed to Al that originated from the γ-γ′-eutectic phase (Figure 28c) and the MC phase (Figure 28d), and was transported into the coarse γ/γ′-phase through the increase in the cumulative strain. Cr, Co, and Mo possess lower activation energies than Al, and their concentrations increased primarily in the γ/γ′-eutectic phase (Figure 28c), also originating from the coarse γ/γ′-phase and the MC phase (Figure 28d).

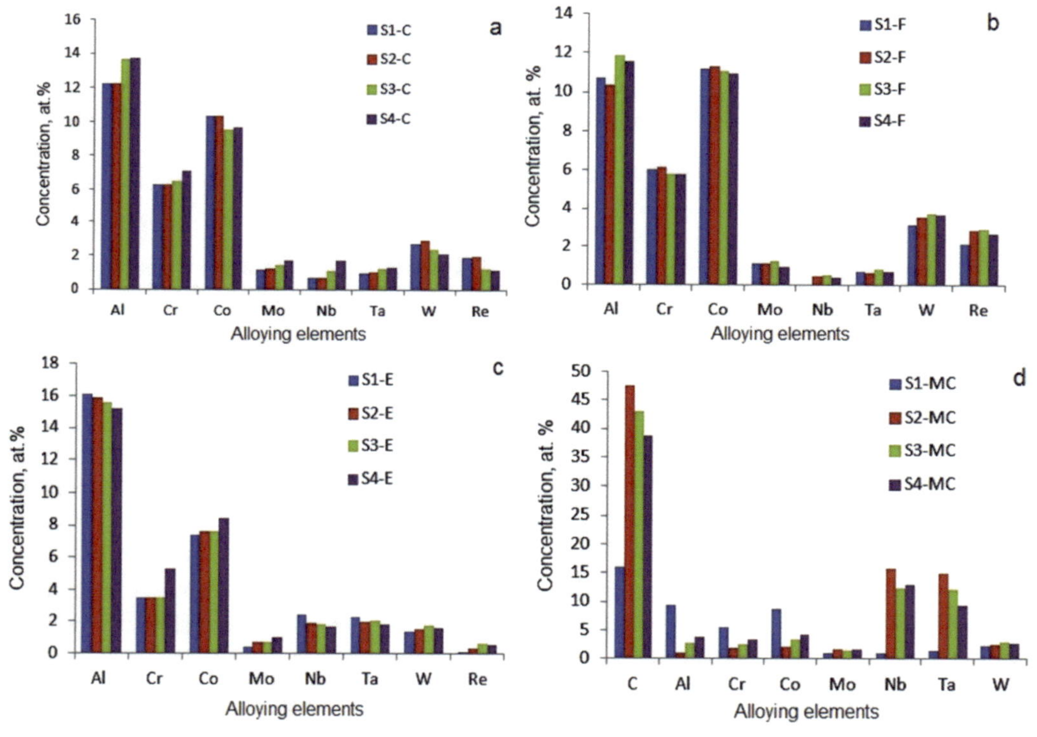

Figure 28. Elements concentration changes in coarse γ/γ′-phase (a), in fine γ/γ′-phase (b), in γ-γ′-eutectics phase pools (c), and in MC phases (d) of samples S1, S2, S3, and S4, respectively.

It is well known that an increase in Re and W content can increase the hardness and elasticity modulus of the γ-phase matrix [48]. The changes in the Nb/Ta concentration in the MC phase during HCV deformation are illustrated in Figure 28d. The Nb/Ta-rich MC phases also contained a large amount of C and very small amounts of Ni, W, and Cr. The MC phase did not contain Re. A detailed investigation of the MC phase revealed that the C, Nb, and Ta concentrations differed considerably and depended on the cumulative strain and the location of the MC phase in the ID region or near γ-γ′-eutectic phase pools (see Figure 12b). Based on

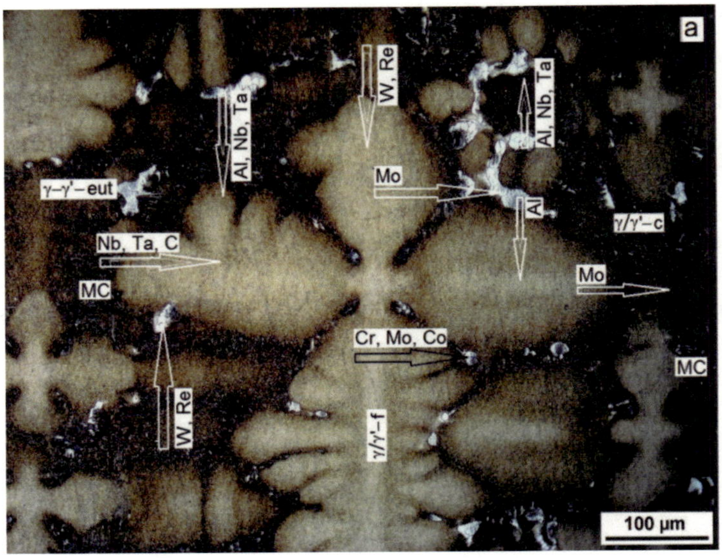

Figure 29. Mine directions of atom diffusion between phases in a cross-section of the primary axis is shown. Adapted from reference [46]

this investigation, the diffusion directions of various elements are shown in the sample cross-sections by the arrows in Figure 29. A large number of energy-dispersive spectrometry investigations demonstrated that Al diffused along the dendrite axis, whereas Re and W diffused from the coarse γ/γ′-phase into the fine γ/γ′-phase and partly into the γ-γ′-eutectic phase pools. Therefore, the Al diffused into the dendrite axis from the γ-γ′-phase eutectic pools. Cr, Mo, and Co diffused into the γ-γ′-phase eutectic pools, whereas the MC, Ta, and Nb diffused into the fine γ/γ′-phase. The elemental concentrations of phases changed as a result of interdiffusion. Orlov [49] previously detected such atomic interdiffusion among different phases and also found that diffusion pores form during the high-temperature heat treatment of superalloys (Figure 29). Such elemental motion at high temperatures changes the chemical compositions and micromechanical properties of the phases. In the present study, because of increased cumulative strain increase (or increased interdiffusion), small pores were formed in the coarse γ/γ′-phase or the interdendritic region of the superalloy. Such pores formed as a result of the high, unbalanced diffusion rate of Al in the Ni-based solid solution. It is well known from previous work [49] that diffusion pores form in metals during heat treatment or during long-term exposure at high temperatures because of the Kirkendall effect. Thus, the interdiffusion of elements during heat treatment as well as during HCV deformation at room temperature has identical mechanisms. As a result of the absence of balanced interdiffusion during HCV deformation, diffusion micropores (Figure 30a) and phase/grain boundaries could be formed (Figure 30b).

Such elemental motions change the chemical composition and micromechanical properties of the phases. At an increased cumulative strain and with the absence of balanced interdiffusion, the formation of small pores and phase boundaries (Figure 30) in the interdendritic region of

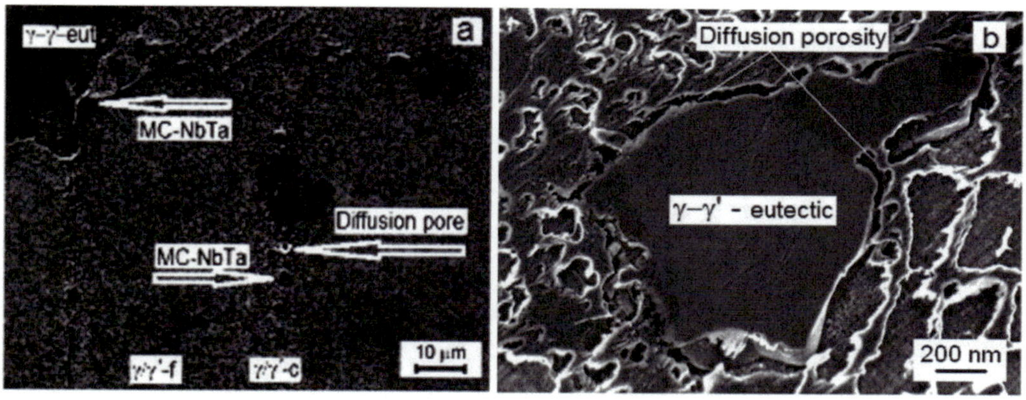

Figure 30. Diffusion micropore (a) and phase/grain boundaries (b) forming in SC as a result of the absence of balanced interdiffusion during HCV deformation.

the SC superalloy can be induced. Such defects form as a result of the high diffusion rate for Al in the Ni-based solid solution [47-49].

4.5. SEM/EDS and XRD analysis of microstructure

The atom interdiffusion between the different phases, which changes their elemental concentration based on the cumulative strain increase. The elemental presence for the γ/γ′-phase (a), γ-γ′-eutectic phase (b), MC phase (c), and Nb/Ta content in MC phase (d) are also illustrated in Figure 31. The changes of elemental content in the phases during HCV deformation have also been previously described (see Figure 28). The coarse and fine γ/γ′-phases (Figure 31a) contain these same elements but in different concentrations (compare a and b graphs in Figure 28). The eutectic phase pools elemental concentration differed when compared with the γ/γ′-phase. The W and Re contents were lowered. The Nb/Ta-rich MC phases (Figure 31c) also contain C and very small amounts of Ni and Cr. A detailed investigation of the MC phase shows that the C, Nb, and Ta concentrations differed greatly and depended on the cumulative strain and location for the MC phase in the SC region (Figure 31c). The measured nanohardness is high (see Figure 22b) in MC phase near eutectic pools and in the MC phase in the coarse γ/γ′-phase. The PDA and SDA regions don't contain MC phase.

The X-ray investigations (Figure 32) of the initial SC superalloy (S1) showed only one large reflection at 74° on the 2-theta scale and a smaller reflection at 117.7° (d = 0.899; not shown) on the 2-theta scale. These results indicate that the material has a single crystalline microstructure with no grain boundaries or structural defects. The reflection at 74° on the 2-theta scale was significantly decreased at a very low increase of strain-stress influence on the microstructure in the SC sample (S2). Increasing the stress amplitude of the sample (S3) allowed X-ray reflection to occur at 50.6° in the 2-theta scale along with smaller reflections from the tantalum carbide (TaC) compound at lower 2-theta values. Increasing the strain-stress loading formed new peaks and increased the intensity at 50.6° (d = 1.797). The XRD shows that Ni3AlC, Mo0.57Re0.43, and W0.15Ni0.85 formed at the maximum stress-strain loading (S4). The XRD

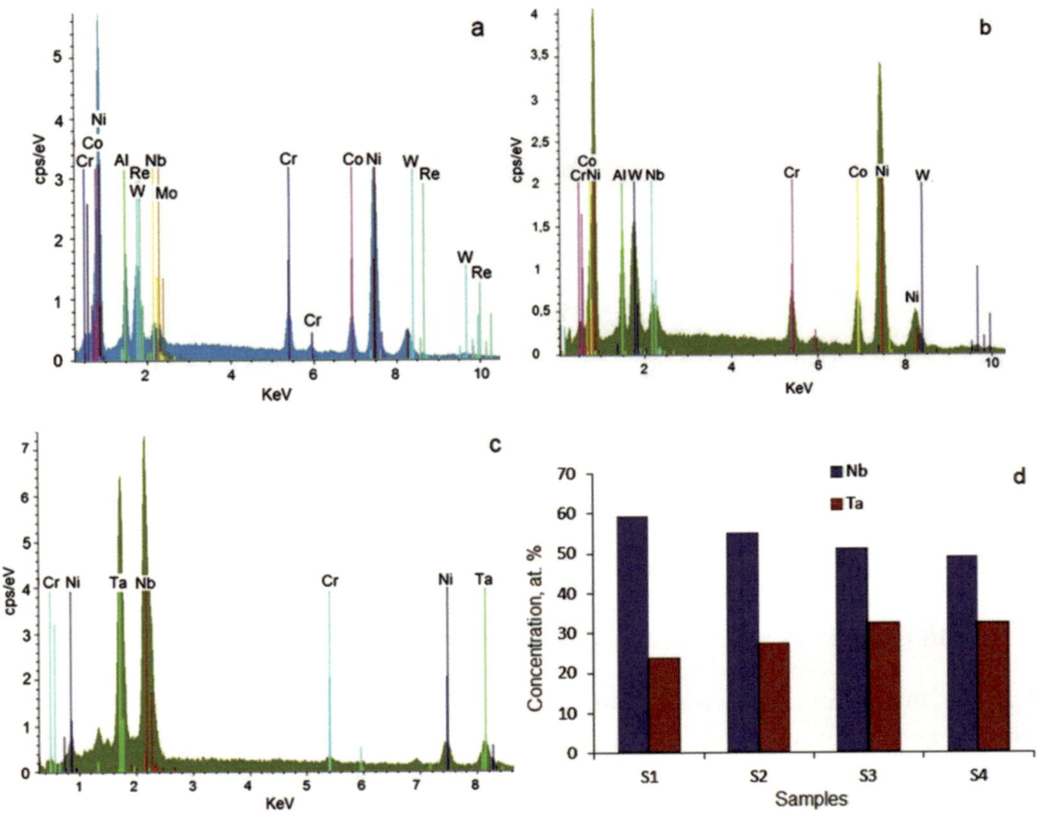

Figure 31. SEM-EDS diagrams of the sample (S4) in the primary dendrite axis or fine γ/γ′-phase (a), in the eutectic γ-γ′-phase region (b), in the MC phase (c), and Nb/Ta content evolution at interdiffusion in the MC phase (d) of samples.

peaks of the formation of a new proof of the formation of microstructural defects such as pores and grain boundaries (see Figure 30). As the deformation defects (micropores and phase boundaries) were formed it means that the material with SC microstructure becomes to material with polycrystalline microstructure.

4.6. Microstructure evolution and fracture mechanism

Results of the different investigations show [28, 44] that prefatigue by HCV deformation of the SC Ni-based superalloy influence the microstructure as well as fracture mechanism changes (compare with as-cast condition). During HCV deformation in the interdendritic region, the γ-γ′-eutectic phase was plastically deformed (Figure 33a). The single intermetallic γ′-phase had a plaid microstructure (Figure 33b), which was driven symmetrically to tensile compression direction of the deformation. The newly formed lines are about 100 nm in width. The MC carbide phase was mainly rifled in the same direction as the lines. The development of rafted structures strongly influences the mobility of the dislocations and twins [44] and then the rate of strain accumulation especially at relatively low applied stress amplitudes. Deformation

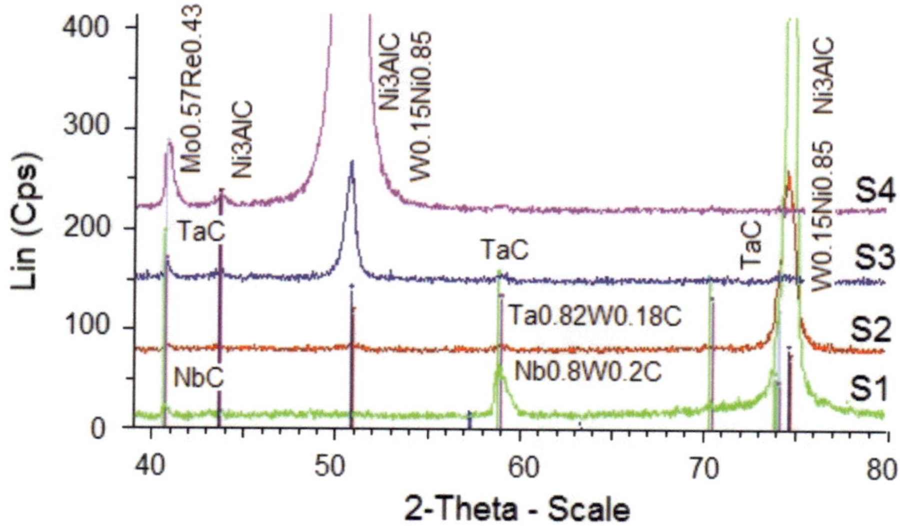

Figure 32. X-ray diffraction diagrams of samples S1, S2, S3, and S4 for different cumulative strains. Adapted from reference [3].

twinning and twinning-related fracture [44] in nickel-based single-crystal superalloys during HCV deformation (Figure 34a) and during thermomechanical fatigue cycling in [6, 7, 51] was studied (Figure 34b). During HCV deformation, a small number of rafts and twins were formed before fracture (Figure 34, a). Such deformation twinning and its related fracture during TMF cycling of Ni-based SC superalloys was studied previously by Zhang et al. [7, 51]. The fracture crack lies by twins formed during TMF cycling (Figure 34b). Analysis of fractured surfaces (after HCV deformation) of tensile specimens show that PDA in the center (Figure 35a) is one of the principal sites for crack initiation in (001) direction, as illustrated in Figure 35b. It is well known that in samples of plastic metals at tension, necking take place with diameter (or cross-section) decrease. In HCV-deformed and tension-fractured samples, necking occurs for each primary dendrite in the core center in the direction of (001). The large number of such longitudinal cracks were formed in fracture surfaces. As a result of interdiffusion, the center part of the dendrite core of PDA has a lowered concentration of Ni and high concentration of Al, and the material has a brittle nature. We believe that such observations have not been shown before. The fracture crack at tension was formed by γ/γ'-phase and by γ-γ'-eutectic phase in the interdendritic region of SC (Figure 36a and b, respectively). As is shown in Figure 36a, the square γ-phase channels were rounded at fracture of the prefatigued SC superalloy. The crack lies mainly by interdendritic zone of secondary dendrites (Figure 37).

As was shown before [10, 11] the damage and fracture of SC Ni-based superalloys are mainly studied by high cycle fatigue testing at elevated temperatures. For testing, smooth and matched specimens are used. In our studies [3, 28] we show that the damage and fracture depend on phases' chemical content as well phases' micromechanical property changes during testing.

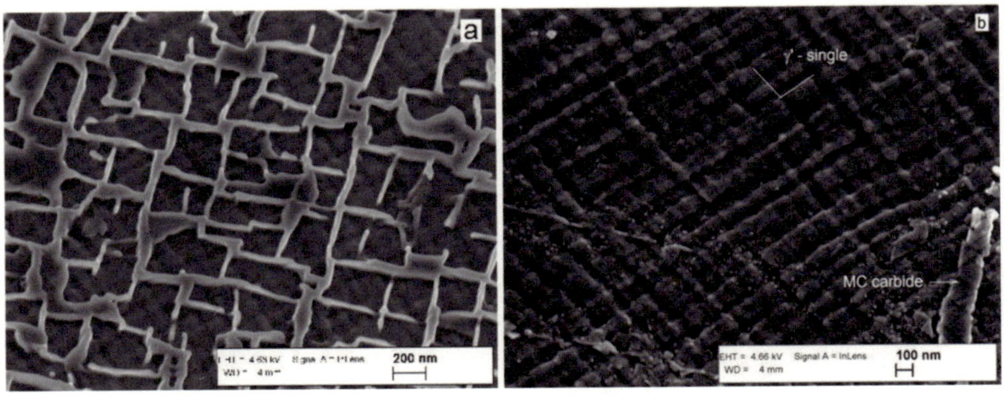

Figure 33. Microstructure of interdendritic zone (a) and plaid microstructure of a single γ′-phase (b) after HCV deformation.

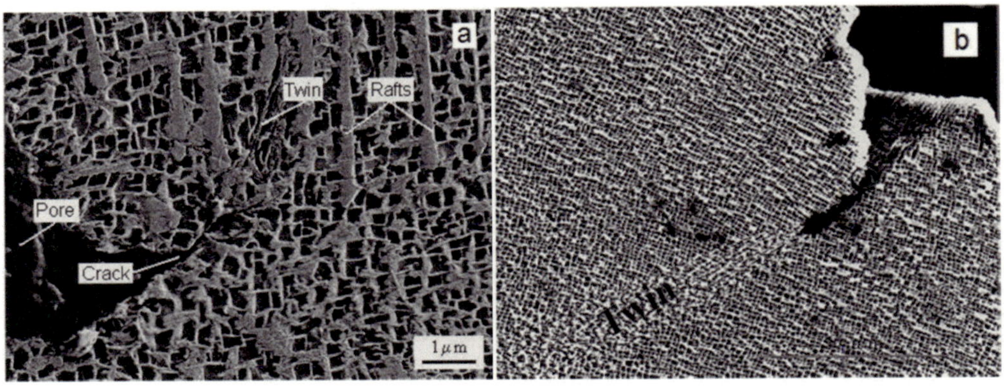

Figure 34. Rafts and twins forming during HCV deformation (a) and deformation twinning and its related fracture during TMF cycling of four nickel-based single-crystal superalloys (b). Adapted from references [28, 51].

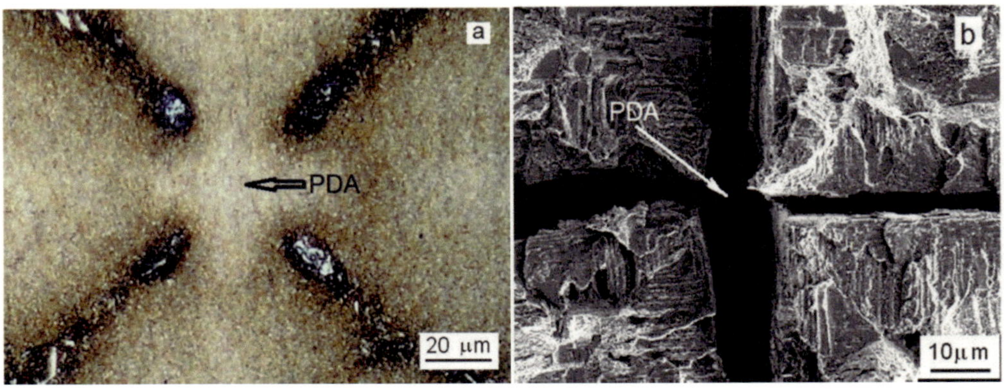

Figure 35. PDA in cross-section (a) and fracture of a preliminarily HCV deformed SC by PDA in (001) direction. Adapted from reference [28]

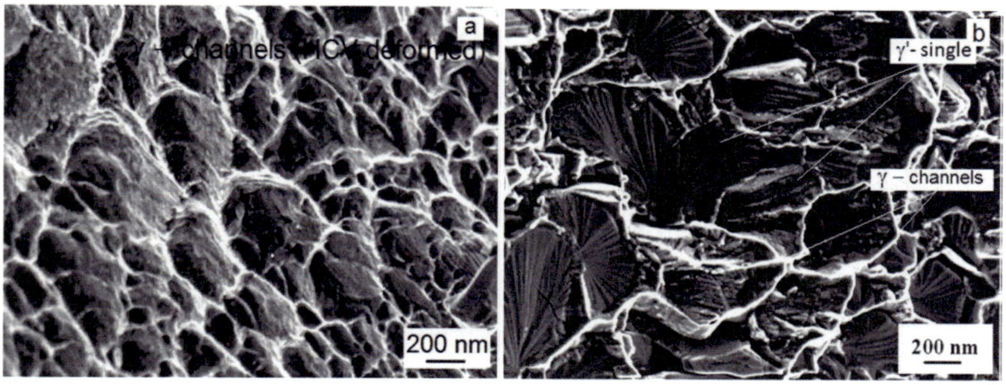

Figure 36. Fracture surface after HCV deformation and tension in γ/γ'-phase (a) and in γ-γ'-eutectic phase pool (b). Adapted from reference [28]

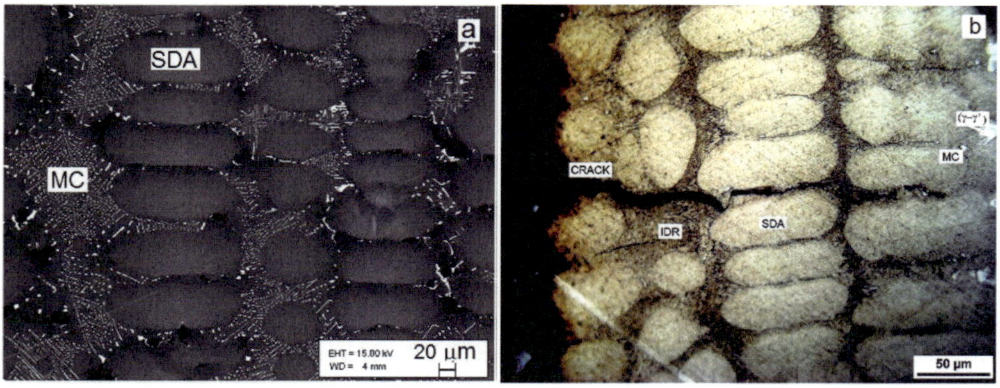

Figure 37. Microstructure of SC in the SDA and interdendritic region with MC(Nb/Ta) arrows (a) and fracture crack at tension lies at the interdendritic region between secondary dendrites. Adapted from reference [28].

5. Conclusions and summary

In this chapter, the phase's equilibrium evolution in single-crystal nickel-based superalloys via chemical condition and micromechanical property changes during hard cyclic viscoplastic deformation at room temperature is presented as a new material characterization method. The samples' microstructure changes at creep testing, as well as damage and fracture mechanisms for different testing modes, are presented. The experimental observation methods—optical microscopy, scanning electron microscopy, transmission electron microscopy, energy-dispersive spectroscopy, microchemical analyses, X-ray diffraction, hard cyclic viscoplastic deformation, and nanoindentation—were combined to obtain new insights into the phase's chemical composition and micromechanical property characterization. Based on the analysis of experimental results and observations, conclusions could be given as follows:

1. The as-cast microstructure of single-crystal nickel-based superalloys, which contain coarse dendrites in the (001) direction with length up to 4.5 mm, were separated by (γ + γ')-eutectic pools, which significantly reduced the length of the newly formed fine dendrites and, as a result, superalloys with a fine microstructure were formed.
2. The area and number of (γ + γ')-eutectic pools in superalloys increased by more than twofold.
3. The interdiffusion of alloyed elements and atoms among the phases changed their micromechanical properties by changing their chemical compositions.
4. These changes differed by elements and depend on the activation energy of elemental diffusion during HCV deformation.
5. At increased strain amplitudes and number of cycles, the SC superalloy softened, and very small diffusion micropores formed in the interdendritic region.
6. The fracture at tension of damaged materials takes place in the interdendritic region and necking by primary dendrites.
7. The results of this investigation provide realistic information concerning shifts in the phase's equilibria evolution and on the changes in the micromechanical properties of phases in SC Ni-based superalloys.
8. Such changes in the phase equilibrium in the SC Ni-based superalloys were influenced by atomic interdiffusion between different neighbor phases by diffusion paths.

We propose that such an investigation can provide realistic information concerning the shifts in phase equilibria of SC Ni-based superalloys, their viability evolution and properties change as well as their fracture mechanisms at increased cumulative strain during HCV deformation. We believe that some results in this chapter have not been reported before.

Acknowledgements

The author would like to acknowledge support from the Estonian Ministry of Education and Science (Projects IUT 19-29 and EU 7FP ERA.Net RUS STProjects-219). The support and vision of doctors Rainer Traksmaa and Mart Viljus are greatly appreciated.

Author details

Lembit Kommel

Address all correspondence to: lembit.kommel@ttu.ee

Tallinn University of Technology, MTI, Tallinn, Estonia

References

[1] Reed RC. The Superalloys: Fundamentals and Applications. New York: Cambridge University Press; 2006.

[2] Wang XG, Liu JL, Jin T. Sun XF. Tensile behaviors and deformation mechanism of a nickel-base single crystal superalloy at different temperatures. Materials Science and Engineering A 2014; 598 154–161.

[3] Kommel L, Kimmari E, Viljus M, Traksmaa R, Volobueva O, Kommel I. Phases micromechanical properties of Ni-base superalloy measured by nanoindentation. Materials Science (Medžiagotyra) 2012; 18(1) 28–33.

[4] Tang H, Cao D, Yao H, Xie M, Duan R. Fretting failure of an aero engine turbine blade. Engineering Failure Analysis 2009; 16 2004–2008.

[5] Sakamoto M, Harada H, Yokokawa T, Koizumi Y, Kobayashi T, Zhou H, Zhang JX, Miyamoto N. Assessment on the thermo-mechanical fatigue properties of 98 Ni-base single crystal superalloys, In: Reed RC, Green KA, Caron P, Gabb TP, Fahrmann MG, Huron ES, Woodard SA (eds.) Superalloys 2008, TMS (The Minerals, Metals & Materials Society) 2008, 521–526.

[6] Zhang JX, Harada H, Ro Y, Koizumi Y, Kobayashi T. Thermomechanical fatigue mechanism in a modern single crystal nickel base superalloy TMS-82. Acta Materialia 2008; 56(13) 2975–2987.

[7] Zhang JX, Harada H, Koizumi Y, Kobayashi T. Crack appearance of single-crystal nickel-base superalloys after thermomechanical fatigue failure. Scripta Materialia 2009; 61 1105–1108.

[8] Moverare JJ, Johansson S, Reed RC. Deformation and damage mechanism during thermal-mechanical fatigue of a single-crystal superalloy. Acta Materialia 2009; 57 2266–2276.

[9] Zhou H, Ro Y, Harada H, Aoki Y, Arai M. Deformation microstructures after low-cycle fatigue in a fourth-generation Ni-base SC superalloy TMS-138. Materials Science and Engineering A 2004; 381 20–27.

[10] Lukáš P, Kunz L, Svoboda M. High cycle fatigue of superalloy single crystals at high mean stress. Materials Science and Engineering A 2004; 387–389 505–510.

[11] Liu Y, Yu JJ, Xu Y, Sun XF, Guan HR, Hu ZQ. High cycle fatigue behavior of a single crystal superalloy. Materials Science and Engineering A 2007; 454–455 357–366.

[12] Akhtar A, Hegde S, Reed RC. The oxidation of single-crystal nickel-based superalloys. JOM 2006; 58(1) 37–42.

[13] Kawagishi K, Harada H, Sato A, Sato A, Kobayashi T. The oxidation properties of fourth generation single-crystal nickel-based superalloy. JOM 2006; 58(1): 43–46.

[14] Epishin A, Link T, Brückner U, Portella PD. Kinetics of the topological inversion of the γ/γ'-microstructure during creep of a nickel-based superalloy. Acta Materialia 2001; 49 4017–4023.

[15] Cormier J, Cailletaud G. Constitutive modeling of the creep behavior of single crystal superalloys under non-isothermal conditions inducing phase transformation. Materials Science and Engineering A 2010; 527 6300–6312.

[16] Chatterjee D, Hazari N, Das N, Mitra R. Microstructure and creep behavior of DMS4-type nickel based superalloy single crystals with orientations near (001) and (011). Materials Science and Engineering A 2010; 528 604–613.

[17] Pierce CJ, Palazotto AN, Rosenberger AH. Creep and fatigue interaction in the PWA1484 single crystal nickel-base alloy. Materials Science and Engineering A 2010; 527 7484–7489.

[18] Cutler ER, Wasson AJ, Fuchs, GE. Effect of minor alloying additions on the solidification of single-crystal Ni-base superalloy. Journal of Crystal Growth 2009; 311 3753–3760.

[19] Zeng Q, Ma SW, Zheng YR, Liu SZ, Zhai T. A study of Re and Al diffusion in Ni. Journal of Alloys and Compounds 2009; 480 987–990.

[20] Wilson BC, Cutler ER, Fuchs GE. Effect of solidification parameters on the microstructures and properties of CMSX-10. Materials Science and Engineering A 2008; 479 356–364.

[21] Kim SH, Kim JM, Lee HJ, Son SD, Lee JH, Seo SM, Jo CY. Effect of thermal gradient on solidification microstructure in the Ni-base single crystal superalloy CMSX10. Defect and Diffusion Forum 2008; 273–276 361–366.

[22] Liu L, Hang T, Zou M, Zhang W, Zhang J, Fu H. The effects of withdrawal and melt overheating histories on the microstructure of a nickel-based single crystal superalloy. In: Roger CR, Green KA, Caron P, Gabb TP, Fahrmann MG, Huron ES, Woodard SA. (Eds.) TMS Superalloys-2008 287–293.

[23] El-Bagoury N, Nofal A. Microstructure of an experimental Ni base superalloy under various casting conditions. Materials Science and Engineering A 2010; 527 7793–7800.

[24] Reed RC, Cox DC, Rae CMF. Damage accumulation during creep deformation of a single crystal superalloy at 1150°C. Materials Science and Engineering A 2007; 448(1–2) 88–96.

[25] Ugaste Ű, Priimets J, Laas T. Dependence of diffusion paths on thermodynamic factors in ternary systems. Defect and Diffusion Forum 2008; 277 119–124. DOI: 10.4028/www.scientific.net/DDF.277.119.

[26] Gebura M, Lapin J. Microsegregation induced inhomogeneity of coarsening of γ' precipitates in a nickel-based single crystal superalloy. Defect and Diffusion Forum 2010; 297–301 826–831. DOI:10.4028/www.scientific.net/DDF.297-301.826.

[27] Tiwari GP, Mehrotra RS. Diffusion and melting. Defect and Diffusion Forum 2008; 279 23–37. DOI:10.4028/www.scientific.net/DDF.279.23.

[28] Kommel LA, Straumal BB. Diffusion in SC Ni-base superalloy under viscoplastic deformation. Defect and Diffusion Forum 2010; 297–301 1340–1345. DOI:10.4028/www.scientific.net/DDF.297–301.1340.

[29] Liu LR, Jin T, Zhao NR, Sun XF, Guan HR, Hu ZQ. Formation of carbides and their effects on stress rupture of a Ni-base single crystal superalloy. Materials Science and Engineering A 2003; 361 191–197.

[30] Zhang J, Li B, Zou M, Wang C, Liu L, Fu H. Microstructure and stress rupture property of Ni-based monocrystal superalloy with melt superheating treatment. Journal of Alloys and Compounds 2009; 484 753–756.

[31] Mughrabi H. Microstructural aspects of high temperature deformation of monocrystalline nickel base superalloys: some open problems. Materials Science and Technology 2009; 25(2) 191–204.

[32] Graverend J-B, Cormier J, Jouiad M, Gallerneau F, Paulmier P, Hamon F. Effect of fine γ' precipitation on non-isothermal creep and creep-fatigue behavior of nickel base superalloy MC2. Materials Science and Engineering A 2010; 527 5295–5302.

[33] Rae CMF, Reed RC. Primary creep in single crystal superalloys: Origins, mechanisms and effects. Acta Materialia 2007; 55 1067–1081.

[34] Huda Z. Metallurgical failure analysis for a blade failed in a gas-turbine engine of a power plant. Materials and Design 2009; 30 3121–3125.

[35] Brückner U, Epishin A, Link T, Fedelich B, Portella PD. Dendritic stresses in nickel-base superalloy. Materials Science Forum 2005; 490–491 497–502.

[36] Huang M, Zhao L, Tong J. Discrete dislocation dynamic modelling of mechanical deformation of nickel-based single crystal superalloys. International Journal of Plasticity 2012; 28 141–158.

[37] Yashiro K, Kurose F, Nakashima Y, Kubo K, Tomita Y, Zbib HM. Discrete dislocation dynamic simulation of cutting of γ' precipitate and interfacial dislocation network in Ni-based superalloy. International Journal of Plasticity 2006; 22 713–723.

[38] Shenoy MM, McDowell DL, Neu RW. Transversely isotropic viscoplasticity model for a directionally solidified Ni-base superalloy. International Journal of Plasticity 2006; 22 2301–2326.

[39] Kommel L. Viscoelastic behavior of a single-crystal nickel–base superalloy. Materials Science (Medžiagotyra) 2009; 15(2) 123-128.

[40] Takagi H, Fujiwara M, Kakehi K. Measuring Young's modulus of Ni-based superalloy single crystals at elevated temperatures through microindentation. Materials Science and Engineering A 2004; 387–389 348–351.

[41] Durst K, Göken M. Micromechanical characterisation of the influence of rhenium on the mechanical properties in nickel-base superalloys. Materials Science and Engineering A 2004; 387–389 312–316.

[42] Schöberl T, Gupta HS, Fratzl P. Measurements of mechanical properties in Ni-base superalloys using nanoindentation and atomic force microscopy. Materials Science and Engineering A 2003; 363 211–220.

[43] Neumeier S, Dinkel M. Pyczak F. Göken M. Nanoindentation and XRD investigations of single crystalline Ni-Ge brazed nickel-base superalloys PWA 1483 and René N5. Materials Science and Engineering A 2011; 528 815–822.

[44] Kommel L. New technique for characterization of microstructure of the nickel-base superalloy. In: Proceedings of the 7th International DAAAM Baltic Conference, Industrial Engineering, 22–24 April 2010, Tallinn, Estonia.

[45] Epishin A, Link T, Klingelhöffer H, Fedelich B, Brückner U, Portella PD. New technique for characterization of microstructural degradation under creep: Application to the nickel–base superalloy CMSX-4. Materials Science and Engineering A 2009; 510–511 262–265.

[46] Kommel L. Effect of hard cyclic viscoplastic deformation on phases chemical composition and micromechanical properties evolution in single crystal Ni-base superalloy. Acta Physica Polonica (In print)

[47] Whitesell HS, Li L, Overfelt RA. Influence of solidification variables on the dendrite arm spacing of Ni-based superalloy. Metallurgical and Materials Transactions 2000; 31B 546–551.

[48] Razumovskii IM, Ruban AV, Razumovskiy VI, Logunov AV, Larionov VN, Ospennikova OG, Poklad VA, Johansson B. New generation of Ni-based superalloys designed on the basis of first-principles calculations. Materials Science and Engineering A 2008; 497 18–24.

[49] Orlov MP. The mechanism of porous forming and viability of monocrystallic turbine blades. Deformation and Fracture of Materials 2008; 6 43–48 (in Russian).

[50] Dirand L, Cormier J, Jacques A, Chateau-Cornu J-P, Schenk T, Ferry O, Bastie P. Measurement of the effective γ/γ' lattice mismatch during high temperature creep of Ni-based single crystal superalloy. Materials Characterization. 2013; 77 32–46.

[51] Sun F, Zhang J, Harada H. Deformation twinning and twinning-related fracture in nickel-base single-crystal superalloys during thermomechanical fatigue cycling. Acta Materialia 2014; 67 45-57.

Chapter 9

Modeling and Simulation of Shape Memory Alloys using Microplane Model

Reza Mehrabi

Additional information is available at the end of the chapter

http://dx.doi.org/10.5772/61124

Abstract

In this chapter, a three-dimensional phenomenological constitutive model for the simulation of shape memory alloys is introduced. The proposed macromechanical model is based on microplane theory. Microplane approach is chosen to have limited material parameters in that all of those are measurable by simple tests. User material subroutine is developed to implement the proposed model in a commercial finite element package. NiTi hollow tube specimens are under various loading conditions in order to experimentally study the superelastic response of shape memory alloys. Comparing experimental data with numerical results in simple tension and pure torsion as well as proportional and nonproportional tension-torsion loadings demonstrates the capability of proposed model in constitutive modeling of shape memory alloys.

Keywords: Shape memory alloy, phenomenological model, microplane, proportional, nonproportional, experiment

1. Introduction

Shape memory alloys (SMAs) can recover their original shape when subjected to thermomechanical loading. If the original shape is remembered under external load, it is in superelastic form, and when the original shape is remembered under thermal load, it is in shape memory effect form. Commercial applications of these materials get more attention in recent years. Some of these unique properties are including biocompatibility, good mechanical properties (very similar to the different parts of human body), hysteresis damping, excellent fatigue properties in cyclic loads, and strain hardening. These characteristics make them attractive for diverse fields of application such as biomedical tools (blood clot filters, vascular stents, orthodontic arches), aerospace applications (space shuttle, morphing aircraft, hydraulic fittings and couplings for airplanes), automotive (engines and actuators for smart systems, thermostats), robotic (human robots, artificial muscle), eyeglass frames, cellular phone antennae, coffee maker, and so on [1-9].

In order to design new application of shape memory alloys, basic understanding behavior of these materials under different conditions is essential. Different aspects of SMAs are identified with constitutive modeling. Most models are based on thermodynamics using free energy formulation. However, some models are based on micromechanics concepts or macromechanics model [10]. Micromechanic models deal with crystallographic texture in microscale response of SMAs [11-17]. These models are used to model the phase transformation and grains propagation and consequently are appropriate for modeling the response of SMAs at the microscale [18]. In order to study SMAs polycrystalline, crystallographic texture is a key property. A micromechanic model for single crystals is used to study polycrystal structures [19]. In the micromechanics models, a volume fraction coefficient is defined for variants, and transformation strain is obtained from martensite variants by averaging procedure.

Even though micromechanical models reflect microscopic physical nature of SMAs, they are not suitable for finite element implementation and structural analysis. Therefore, macromechanical models based on phenomenological findings are proposed. These models are efficient in modeling the thermomechanical behavior of SMAs in different engineering applications. In a macromechanic model, macro-scale response of SMAs is considered [20-24]. Phenomenological models belong to a class of macromechanic models that are defined by macroscopic energy functions. These models are calibrated from material characterization and experimental data that depend on the internal variables.

Macrophenomenological models are categorized to one- and three-dimensional models. One-dimensional models are used for the simulation of wire and bar samples under uniaxial loads, while 3-D models can be used for complicated devices under complex loadings. Since most SMA devises are used in multidimensional or under complex loadings, one-dimensional models should extend to three-dimensional models. Three-dimensional constitutive modeling is an important step in analysis and design of SMAs in different industries. Some of the existing macro phenomenological models with specifications and limitations are summarized in Table 1.

Group	Formulation	Characteristics	Limitation
Tanaka [25]	Helmholtz free energy	(1) 1-D model (2) Exponential phase transformation equation	3-D, shape memory
Liang and Rogers [26]	Helmholtz free energy	(1) 1-D model, extension of the Tanaka model (2) Cosine phase transformation equation	Shape memory

Group	Formulation	Characteristics	Limitation
Brinson [27]	Helmholtz free energy	(1) 1-D model, martensite volume fraction separate into two parts induced by stress and temperature (2) Cosine phase transformation equation (3) Different elastic moduli for austenite and martensite	3-D
Auricchio et al. [22, 28]	Based on phase diagram	(1) 1-D and 3-D model (2) Finite-strain regime, rate-dependent behavior (3) Asymmetry behavior and numerical implementation	Shape memory
Lagoudas et al. [2, 20, 29-31]	Gibbs free energy	(1) 3-D model, exponential phase transformation equation (2) Tension-compression asymmetry behavior (3) Nonassociated flow rule during reverse transformation	Depend on the specific alloy
Panico and Brinson [32]	Helmholtz free energy	(1) 3-D model, extension of the Lexcellent model [33, 34] (2) Variant reorientation modeling under nonproportional loadings (3) Computational modeling of porous SMAs	Depend on the specific alloy
Helm and Haupt [35]	Free energy	(1) 3-D model based on continuum thermodynamics framework (2) Simulation of nonproportional loading (3) Finite strain regime	Depend on the specific alloy
Oliveira et al. [36]	Helmholtz free energy	(1) 3-D model (2) Based on Fremond's method (3) Asymmetric behavior and based on plasticity	Unmeasurable material parameter
Brocca et al. [37]	Microplane theory	(1) 3-D model (2) Constant module for different phases (3) Simulation of nonproportional loading	Fundamental concept need to be revised
Arghavani et al. [21, 38]	Helmholtz free energy	(1) 3-D model (2) Finite deformations (3) Martensite volume fraction is a scalar parameter and variant orientation is a tensor	Unmeasurable material parameter
Saleeb et al. [23, 39]	Gibbs free energy	(1) 3-D model (2) Deviation from normality and reorientation under nonproportional loading (3) Cyclic behavior	Unmeasurable material parameter
Mehrabi et al. [40, 41]	Gibbs free energy	(1) 3-D model, microplane formulation	Under investigation

Group	Formulation	Characteristics	Limitation
		(2) Anisotropic behavior under nonproportional loadings (3) Tension-compression asymmetry and numerical implementation	
Patoor et al. [15, 39, 42]	Gibbs free energy	(1) 3-D model (2) Tension-compression asymmetry (3) Cyclic behavior and nonproportional loading	Unmeasurable material parameter
Zaki et al. [43-46]	Helmholtz free energy	(1) 3-D model (2) cyclic and asymmetry behavior (3) Capture plastic deformations	Unmeasurable material parameter

Table 1. Macromechanical models

One-dimensional constitutive model was proposed by Tanaka [47] based on exponential phase transformation equation. Liang and Rogers [26] proposed a one-dimensional phase transformation based on cosine type. Martensite volume fraction suggested by Liang and Rogers is modified by Brinson [27] into two different parts. These two fractions included martensite volume fraction induced by stress and temperature. This assumption was done to distinct superelastic behavior from shape memory effect. Boyd and Lagoudas [30] had developed 1-D model to 3-D constitutive model. For this development, they used effective stress and strain in phase transformation equations. A three-dimensional constitutive model includes twinned martensite and detwinned martensite. In addition, multidimensional models could predict behavior of shape memory effect as well as superelastic behavior. A 3-D model based on the phase transformation equation of exponential type was proposed by Peng et al. [48]. They used classical plastic theory and defined equivalent stress and strain for modeling SMA response. Reali et al. [22] proposed a 3-D constitutive model, which is capable of simulating superelastic and shape memory behavior. Proposed model is developed within the framework of thermodynamics by defining a scalar and a tensorial internal variable [21]. The 3-D phenomenological model demonstrates the ability of developed model in proportional and nonproportional loadings. In most existing 3-D phenomenological constitutive models, some internal variables are necessary to be calibrated. In calibration process, most models estimate various material parameters in which some of those are not simply measurable by experimental tests. Among various phenomenological models, the microplane model is utilized due to its simplicity and the limited material parameters needed for calibration.

The behavior of some quasi-brittle materials, such as concrete, soil, and stiff foams, is studied using the microplane method [49-52]. A one-dimensional phase transformation model using microsphere formulation was proposed by Ostwald et al. [53] to simulate the polycrystalline materials. The three-dimensional model based on the microplane model was proposed by Brocca et al. [37]. In the microplane model, all material parameters can be determined from uniaxial tension tests at different temperatures. The microplane model considers 1-D equations for some directions on arbitrary plane and is extended to 3-D model using homogenization process [54]. Mehrabi et al. [40, 41, 55] developed this idea within the framework of thermodynamics and proved the capability of the proposed model under multiaxial loadings. Some unique characteristics of SMAs such as deviation from normality in nonproportional loading, anisotropic behavior, and tension-compression asymmetry behavior of SMAs were studied by Mehrabi et al. [56, 57] using the microplane approach.

The proposed microplane model is implemented in a standard commercial finite element package. Some experimental studies under tension, torsion, proportional, and nonproportional loadings have been performed on SMA hollow tubes to assess the proposed model [58]. The numerical results extracted from the microplane model compared with experimental data demonstrate the ability of the microplane approach.

2. Phenomenological constitutive modeling

In this section, phenomenological model based on microplane theory is used to describe SMAs material behavior in a simple way. The general definition of the microplane approach is that the 1-D constitutive law is defined for associated normal and tangential stress/strain components on any microplane at each material point. The generalization of 1-D equation to 3-D model is done by homogenization process. In microplane formulation, strain tensor is in closed form of stress tensor. Mehrabi and co-workers have done a thorough research on 3-D phenomenological model based on microplane theory to demonstrate their model features [40, 41, 55, 57, 59, 60]. The three main steps of the microplane model are summarized in Figure 1.

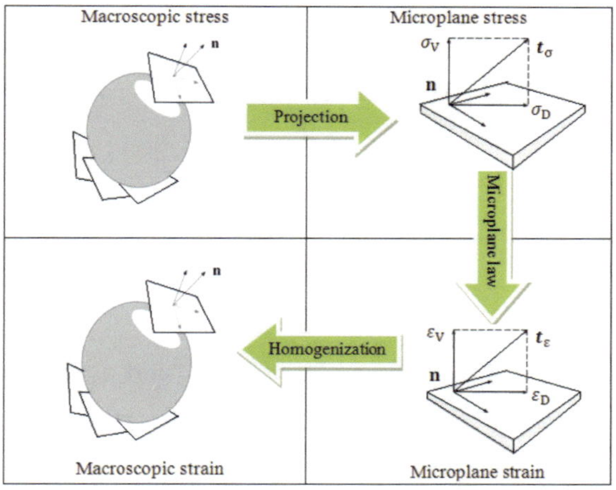

Figure 1. General schematic of the microplane model based on the volumetric-deviatoric split.

For any plane on the shape memory alloys, microscopic Gibbs free energy (G^{mic}) is defined as

$$G^{mic} = \hat{G}^{mic}\left(\sigma_V, \boldsymbol{\sigma}_D, T, \xi\right) \tag{1}$$

Here, σ_V and $\sigma_D$are stress components on each microplane, T is temperature, and ξ is martensite volume fraction, which is defined as internal variable [27, 61].

Macroscopic strain tensor is defined as [41]

$$\begin{aligned}\varepsilon &= -\rho\frac{\partial G}{\partial \boldsymbol{\sigma}} = -\rho\frac{3}{2\pi}\int_{\Omega}\frac{\partial G^{\text{mic}}}{\partial \boldsymbol{\sigma}}\text{d}\Omega = \\ &= \frac{3}{2\pi}\int_{\Omega}\left[-\rho\frac{\partial G^{\text{mic}}}{\partial \sigma_{\text{V}}}\frac{\partial \sigma_V}{\partial \boldsymbol{\sigma}} - \rho\frac{\partial G^{\text{mic}}}{\partial \boldsymbol{\sigma}_{\text{D}}}\frac{\partial \boldsymbol{\sigma}_D}{\partial \boldsymbol{\sigma}}\right]\text{d}\Omega = \\ &= \frac{3}{2\pi}\int_{\Omega}\left(\varepsilon_{\text{V}}\mathbf{V} + \mathbf{Dev}^{\text{T}}.\varepsilon_{\text{D}}\right)d\Omega\end{aligned} \tag{2}$$

These integrals calculated on different orientations of hemisphere at a material point. Projection tensor ***Dev*** and the transpose of the deviatoric projection tensor ***Dev***^{T} are defined as [62]

$$\mathbf{Dev} := \boldsymbol{n}.\mathbf{I}^{\text{dev}}, \mathbf{Dev}^{\text{T}} := \mathbf{I}^{\text{dev}}.\boldsymbol{n} \tag{3}$$

in which $\boldsymbol{n}$ represents the unit normal vector on the plane and $\boldsymbol{I}^{\text{dev}}$ is deviatoric projection tensor (forth-order identity tensor) that is defined as follows

$$\mathbf{I}^{\text{dev}} = \frac{3}{2\pi}\int_{\Omega}\mathbf{Dev}^{\text{T}}.\mathbf{Dev}d\Omega \tag{4}$$

The local 1-D constitutive laws in the volumetric and deviatoric components of the strain for SMAs are defined as

$$\varepsilon_{\text{V}} = \frac{\sigma_{\text{V}}}{E_{\text{V}}^{0}},\ \boldsymbol{\varepsilon}_{\text{D}} = \frac{\boldsymbol{\sigma}_{\text{D}}}{E_{\text{D}}^{0}} + \mathbf{R}\varepsilon^{*}\xi \tag{5}$$

where ε^{*} is the axial maximum recoverable strain, $E_{\text{V}}^{0}$and E_{D}^{0} are the local linear elastic modulus, which are a function of the global elastic constants. Transformation strain is only initiated in deviatoric direction of microplanes.

Here a standard procedure in the microplane model [57] is used to generalize the 1-D equations to 3-D formulation. Macroscopic strain tensor for shape memory alloys is calculated as

$$\varepsilon = \varepsilon^{e} + \varepsilon^{tr} = \frac{3}{2\pi}\int_{\Omega}\left(\frac{1-2\upsilon}{E(\xi)}\mathbf{V}\otimes\mathbf{V}:\boldsymbol{\sigma} + \frac{1+\upsilon}{E(\xi)}\mathbf{Dev}^{\text{T}}.\mathbf{Dev}:\boldsymbol{\sigma}\right)d\Omega + \frac{3}{2\pi}\int_{\Omega}\varepsilon^{*}\xi\mathbf{Dev}^{\text{T}}.\mathbf{R}\,d\Omega \tag{6}$$

where R is a vector that is defined in reference [41]. In order to numerically calculate these integrals, an integration technique consisting of 21 Gaussian integration points is utilized [63].

3. Experimental study

In order to have a robust constitutive modeling, an experimental study of polycrystalline SMAs gets more attention in recent years. Since some SMA devices experience complex loading paths, investigation of material behavior under multiaxial loadings is essential. Some of the experimental findings are summarized in Table 2. In this table, some of the famous groups and loading schemes on the specific materials are introduced. As it is shown, most experimental studies are on the NiTi materials and copper based SMAs.

Group	Specification	Material
Sittner et al. [64, 65]	Tension-Torsion	CuAlZnMn
Jacobus et al. [66, 67]	Triaxial stress	CuZnAl, NiTi
Tokuda et al. [68, 69]	Tension-Torsion	CuAlZnMn
Lim and McDowell [70]	Tension-Torsion	NiTi
McNaney et al. [71]	Tension-Torsion	NiTi
Grabe and Bruhns [72, 73]	Tension-Torsion	NiTi
S. Arbab-Chirani and C. Lexcellent [74, 75]	Tension-Pressure-Torsion	CuAlBe, NiTi
Wang et al. [76, 77]	Tension-Torsion	NiTi
Reedlunn et al. [78]	Tension-Pressure-Bending	NiTi
Mehrabi et al. [55-58]	Tension-Torsion	NiTi

Table 2. Experimental study

A vast experimental study of the NiTi hollow tubes under uniaxial tension, pure torsion, and proportional and nonproportional tension-torsion were done by the author in the Dynamic and Smart Systems Laboratory at the University of Toledo, USA. The Johnson Matthey provided NiTi tube specimens, and the experimental tests were performed using BOSE ElectroForce machine. All mechanical tests were performed at room temperature, and in order to have an isothermal condition, the loading rate was below $10^{-3}s^{-1}$ [76]. The experimental results are compared with numerical findings to show the capability of the proposed approach in the next section.

4. Numerical simulation

In order to use microplane approach to simulate real SMA devices, the proposed model is implemented and developed into the FE code. The computational algorithm is outlined in Table 3.

1. Import the strain increment and the stress evaluated from ABAQUS
2. Check for transformation according to the phase diagram and compute the transformation strain if necessary
3. Compute the elastic strain
4. Compute the total strain
5. Compute the Jacobian matrix
6. Compute the incremental stress tensor
7. Update stress
8. End the program

Table 3. Algorithm for implementation of constitutive modeling of SMA

Tensile and torsional tests are conducted on the NiTi tubes to investigate the capability of the microplane approach in capturing the behavior of SMAs. The material parameters calibrated for the microplane model are listed in Table 4 [55].

Symbols	Values	Units
E_A	20,000	MPa
E_M	13,300	MPa
$\upsilon_A = \upsilon_M$	0.33	
T_f^M	-32	°C
T_s^M	-15	°C
T_s^A	-5	°C
T_f^A	15	°C
σ_s^{cr}	20	MPa
σ_f^{cr}	100	MPa
C_M	6	MPa / °C
C_A	8.2	MPa / °C
ε^*	0.038	

Table 4. Material properties

http://dx.doi.org/10.5772/61124

Figures 2 and 3 represent comparison between the axial stress-strain responses of the microplane model and the experimental results as well as shear stress-strain response at room temperature. These comparisons confirm the fact that material parameter calibration process is done as well. Calibrated material parameters are constant during numerical study of proportional and nonproportional loadings.

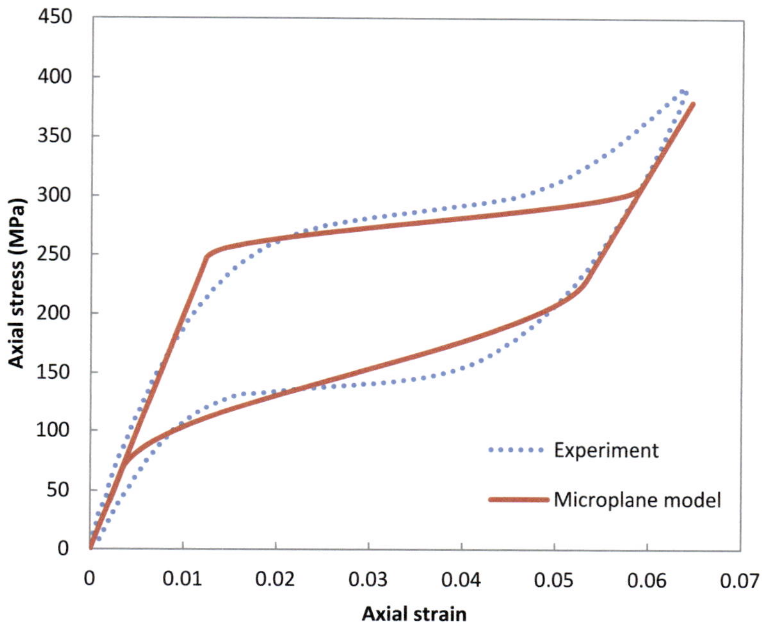

Figure 2. Comparison of the microplane model with experimental result [55].

To demonstrate other aspects of the microplane model, proportional tension-torsion loading is experimentally performed. Proportional loading path is shown in Figure 4(a). According to this loading path, axial stress and shear stress are increasing in step 1 and are decreasing to zero during step 2. The experimental findings are compared with the numerical results for axial stress-strain and shear stress-strain in Figures 4(b) and (c). The studied proportional loading demonstrates the capability of the proposed model.

In order to show the capability of the proposed approach in multiaxial loading, one complex loading path is considered here. In Figure 5, nonproportional tension-torsion loading path is shown. At first, shear stress increases while axial stress is zero. During step 2, shear stress is constant, and axial stress increases. Then, shear stress and axial stress are recovered to zero, respectively. Experimental results are compared with microplane numerical results in Figure 6. Comparison of results shows that the proposed model has good agreement with experimental results in both axial stress-strain and shear strain-axial strain. It is obvious that a discrepancy between experimental results and numerical results in the shear stress-strain curve is found. As the proposed model could predict general behavior of SMAs in different loadings, this negligible discrepancy is acceptable.

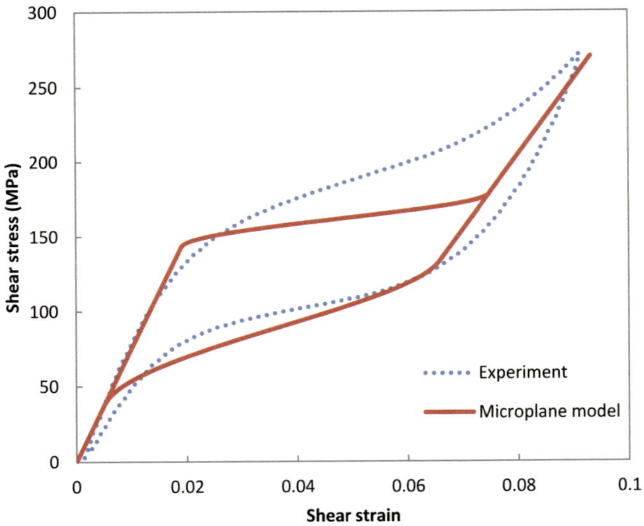

Figure 3. Comparison of the microplane model with experimental result [55].

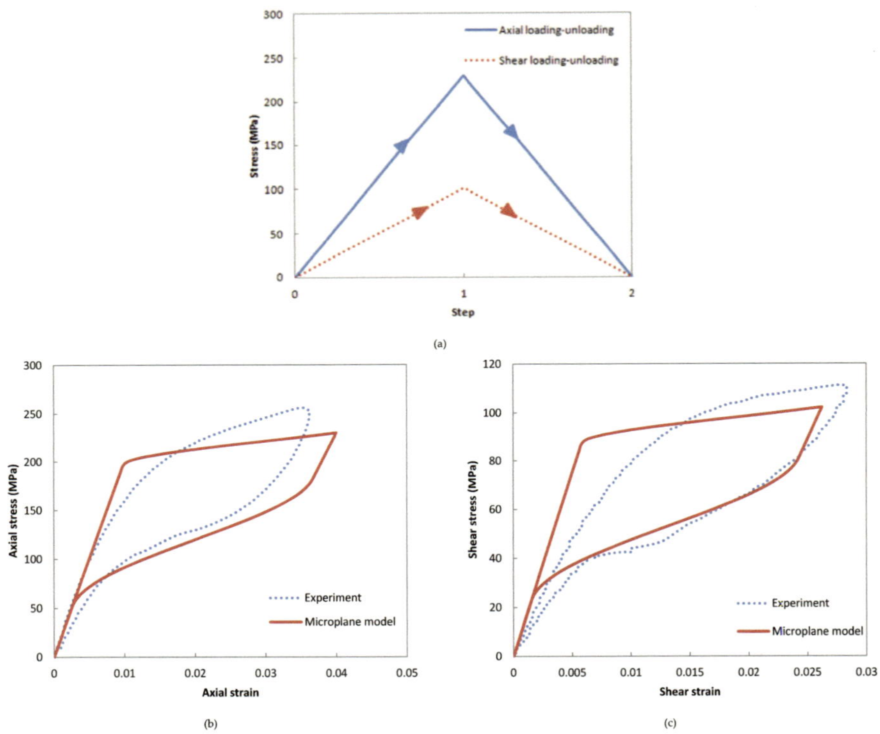

Figure 4. Comparison of the microplane model with experimental result in proportional loading: (a) proportional loading path, (b) axial stress-strain, (c) shear stress-strain [55].

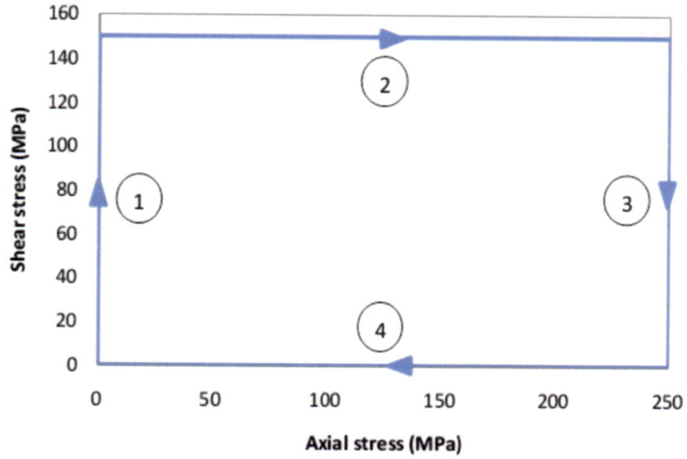

Figure 5. Nonproportional loading path.

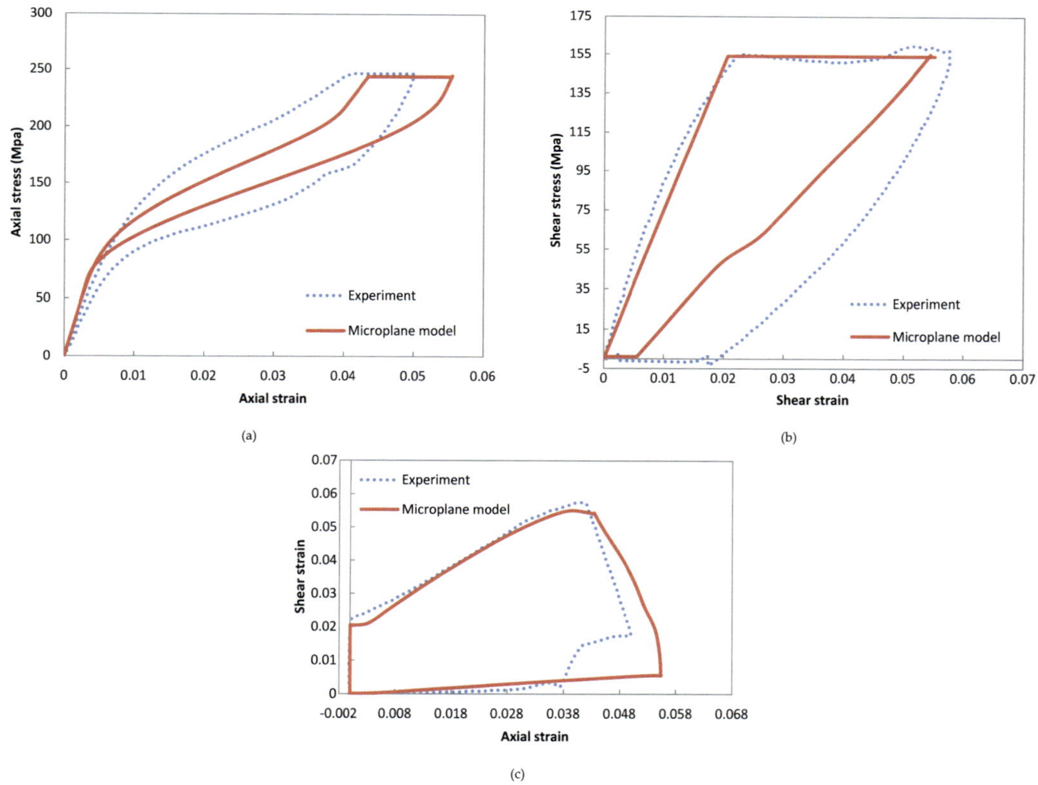

Figure 6. Comparison of the microplane model with experimental result in nonproportional loading: (a) axial stress-strain, (b) shear stress-strain, (c) axial strain-shear strain [57].

In recent years, some biomedical applications [5] such as stent [79], catheters [80], muscle [81], artificial muscles [82], and shape memory implants [83] are produced by SMAs. Therefore, the simulation of biomedical devices using 3-D finite element method [84] is an interesting topic that leads to future works.

5. Conclusion

Constitutive modeling of shape memory alloys (SMAs) is a key property that leads researchers to find new engineering applications. Phenomenological modeling in macroscopic frame is an appropriate way for modeling the thermomechanical response of SMAs. One of the unique constitutive models based on the microplane model is utilized to investigate behavior of SMAs. Material parameters defined in the proposed model are limited and are calibrated with simple experimental tests. The proposed model is developed to implement and analyze in a finite element package. Some multiaxial loadings as proportional and nonproportional loadings are investigated with constitutive model. Numerical results in comparison of experimental findings show the microplane approach ability in simulation of SMAs behavior.

Author details

Reza Mehrabi

Address all correspondence to: r.mehrabi@vru.ac.ir

Department of Mechanical Engineering, Vali-e-Asr University of Rafsanjan, Rafsanjan, Iran

References

[1] Anson, T., *Shape memory alloys—medical applications.* Materials World, 1999. 7 (20): p. 745-747.

[2] Lagoudas, D.C., *Shape memory alloys: modeling and engineering applications.* 2008: Springer.

[3] Lagoudas, D.C. and D.J. Hartl, *Aerospace applications of shape memory alloys.* Proceedings of the Institution of Mechanical Engineers. Part G—Journal of Aerospace Engineering, 2007. 221(G4): p. 535-552.

[4] Mohd Jani, J., et al., *A review of shape memory alloy research, applications and opportunities.* Materials and Design, 2014. 56: p. 1078-1113.

[5] Petrini, L. and F. Migliavacca, *Biomedical applications of shape memory alloys.* Journal of Metallurgy, 2011. 2011.

[6] Song, G., N. Ma, and H.-N. Li, *Applications of shape memory alloys in civil structures.* Engineering Structures, 2006. 28(9): p. 1266-1274.

[7] Van Humbeeck, J., *Non-medical applications of shape memory alloys.* Materials Science and Engineering: A, 1999. 273: p. 134-148.

[8] Webb, G., et al., *Adaptive control of shape memory alloy actuators for underwater biomimetic applications.* AIAA Journal, 2000. 38(2): p. 325-334.

[9] Barbarino, S., et al., *A review on shape memory alloys with applications to morphing aircraft.* Smart Materials and Structures, 2014. 23(6): p. 063001.

[10] Brinson, L.C., I. Schmidt, and R. Lammering, *Stress-induced transformation behavior of a polycrystalline NiTi shape memory alloy: micro and macromechanical investigations via in situ optical microscopy.* Journal of the Mechanics and Physics of Solids, 2004. 52(7): p. 1549-1571.

[11] Entchev, P.B. and D.C. Lagoudas, *Modeling porous shape memory alloys using micromechanical averaging techniques.* Mechanics of Materials, 2002. 34(1): p. 1-24.

[12] Bhattacharya, K. and A. Sadjadpour, *A micromechanics inspired constitutive model for shape-memory alloys: the one-dimensional case.* Smart Materials and Structures, 2007. 16(1): p. S51-S62.

[13] Gao, X.J., M.S. Huang, and L.C. Brinson, *A multivariant micromechanical model for SMAs Part 1. Crystallographic issues for single crystal model.* International Journal of Plasticity, 2000. 16(10-11): p. 1345-1369.

[14] Levitas, V.I. and I.B. Ozsoy, *Micromechanical modeling of stress-induced phase transformations. Part 1. Thermodynamics and kinetics of coupled interface propagation and reorientation.* International Journal of Plasticity, 2009. 25(2): p. 239-280.

[15] Patoor, E., A. Eberhardt, and M. Berveiller, *Micromechanical modelling of superelasticity in shape memory alloys.* Journal De Physique IV, 1996. 6(C1): p. 277-292.

[16] Sadjadpour, A. and K. Bhattacharya, *A micromechanics-inspired constitutive model for shape-memory alloys.* Smart Materials and Structures, 2007. 16(5): p. 1751-1765.

[17] Thamburaja, P., H. Pan, and F. Chau, *The evolution of microstructure during twinning: constitutive equations, finite-element simulations and experimental verification.* International Journal of Plasticity, 2009. 25(11): p. 2141-2168.

[18] Mirzaeifar, R., et al., *A micromechanical analysis of the coupled thermomechanical superelastic response of textured and untextured polycrystalline NiTi shape memory alloys.* Acta Materialia, 2013. 61(12): p. 4542-4558.

[19] Thamburaja, P. and L. Anand, *Superelastic behavior in tension-torsion of an initially-textured Ti-Ni shape-memory alloy.* International Journal of Plasticity, 2002. 18(11): p. 1607-1617.

[20] Lagoudas, D., et al., *Constitutive model for the numerical analysis of phase transformation in polycrystalline shape memory alloys.* International Journal of Plasticity, 2012. 32: p. 155-183.

[21] Arghavani, J., et al., *A 3-D phenomenological constitutive model for shape memory alloys under multiaxial loadings.* International Journal of Plasticity, 2010. 26(7): p. 976-991.

[22] Reali, A., F. Auricchio, and U. Stefanelli, *A three-dimensional model describing stress-induced solid phase transformation with permanent inelasticity.* International Journal of Plasticity, 2007. 23(2): p. 207-226.

[23] Saleeb, A.F., S.A. Padula, and A. Kumar, *A multi-axial, multimechanism based constitutive model for the comprehensive representation of the evolutionary response of SMAs under general thermomechanical loading conditions.* International Journal of Plasticity, 2011. 27(5): p. 655-687.

[24] Sedlák, P., et al., *Thermomechanical model for NiTi-based shape memory alloys including R-phase and material anisotropy under multi-axial loadings.* International Journal of Plasticity, 2012.

[25] K. Tanaka, S. Kobayashi, and Y. Sato, *Thermomechanics of transformation pseudoelasticity and shape memory effect in alloys.* International Journal of Plasticity, 1986. 2: p. 59-72.

[26] Liang, C. and C. Rogers, *One-dimensional thermomechanical constitutive relations for shape memory materials.* Journal of Intelligent Material Systems and Structures, 1990. 1(2): p. 207-234.

[27] Brinson, L.C., *One-dimensional constitutive behavior of shape memory alloys: thermomechanical derivation with non-constant material functions and redefined martensite internal variable.* Journal of Intelligent Material Systems and Structures, 1993. 4(2): p. 229-242.

[28] Auricchio, F. and E. Sacco, *A one-dimensional model for superelastic shape-memory alloys with different elastic properties between austenite and martensite.* International Journal of Non-Linear Mechanics, 1997. 32(6): p. 1101-1114.

[29] Bekker, A., et al., *Impact induced propagation of phase transformation in a shape memory alloy rod.* International Journal of Plasticity, 2002. 18(11): p. 1447-1479.

[30] Boyd, J.G. and D.C. Lagoudas, *Thermomechanical response of shape-memory composites.* Journal of Intelligent Material Systems and Structures, 1994. 5(3): p. 333-346.

[31] Hartl, D.J., et al. *Phenomenological modeling of induced transformation anisotropy in shape memory alloy actuators. Proceedings of SPIE.* 2012.

[32] Panico, M. and L.C. Brinson, *A three-dimensional phenomenological model for martensite reorientation in shape memory alloys.* Journal of the Mechanics and Physics of Solids, 2007. 55(11): p. 2491-2511.

[33] Bouvet, C., S. Calloch, and C. Lexcellent, *A phenomenological model for a Cu-Al-Be shape memory alloy pseudoelasticity under complex loading.* Journal De Physique IV, 2003. 112: p. 201-204.

[34] Lexcellent, C. and G. Bourbon, *Thermodynamical model of cyclic behaviour of Ti-Ni and Cu-Zn-Al shape memory alloys under isothermal undulated tensile tests.* Mechanics of Materials, 1996. 24(1): p. 59-73.

[35] Helm, D. and P. Haupt, *Shape memory behaviour: modelling within continuum thermomechanics.* International Journal of Solids and Structures, 2003. 40(4): p. 827-849.

[36] Oliveira, S., M. Savi, and A. Kalamkarov, *A three-dimensional constitutive model for shape memory alloys.* Archive of Applied Mechanics, 2010. 80(10): p. 1163-1175.

[37] Brocca, M., L.C. Brinson, and Z. Bazant, *Three-dimensional constitutive model for shape memory alloys based on microplane model.* Journal of the Mechanics and Physics of Solids, 2002. 50(5): p. 1051-1077.

[38] Arghavani, J., et al., *A 3D finite strain phenomenological constitutive model for shape memory alloys considering martensite reorientation.* Continuum Mechanics and Thermodynamics, 2010. 22(5): p. 345-362.

[39] Chemisky, Y., et al., *Constitutive model for shape memory alloys including phase transformation, martensitic reorientation and twins accommodation.* Mechanics of Materials, 2011. 43(7): p. 361-376.

[40] Mehrabi, R. and M. Kadkhodaei, *3D phenomenological constitutive modeling of shape memory alloys based on microplane theory.* Smart Materials and Structures, 2013. 22(2): p. 025017.

[41] Mehrabi, R., M. Kadkhodaei, and M. Elahinia, *A thermodynamically-consistent microplane model for shape memory alloys.* International Journal of Solids and Structures, 2014. 51(14): p. 2666-2675.

[42] Lagoudas, D.C., et al., *Shape memory alloys, Part II: modeling of polycrystals.* Mechanics of Materials, 2006. 38(5-6): p. 430-462.

[43] Morin, C., Z. Moumni, and W. Zaki, *A constitutive model for shape memory alloys accounting for thermomechanical coupling.* International Journal of Plasticity, 2011. 27(5): p. 748-767.

[44] Zaki, W. and Z. Moumni, *A 3D model of the cyclic thermomechanical behavior of shape memory alloys.* Journal of the Mechanics and Physics of Solids, 2007. 55(11): p. 2427-2454.

[45] Zaki, W. and Z. Moumni, *A three-dimensional model of the thermomechanical behavior of shape memory alloys.* Journal of the Mechanics and Physics of Solids, 2007. 55(11): p. 2455-2490.

[46] Zaki, W., *An efficient implementation for a model of martensite reorientation in martensitic shape memory alloys under multiaxial nonproportional loading.* International Journal of Plasticity, 2012. 37: p. 72-94.

[47] Tanaka, K., *Thermomechanical sketch of shape memory effect: one-dimensional tensile behavior.* Res Mechanica: International Journal of Structural Mechanics and Materials Science, 1986. 18(3): p. 251-263.

[48] Peng, X., Y. Yang, and S. Huang, *A comprehensive description for shape memory alloys with a two-phase constitutive model.* International Journal of Solids and Structures, 2001. 38(38-39): p. 6925-6940.

[49] Bazant, Z., *Microplane model for strain controlled inelastic behavior.* Mechanics of engineering materials, 1984. 3: p. 45-59.

[50] Bažant, Z.P. and B.H. Oh. *microplane model for fracture analysis of concrete structures.* In Proceedings of Symposium on the Interaction of Non-nuclear Munitions with Structures. 1983. Washington.

[51] Bažant, Z.P. and P.C. Prat, *Microplane model for brittle plastic material. I: theory.* Journal of Engineering Mechanics, 1988. 114: p. 1672-1688.

[52] Carol, I., M. Jirasek, and Z. Bazant, *A thermodynamically consistent approach to microplane theory. Part I. Free energy and consistent microplane stresses.* International Journal of Solids and Structures, 2001. 38(17): p. 2921-2931.

[53] Ostwald, R., T. Bartel, and A. Menzel, *A computational micro-sphere model applied to the simulation of phase-transformations.* ZAMM-Journal of Applied Mathematics and Mechanics/Zeitschrift für Angewandte Mathematik und Mechanik, 2010. 90(7-8): p. 605-622.

[54] Kadkhodaei, M., et al., *Modeling of shape memory alloys based on microplane theory.* Journal of Intelligent Material Systems and Structures, 2008. 19(5): p. 541-550.

[55] Mehrabi, R., et al., *Microplane modeling of shape memory alloy tubes under tension, torsion, and proportional tension-torsion loading.* Journal of Intelligent Material Systems and Structures, 2014: p. 1045389X14522532.

[56] Mehrabi, R., et al., *Anisotropic behavior of superelastic NiTi shape memory alloys; an experimental investigation and constitutive modeling.* Mechanics of Materials, 2014. 77: p. 110-124.

[57] Mehrabi, R., M. Kadkhodaei, and M. Elahinia, *Constitutive modeling of tension-torsion coupling and tension-compression asymmetry in NiTi shape memory alloys.* Smart Materials and Structures, 2014. 23(7): p. 75021-75035.

[58] Mehrabi, R., et al., *Experimental study of NiTi thin-walled tubes under uniaxial tension, torsion, proportional and non-proportional loadings.* Experimental Mechanics, 2015: p. 1-14.

[59] Mehrabi, R., M. Kadkhodaei, and S. Arbab Chirani, *Microplane modeling of martensite reorientation in shape memory alloys.* ECCOMAS 2012, Austria, 10-14 September, 2012.

[60] Mehrabi, R., M. Kadkhodaie, and M. Elahinia, *Microplane modeling of shape memory alloys in an alternative formulation.* SMASIS2013, Snowbird, Utah, 2013, 2013.

[61] Brinson, L.C. and M.S. Huang, *Simplifications and comparisons of shape memory alloy constitutive models.* Journal of Intelligent Material Systems and Structures, 1996. 7(1): p. 108-114.

[62] Leukart, M. and E. Ramm, *A comparison of damage models formulated on different material scales.* Computational Materials Science, 2003. 28(3): p. 749-762.

[63] Bažant, P. and B. Oh, *Efficient numerical integration on the surface of a sphere.* ZAMM-Journal of Applied Mathematics and Mechanics/Zeitschrift für Angewandte Mathematik und Mechanik, 1986. 66(1): p. 37-49.

[64] Sittner, P., Y. Hara, and M. Tokuda, *Experimental-study on the thermoelastic martensitic-transformation in shape-memory alloy polycrystal induced by combined external forces.* Metallurgical and Materials Transactions A. Physical Metallurgy and Materials Science, 1995. 26(11): p. 2923-2935.

[65] Sittner, P., et al., *On transformation pathways of general stress controlled thermoelastic martensitic transformation in shape memory alloys.* Journal De Physique IV, 1996. 6(C1): p. 357-366.

[66] Gall, K., et al., *Stress-induced martensitic phase transformations in polycrystalline CuZnAl shape memory alloys under different stress states.* Metallurgical and Materials Transactions A, 1998. 29(3): p. 765-773.

[67] Jacobus, K., H. Sehitoglu, and M. Balzer, *Effect of stress state on the stress-induced martensitic transformation in polycrystalline Ni-Ti alloy.* Metallurgical and Materials Transactions A. Physical Metallurgy and Materials Science, 1996. 27(10): p. 3066-3073.

[68] Tokuda, M., et al., *Multi-axial constitutive equations of polycrystalline shape memory alloy —(Experimental background).* JSME International Journal Series A. Solid Mechanics and Material Engineering, 2002. 45(2): p. 276-281.

[69] Tokuda, M., et al., *Thermomechanical behavior of shape memory alloy under complex loading conditions.* International Journal of Plasticity, 1999. 15(2): p. 223-239.

[70] Lim, T.J. and D.L. McDowell, *Mechanical behavior of an Ni-Ti shape memory alloy under axial-torsional proportional and nonproportional loading.* Journal of Engineering Materials and Technology-Transactions of the Asme, 1999. 121(1): p. 9-18.

[71] McNaney, J.M., et al., *An experimental study of the superelastic effect in a shape-memory Nitinol alloy under biaxial loading.* Mechanics of Materials, 2003. 35(10): p. 969-986.

[72] Bruhns, O.T. and C. Grabe, *Path dependence and multiaxial behavior of a polycrystalline NiTi alloy within the pseudoelastic and pseudoplastic temperature regimes.* International Journal of Plasticity, 2009. 25(3): p. 513-545.

[73] Grabe, C. and O.T. Bruhns, *Tension/torsion tests of pseudoelastic, polycrystalline NiTi shape memory alloys under temperature control.* Materials Science and Engineering A. Structural Materials Properties Microstructure and Processing, 2008. 481: p. 109-113.

[74] Lexcellent, C., et al., *Experimental and numerical determinations of the initial surface of phase transformation under biaxial loading in some polycrystalline shape-memory alloys.* Journal of the Mechanics and Physics of Solids, 2002. 50(12): p. 2717-2735.

[75] Saint-Sulpice, L., S.A. Chirani, and S. Calloch, *A 3D super-elastic model for shape memory alloys taking into account progressive strain under cyclic loadings.* Mechanics of Materials, 2009. 41(1): p. 12-26.

[76] Wang, X., et al., *An experimental study of the superelastic behavior in NiTi shape memory alloys under biaxial proportional and non-proportional cyclic loadings.* Mechanics of Materials, 2010. 42(3): p. 365-373.

[77] Wang, Y., Z. Yue, and J. Wang, *Experimental and numerical study of the superelastic behaviour on NiTi thin-walled tube under biaxial loading.* Computational Materials Science, 2007. 40(2): p. 246-254.

[78] Reedlunn, B., et al., *Tension, compression, and bending of superelastic shape memory alloy tubes.* Journal of the Mechanics and Physics of Solids, 2013.

[79] Thériault, P., et al., *Finite element modeling of a progressively expanding shape memory stent.* Journal of biomechanics, 2006. 39(15): p. 2837-2844.

[80] Migliavacca, F., et al., *Stainless and shape memory alloy coronary stents: a computational study on the interaction with the vascular wall.* Biomechanics and Modeling in Mechanobiology, 2004. 2(4): p. 205-217.

[81] Stirling, L., et al., *Applicability of shape memory alloy wire for an active, soft orthotic.* Journal of Materials Engineering and Performance, 2011. 20(4-5): p. 658-662.

[82] Shiraishi, Y., et al. *Development of an artificial myocardium using a covalent shape-memory alloy fiber and its cardiovascular diagnostic response.* In *Engineering in Medicine and Biology Society, 2005. 27th Annual International* Conference of the IEEE-EMBS, 2005. IEEE.

[83] Kim, Y.-S., H.-Y. Zhang, and C. Illustrations, Shape Memory Implant. Dynamic Reconstruction of the Spine, 2006: p. 292.

[84] Mehrabi, R., KaramoozRavari, M.R., *Superelastic Simulation of SMA Helical Springs.* Smart Structures and Systems, An International Journal, 2015. 16 (1): p.183-194.

Chapter 10

Comparison of Additive Technologies for Gradient Aerospace Part Fabrication from Nickel-Based Superalloys

Igor V. Shishkovsky, Aleksey P. Nazarov, Dmitry V. Kotoban and Nina G. Kakovkina

Additional information is available at the end of the chapter

http://dx.doi.org/10.5772/61121

Abstract

In our papers, the laser beam-aided control of the self-propagating high-temperature synthesis in Ni–Al systems for the layerwise manufacture of three-dimensional (3D) parts was offered and experimentally realized. As for the laser in situ synthesis of NiAl and Ni_3Al intermetallides and their layerwise laser cladding without any visible cracks and pores, it was successfully performed later on. The present chapter is dedicated to the comparison of optimal conditions for the selective laser melting and laser direct metal deposition processes of the nickel-based powders and fabrication of a full-density, functionally graded, and crack-free structures on the maximum deposition rate for technological applications. The effects of laser parameters on the phase composition and microstructure of the resulting intermetallic samples will be discussed.

The possibility of controlling to change the hardness of the gradient structures from layer to layer by changing of the powder composition and by using the reinforced intermetallic inclusions into superalloy matrix widens the range of possible applications of 3D parts in aerospace and nuclear industries. Comparing different methods of additive manufacturing reveals their advantages and disadvantages for making large samples, scalability, and customizability, finding ways to control the distribution of residual stresses and specified grain-growth direction for the fabrication of more functional and high-precision samples.

Keywords: Selective laser melting (SLM), direct metal deposition (DMD), nickel-based superalloy, laser-controlled reaction synthesis, functional graded (FG) intermetallic structures, nickel aluminide

1. Introduction

The range of use of nickel superalloys is diverse and covers gas turbines of air, sea, and road transport and industrial turbines for electro- or gas-pumping stations, rocket motors, automatic spacecrafts, and nuclear reactors. The basic units of the turbines where nickel-based superalloys could be used are combustion chambers, guide blades in nozzles, rotor blades, and turbine disks. It is known that with the temperature rise by every 100° at the turbine inlet, the increase in efficiency is about 3–4%. Therefore, the determination to use these heat-resistant nickel superalloys is reasonable and well grounded [42].

In addition to heat resistance, the materials for the turbine blades are required to be creep resistive to possess lasting plasticity, resistance to gas corrosion and oxidation, high strength, fluidity, and viscosity. This is why the nickel superalloys are mainly used for turbine blades and disks.

Five mechanisms are known for the strengthening of superalloys. They are solid solution, dispersion, grain boundary, deformation, and textural strengthening [3, 42]. The first three ones depend on the alloy nature. In nickel-based alloys, strengthening occurs essentially by the dispersive mechanism due to the γ $Ni_3(Al, Ti)$ phase release. The phase with the $L1_2$–Ni_3Al superlattice is the basis for the fabrication of the promising superalloys of a new generation on the nickel base, owing to the observed anomalous temperature dependence of its mechanical properties. This anomaly is manifested in the form of the yield point increase with the temperature growth within a certain temperature interval. It should be mentioned that such behavior is observed exclusively in alloys with a long-range atomic order, i.e., in superlattice alloys.

Depending on the temperature range necessary for the availability of the γ-solution suitable for the hot deformation of the matrix, the following nickel superalloys are distinguished: deformed, hardly deformed, and nondeformable alloys. There exist several basic groups of the alloying elements for the nickel superalloys:

- Carbide-generating (Cr, Mo, Nb, Hf, and Ta of etc) and oxide-generating (Cr and Al), the latter protect an article against corrosion
- Grain boundary (B, C, and Zr)
- γ precipitates for the (Al, Ti, Hf, and Ta) phases coherent to the austenitic matrix and ensuring the dispersion strengthening. Studies show that their percentage must not exceed

24–26%. The shape of the γ phase particles (spherical, cubic, and lamellar) depends on the nonconformity degree to the lattice parameter of matrix and temperature regimens of synthesis. It influences the mechanical properties of the article.

- Finally, austenitic γ-matrix additives (Co, Fe, Cr, and Mo) are known, which form an fcc lattice with nickel and ensure the solid solution strengthening.

Meanwhile, the formation of σ, μ, and Laves phases resulting in the decrease of the nickel superalloys strength and plasticity is extremely undesirable. It is considered that the basic task for the achievement of the nickel alloy's lasting strength is the ability to control the grain size and orientation, with respect to the size of the article (component). Ensuring the primary grain growth by the oriented crystallization and, if possible, creating single-crystal articles are considered in recent years to be the main line of studies under this technology.

1.1. The nickel superalloy using at the additive technologies

Machining of Ni superalloys by conventional methods is difficult due to rapid work hardening. Therefore, powder metallurgy methods, such as conventional casting, powder sintering, and/or self-propagating high-temperature synthesis (SHS) are not allowed to fabricate Ni superalloy-based functional parts directly.

Recently, additive manufacturing (AM) has been applied to produce Ni superalloy-based parts from corresponded powders. The AM is able to make three-dimensional objects from computer-aided design (CAD) models. Studies in ATs date back to the mid-1980s. Since then, there has been a rapid advance in its development. Actually, selective laser sintering/melting (SLS/M) (DTM and 3D Systems), laser engineered net shaping (LENS) (Sandia Lab), direct metal deposition (DMD) (POM and Optomec), and 3D laser cladding (LC) or 3D laser welding (LW) technologies could be utilized as metal powders and to produce functional parts.

In the case of the SLS/M technology, a layer-by-layer fabrication is realized by powder delivering into the bed substrate. Traditionally, a powder fraction ranging from 20 to 60 μm and a laser beam diameter ~50–70 μm that allows the manufacture of more precise parts and tools are used. In the cases of the very similar LENS, DMD, and 3D LC/LW methods, a coaxial nozzle with the laser beam delivery system, the multichannel powder feeding system, and the numerically controlled multiaxis table are carried out. The particle flow through the nozzle demands particles' sizes to be of more than hundred micron and the beam diameter increasing to some centimeters, so modest precision and increasing productivity could be achieved. Hence, these methods are more suitable for the repair of tools and articles. An electron beam melting (EBM) process is occupying a separate place; i.e., it conducts into vacuum and is characterized to an oxidation and/or nitration absence and good productivity and precision. The 3D LW of a nickel alloy-based wire could be boosted to greater productivity but very modest precision. On the whole, the necessity of AT application is connected with the heightened requirements to the physical and mechanical characteristics of construction materials used in aerospace technologies, and also with the wide prospects opened by the AT for the fabrication of ready-to-serve articles.

The selective laser melting (SLM) of an Inconel 625 (Ni–Cr 21.5%, Mo 9.0%, Nb and Ta 3.6% in wt.%) superalloy using an Nd:YAG pulsed laser to produce thin wall parts with an emphasis on attaining parts with minimum top surface and side surface roughness was obtained [23]. A sample with 9-mm top surface roughness and 10-mm side roughness was produced. A significant fatigue crack propagation in laser rapid manufactured Inconel 625 structures was reported by [7], and the cracks were observed along the growth direction of the specimens, which was predominantly along the [111] plane. A comparison of the SLM and EBM processes for the Inconel 625 superalloy is given at paper by [24]. They identified γ"-Ni_3Nb bct platelets coincident with the NiCr fcc matrix [111] planes. Similar precipitation and grain orientations (textures) were observed by for the SLM fabricated Inconel 718 components too, although the γ" precipitate platelets were coincident with the NiCr fcc matrix [001] planes rather than the [111] planes. Porous structures of the Inconel-625 by new cross-thin-wall strategy were fabricated at study of [30]. It was found that the yield strength of the fabricated structures followed the power law and decreased from 423 ± 8 MPa for 2.63 ± 0.14% porosity to 226 ± 6.8 MPa for 11.57 ± 0.52% porosity.

An increase in fatigue resistance after the LC process in comparison with wrought and investment cast Inconel 625 was discussed by [44]. The addition of Cr_3C_2 ceramic particles into the Inconel 625 alloy deposited onto a ferrite steel substrate by the LC was analyzed by [45]. As a result, the hard precipitates in the coating microstructure lead to hardness increasing.

[47, 48] described the effect of Al_2O_3 and CeO_2 nanoparticle inclusions into a Ni-based superalloy GH4033 (Ni–Cr 22%; Ti 2.8%; Al 1%; C 0.08%; Fe <1%; Cu, Pb, Bi, Sn, Sb, and As <0.01% in wt.%) during the LC process. The results show that the interface grains, after adding proper nano-Al_2O_3 (1% by mass), grow from epitaxial to nonepitaxial shape gradually, and the columnar dendrites become thinner and denser with cellular shape. Moreover, the dispersive nano-Al_2O_3 particles mainly distribute around cellular substructure and grain boundaries, which prevent the diffusion of alloying elements and restrains the formation of new phase. The addition of 2.0 wt.% nano-CeO_2p showed the most significant improvement effects into the laser-cladded NiCoCrAlY coatings. After adding nano-CeO_2p, an improvement of the microhardness and microstructure uniformity on the cross section and the thermal shock resistance as well were remarked. Results indicate [49] that the hot corrosion resistance of the coatings with nanoparticles is better than that of the one without nanoparticles, among which the one with nano-CeO_2 presented the best hot corrosion resistance. Another effect was observed that the frictional coefficient of the coatings increases and presents the decrease trend with the increase of sliding distance after adding nanoparticles. Moreover [50], the wear rate of the coatings with SiC nanoparticles is only 34.0–64.5% of the coating without nanoparticles.

[60, 6] developed laser induction hybrid rapid cladding process (LIHRC) for the NiCrAlY powder. The preliminary mechanical cryomilling induced the formation of γ'-Ni_3Al and the dissolution of β-NiAl in cryomilled NiCrAlY powder, which in turn was only composed of γ/γ' (γ: Ni, Cr-rich phase). The oxide products formed on the surface of cryomilled and non-cryomilled coatings were predominantly composed of α-Al_2O_3, Cr_2O_3, $NiCr_2O_4$, and $AlYO_3$, but the mechanical cryomilling a significant improves the oxidation resistance of the NiCrAlY coating by the LIHRC. [43] reported about the SLM process optimization into the NiCr alloy.

The authors observed an unusual growth direction-oriented columnar microstructure of [100] texture (corresponding to the [200] plane). Furthermore, columnar grain growth crossing the melt pools was revealed during the SLM process, and this growth was increasing while the laser scanning speed was decreasing.

The results showed that as the laser scanning speed increased, microhardness decreased at the horizontal surface but increased at the vertical surface, and an increase in yield strength and ultimate tensile strength was observed.

[16] produced a Ni-based superalloy Rene-41 (Ni–Cr 20%, Mo 10.5%, Co 12%, Fe 5%, Ti 3%, Al 1.8%, C 0.12% in wt.%) parts by the DMD process. The columnar grain growth was fixed with well-oriented cellular dendrites and with a primary arm spacing of approximately 35 μm. The additional solution aging at 1065 °C for 4 h and followed by air quenching and aged at 760 °C for 16 h followed by air cooling for the Rene-41 superalloy led to the extensive precipitation of γ, which resulted in high hardness and ductility, and the subsequent alignment and coalescence of γ precipitates induced low-strength [16, 58] reported about crack healing in the Rene88DT superalloy prepared by a laser solid forming after an HIP procedure. [20] reported about good laser weldability of the PM1000 superalloy (Ni–20 Cr, 3Fe–0.5Ti, 0.3Al–$0.6Y_2O_3$ wt.%) account of a dispersion strengthened of yttrium oxide particles.

The Waspaloy (19.5% Ni–Cr, 13.5% Co, 4.3% Mo, 3.0% Ti, 1.3% Al, 0.1% C in wt.%) is an established nickel-based superalloy widely used in the manufacture of compressor discs and combustor cans for the aeroengine gas turbines. [10] developed the DMD process with wire deposit for the Waspaloy. A columnar-dendritic solidification structure with the Ni_3(Al, Ti) precipitate phase (γ) forms with the dendrites growing approximately in the plane of the wall and at an angle of around 30° to the build direction. Mumtaz et al. (2012) developed the SLM process of high-density specimens from the Waspaloy. The laser parameters were determined for a high-power pulsed Nd:YAG laser.

The nickel-based niobium-modified superalloy Inconel 718 (Ni–Cr 19.0%, Fe 18.5%, Nb 5.0%, Mo 3.0% in wt.%) is considered as an important material in aeronautic, astronautic, and nuclear industries in virtue of its excellent high-temperature yield strength, anti oxygenic property, hot corrosion resistance, fatigue resistance, and rupture resistance. [2] used the Inconel 718 powder to fabricate cylinder by the SLM method in an argon and nitrogen environment. As-fabricated cylinders were oriented in the build direction (*z*-axis) and perpendicular to the build direction (*x*-axis) exhibited columnar grains and arrays of γ" (body-centered tetragonal) Ni_3Nb oblate ellipsoidal precipitates oriented in a strong [200] texture. [11] discussed a high-temperature oxidation property in the INT-718 superalloy parts during the SLM process. They determined that the oxidation process was controlled by the outward diffusion of oxide-forming elements and in ward penetration of oxygen. Depending on the applied laser energy density, the coarsened columnar dendrites, clustered dendrites, and slender and uniformly distributed columnar dendrites could form [12]. The optimally prepared fully dense Inconel 718 parts had a uniform microhardness distribution with a mean value of 395.8 $HV_{0.2}$, a considerably low-friction coefficient of 0.36, and a reduced wear rate of 4.64×10^{-4} mm^3/N m in sliding wear tests. [19] developed a laser solid forming (LSF) for the Inconel 718 superalloy and ensured that the predominant γ columnar dendrites directionally growing along the deposition direction. Primary dendrite arm space measurement revealed spaces of 11.5, 17.5, and 38.0 mm at the bottom, the middle, and the top section of the LSF sample, respectively, in accordance with the Kurz–Giovanola–Trivedi dendrite growth model.

A multiscale finite element model and stochastic analysis was developed to simulate the evolution of the microstructure of the Nb-bearing nickel-based superalloy (INT 718) during a laser additive manufacturing solidification by [26]. The simulations show that a small equiaxed dendrite arm spacing under a high cooling rate and a low-temperature gradient-to-growth rate (G/R) ratio is beneficial in forming discrete Laves phase particles. Therefore, the improvement of hot cracking resistance by controlling the morphology of the Laves phase particles is possible via analyzing the cracking patterns to numerical analyzed. Safdar et al. (2014) developed detailed CFD models that described melt pool flows dominated by Marangoni and buoyancy-driven convection and taken into account an anisotropic-enhanced thermal conductivity during the DMD process for the Inconel 718. For the repair of the nickel-based turbine blades without hot cracking, [55] recommended a laser powder deposition (LPD) process. The 3D finite element model, including parameters of overlapping and bead geometry, has been developed to simulate the multibead pulse LPD process applied to the Ni-based SX turbine blade repair, using the ANSYS code and element reactivation technology.

[55] reported optimal parameters for the DMD of the Inconel 718. With optimal parameters of laser treatment applied, the authors obtained a directional solidification microstructure with an average distance of 5–10 μm between the columnar crystallites. Between the columnar crystal trunk and the columnar crystallite, the microcomposition segregation was observed. Nb, Mo, and Ti concentrated in the crystal trunk. The segregation was sufficiently lower after heat treatment, and segregation ratios were about 1.

[56] precisely developed the DMD process for the filler wire from the Inconel 718 in the Ar environment. The fracture surfaces revealed the presence of both Al_2O_3 and Cr_2O_3 films, although the latter was reasoned to be the main oxide in the IN718. The exposed surface of the oxide film on the fracture surface has poor wetting with the metal and thus may nucleate some intermetallic compounds, such as the Laves, Ni_3Nb-d, Nb-rich MC, and c0 compounds.

[4] obtained the LC process-induced microstructural characteristics for the INT-738 (Ni–Cr 16%, Co 8.3%, Ti 3.38%, Al 3.4%, W 2.6%, Ta 1.7%, Mo 1.7% in wt.%) superalloy with coarse columnar grains in the range of about 56–158 μm in diameter, and the secondary dendritic arm spacing is about 1.7 μm. The IN-738 alloy after the LC was a supersaturated γ solid solution, but any precipitation of γ particles from the γ matrix was effectively suppressed. [21] reported about liquation cracks in the IN738LC alloy after the LW which were associated with grain boundaries constituents such as γ–γ' eutectic, MC carbide, Cr–Mo boride, and Ni–Zr intermetallic. [5] informed about crack-free laser welds of the INT-738 nickel-based superalloy under preheating at approximately 800°C. [59] informed about minimization of the boundary liquation and interface cracking styles into the Inconel 738 after the DMD. With an extra high-temperature gradient and cooling rate during the deposition process, laser deposition is able to produce directional solidification structure and to control the microstructure.

The LW of the cast Ni-based superalloy K418 (Ni–Cr 12.54%, Mo 4.59%, Al 5.81%, Nb 2.35%, Ti 0.97%, Fe 0.17%, C 0.13%, B 0.13% in wt.%) turbo disk and alloy steel 42CrMo shaft were conducted in [18]. It was remarked that the structure of the laser-welded seam was extremely heterogeneous and consisted mainly of FeCr0.29Ni0.16C0.06 austenite in a form of solid solution dendrites with inclusions of Ni_3Al γ phase and Laves particles, as well as with MC globular or needle-shaped carbides distributed by boundaries between the dendrites.

[29] reported about the LW of the cast nickel-based superalloy K418. Microstructures consisted mainly of austenite solid solution dendrites but also had fine-dispersed $Ni_3(Al, Ti)$ γ phase, MC needle-shaped carbides, and Nb, Ti, and Mo-enriched particles distributed in regions between the dendrites. The microcracks were caused by the liquation of low melting point eutectics in the heat-affected zone (HAZ) and at the grain boundary.

The γ precipitation in the primary γ' of the new Ni–Co-base disc superalloy, TMW-4M3 (Ni–Co 25%, Cr 13.5%, Mo 2.8%, Ti 6.2%, Al 2.3%, W 1.2%, C, B, Zr ~0.02–0.03% in wt.%), was studied by [51]. The size of these tiny particles was unlikely to change after a two-step sagging treatment, but a significant growth was detected after creep rupture at 725°C under various stresses. [14] developed a homogenized, activation energy-based crystal plasticity model for the single-crystal Ni-based superalloys that can be implemented in simulations of polycrystalline aggregates. [57] conducted the LENS process optimization for the 84Ni14.4Cu1.6Sn powder alloy. Thick-wall parts with thickness ranged from 20 to 25 mm were fabricated.

Acharya et al. (2014) presented a comprehensive thermal, fluid flow, and solidification model that can predict the temperature distribution and flow characteristics for the processing of the CMSX-4 (Ni–Cr 6.5%, Co 9.6%, W 6.4%, Ta 6.5%, Al 5.6%, Re 3%, Ti 1%, Mo 0.6%, Hf 0.1% in wt.%) alloy powder through the scanning laser epitaxy (SLE) process. The fabrication of equiaxed, directionally solidified, and single-crystal (SX) deposits of the nickel-based superalloys was successfully using a fast-scanning laser beam. Under temperature gradients at the leading and trailing edges of blades are the order 2.9×10^5 and 10^4 K/m, respectively. In the empirical values for several microstructural characteristics such as the primary dendrite arm spacing of 10–30 μm in the deposit region, unsteadiness of the columnar-to-equiaxed transition criterion value was found, and the oriented-to-misoriented transition criterion is obtained.

Last decade, specific types of studies were devoted to combination of the self-propagated high-temperature synthesis (SHS) and rapid prototyping approaches, which allow synthesized and 3D part fabricated in situ nickel aluminide phases [13, 28, 31, 32, 35, 37]. The experimental parameters controlling the ignition step such as an ignition time, a width of the exothermic reaction zone, and an adiabatic temperature were calculated as a function of initial stoichiometry for different Ni_xAl_y phases. The increase in Al powder content resulted in the rise in adiabatic temperature and the morphology change of nickel aluminide compounds from needlelike to blocky.

[37] and later [32] showed that the laser reaction sintering of the NiAl consists of a complex reaction with several sequential steps. First, the Al-rich compounds $NiAl_3$ and Ni_2Al_3 are formed. Subsequently, these phases react with Ni to form NiAl. In the stage of bulk combustion, Ni continuously dissolved into liquid Al(Ni) solution and transformed the NiAl phase

and a lot of heat. Finally, the Ni reacted with NiAl phase and formed the terminal product–Ni_3Al phase.

Despite the mentioned above serious success in the field of the complex functional articles fabrication out of prepared nickel superalloys, it is still actual to develop new classes of solid, nonbrittle, corrosion- and wear-resistant, and chiefly heat-resistant materials, for instance, on the basis of NixAly phases. There also exists an actual need to generate gradient and laminar structures of the nickel superalloys reinforced by the nickel aluminides, which can be used for production of components and elements for the rocket–space technology. Some approaches to the layered synthesis of intermetallide phases of the NixAly type and gradient of mechanical properties in the obtained laminar structures will be reviewed below. The major task of the work is to develop the basis for the manufacturing of laminar materials with predetermined properties by the different AT approach conditions and to compare conditions required for this.

2. Experimental scheme layerwise fabrication of the FG nickel aluminide's structures

Earlier, we experimentally approved the layerwise SLM and DMD processes in the Ni + Al system [9, 13, 35, 41] and theoretically substantiated [53] the approach, where the strengthened intermetallides are created directly within the laser-assisted AM process in the Ni metallic matrix due to the synthesis reaction.

The LI initiates a chemical reaction between the particles in the powder mixture of stoichiometric composition (Ni + Al, Ni + Ti, Ni + Ti + Al), and this leads to intermetallic phase composition. Moreover, deviation from stoichiometric ratios causes the aggregation of this excess on the matrix of nickel superalloy.

Earlier, we remarked [39, 40] that the LI-controlled intermetallide synthesis reaction is advantageous because it provides uniform and fine distribution of the inclusions and makes the superalloy matrix more stable, while an additional input of energy from exothermic reaction allows using less energy-requiring laser sources.

In addition to activating the synthesis reaction [37, 39, 53], LI also accelerates the directed crystallization and customizable controls the microstructure's properties.

Controlling the thickness of the powdered layer (i.e., the reaction zone volume) during the additive manufacturing process (compare the HAZ under the SMS vs. the DMD processes) can affect the character and direction of the exothermic synthesis reaction and activate the scalable process of directed solidification. Furthermore, the laser power and scanning speed also affect the melting-crystallization conditions.

Functionally graded structures (FGS) and FG articles fabricated by applying complex and dissimilar materials ensure the specific properties of the final product. The manufacturing of 3D FG objects by the 3D LC and/or LDMD is one of the most promising techniques capable of meeting various industrial challenges [27, 36, 41]. This approach permits new freedoms in design and manufacturing, thus allowing, for example, to create an object with the desired shape, internal structure, and engineering composition, including the appropriate physic-mechanical properties, within a single-step fabrication process.

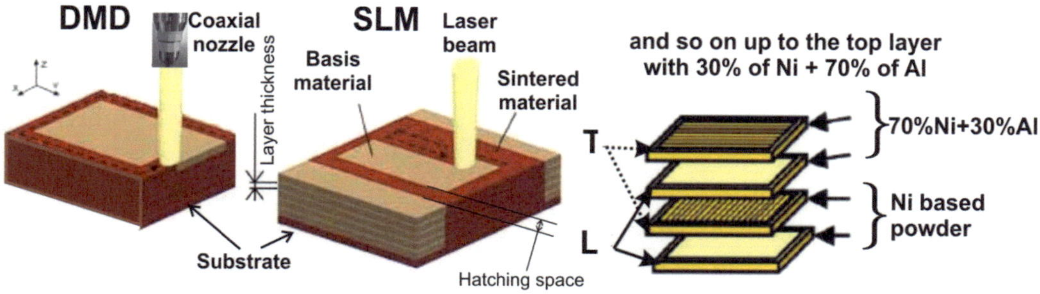

Figure 1. Schematic of the multicomponent-graded structure fabrication by the DMD and SLM processes. Longitudinal —L; transversal—T.

The next powders were used for the FGS fabrication via the LDMD. The aluminum powder had 99 wt.% of Al (TLS Technik GmbH&Co.). The NiCr superalloy Diamalloy 1005 (Sulzer Metco Co.) was used as the Ni powder, which had the following chemical composition: Cr—21.5, Fe—2.5, Mo—9, Nb 3.7 wt.%, bal.–Ni. The powder particles were mainly spherical with the size of ~80–100 μm for 95% of them. The steel substrates of a square shape with a width of 65 and 5 mm height were used. The Diamalloy is equivalent to the Inconel 525 alloy, which was studied earlier in the DMD process [6].

The LDMD method for the FGS fabrication used in the present study is schematically presented in Fig. 1 and was proposed us early [38]. The hatching distance was 2 mm, the layer thickness was ~1 mm, and the powder feeding rate was ~10 g/min. The layers were made out of Ni-based (Diamaloy) and Al powders on a related substrate by the following strategy: the first two layers were of pure NiCr, the next two consisted of 70% NiCr + 30% Al, the third couple of layers were of 50% NiCr + 50% Al, and lastly the upper 7th and 8th layers had the ratio of 30% of NiCr + 70% of Al. Each second layer was formed on the bottom layer after its turning by 90°. For the Fe–Al system, this scheme was successfully approbated at [39, 40]. Argon was the carrying gas. Laser scanning speed was 500 mm/min, laser power varied within the range of 800–1200 W, and laser beam spot diameter was 3 mm. The first channel of the feeder with the Diamaloy powder had a gas flow rate of approximately 20 l/min, while the second one with the Al powder was ~10 l/min.

The following powders were used for the SLM process of FGS fabrication (Fig. 1) in the Ni–Al system. The aluminum powder used was AMDRY 355 (Al + Si 12 wt.%, –45 μm) and nickel powder was Metco 56C-NS (Ni > 99.5 wt.% –75 + 10 μm), which were supplied by Sulzer Metco, GmbH. The scheme of alteration for Ni- and Al-based powders (Ni + Al = 3:1; 1:1; 1:3 wt.%) during the process using SLM approach (Fig. 1) was analogical to the one mentioned above for DMD process.

Granulomorphological analysis of the micron sized powders was carried out by the optical granulomorphometer ALPAGA 500 NANO. The SLM process was performed on the Concept M3 setup (Germany), and the DMD process was realized with the aid of the HAAS 2006D (Nd $^{+3}$:YAG, 4000 W, cw) with the laser beam delivery system, powder feeding system, coaxial nozzle, and numerically controlled five-axes table (Fig. 1). The FGS results after DMD and SLM processes were compared with each other and with the SLM process of the Ni85Al15 nickel superalloy powder produced by the JSC Polema Ltd. (Tula, RF). The Ni85Al15 powder was represented by a 20- to 63-μm fraction and containing >95 wt.% of intermetallic Ni_3Al phase. Results of the LAM processes are shown in Fig. 2.

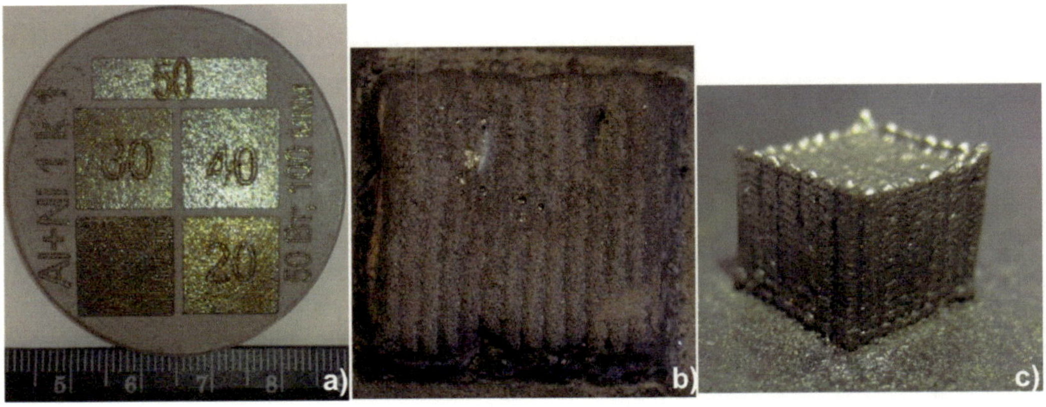

Figure 2. Appearance of the SLM and DMD processes: (a) SLM in Al + Ni = 1:1 system, three layers, hatch distance 0.1 mm; (b) FGS in Ni–Al after DMD; (c) DMD of cube (5 mm^3) from the Ni85Al15 powder.

3. Laser DMD of FG layered nickel aluminides

The optical metallography after FGS fabrication via DMD process is presented in Fig. 3.

The photographs are selected in order to show the characteristic microstructures based on the lower (a1, a2), middle (b, b1), and upper (c, c1) parts of the FGS, i.e., where the proportions of the powdered Ni + Al ≈ 3:1; 1:1 and 1:3 by weight ratios were comprised. Under large magnification, these layers are presented on the inserts a1 and a2 for bottom, b1 for panel b, and c1 for panel c. In the lower layers in the photo (Fig. 3a1), a columnar dendrite pattern was observed, which characterized the high-speed quenching of the NiCr superalloy practically without Al additive participation (compare with [6]). The dendrite growth direction (Fig. 3a1) is stipulated for maximum heat dissipation to the massive substrate. We connect superfluous crackability in Figs. 3a1 and a2 (shown by arrows) with the nonoptimal selection of the increment height from one layer to the next. At the top layers (Figs. 3a1 and b), the microstructure radically changes. In the middle (bottom of Fig. 3a2 or Figs. 3b–b1), there is a cellular and rosette microstructure, which can testify to the equivalence of the heat dissipation speeds in different directions from this area. The upper layers (Figs. 3c–c1) have a needle-shaped

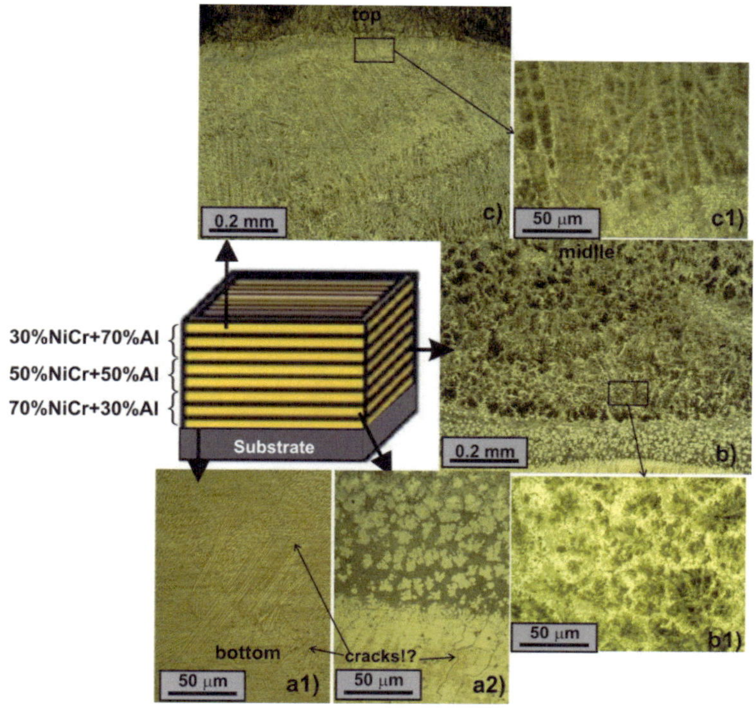

Figure 3. OM micrographs showing the typical microstructures of 3D laser clad coating of NiCr–Al multilayer system: (a) bottom layers ~2–4 mm from substrate; (b) middle layers ~4–6 mm from substrate; (c) top layers ~6–8 mm from substrate.

microstructure with the impregnations of intermetallide phases. At the middle layers, we also observed the triple eutectic structure, which consists of a γ-solid solution on the nickel basis, an α-solid solution on the chromium basis, and also a γ-solid solution on the basis of the Ni_3Al intermetallic compound in the middle and NiAl intermetallide phase above. By the boundaries of the ($\alpha + \gamma + \gamma$) triple eutectic colonies, the two-component eutectic is arranged, which includes the γ-solid solution on the Ni basis and $AlCr_2$ intermetallide phase.

The results of a microhardness measuring are shown in Fig. 4. If the microhardness is equal to 500–550 $HV_{0.1}$ near the base, then it unevenly grows and reaches up to 650–750 $HV_{0.1}$. We connect the separate dips in the microhardness value in the bottom layers (up to 2 mm from the substrate; Fig. 4) with the superfluous cracks, and in the top layers with the indentor entry into the Diamalloy 1005 matrix. On the whole, the measured microhardness values considerably exceed other researchers' data (200–250 $HV_{0.1}$ for the Inconel 625 after the DMD process by [6]).

The XRD pattern of a transverse section after the DMD in the NiCr–Al system is shown in Fig. 5. We can draw the following basic conclusions. After the FGS fabrication, the free nickel and aluminum are practically absent, i.e., it has completely interacted with the Ni_3Al and NiAl intermetallic phase formation. This distinguishes our results from the data of [6], which fixed both the free nickel, γ-, and δ-Ni_3Nb phases under similar DMD regimens with the subsequent

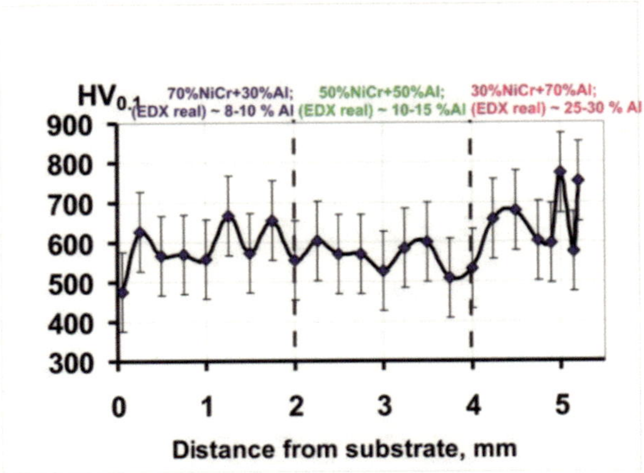

Figure 4. Microhardness distributions of NiCr–Al FG multilayer system.

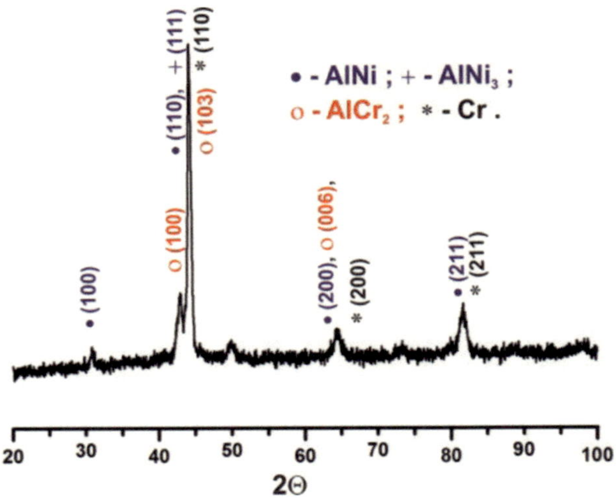

Figure 5. X-ray diffraction patterns of NiCr–Al DMD multilayer system.

annealing. Since the intensity lines of free chromium practically coincide with the intensity lines for the intermetallic compound-AlNi, asserting unambiguously about the free Cr presence is impossible. Meanwhile, the presence of intermetallic phase of $AlCr_2$ is possible.

Fig. 6 shows the SEM data in the bottom, middle, and top layers of the FGS after the layerwise DMD in the NiCr–Al system. For clearness, Figs. 6b–c are similar to Figs. 3b–c, and inserts in Figs. 6b1–c1 are similar to inserts in Figs. 3b1–c1. It is evident that its microelement composition (S1 and S2) practically repeats the initial Diamalloy 1005 composition plus 3.37% wt of Al (see Table 1). By the element relationship in the S3 and S4 areas (Fig. 6b), we have the nickel superalloy matrix depleted by Al. The existence of the metastable Ni_5Al_3 intermetallic phase is possible, on the boundaries of which free Cr extraction has been observed (see Fig. 6b1 and

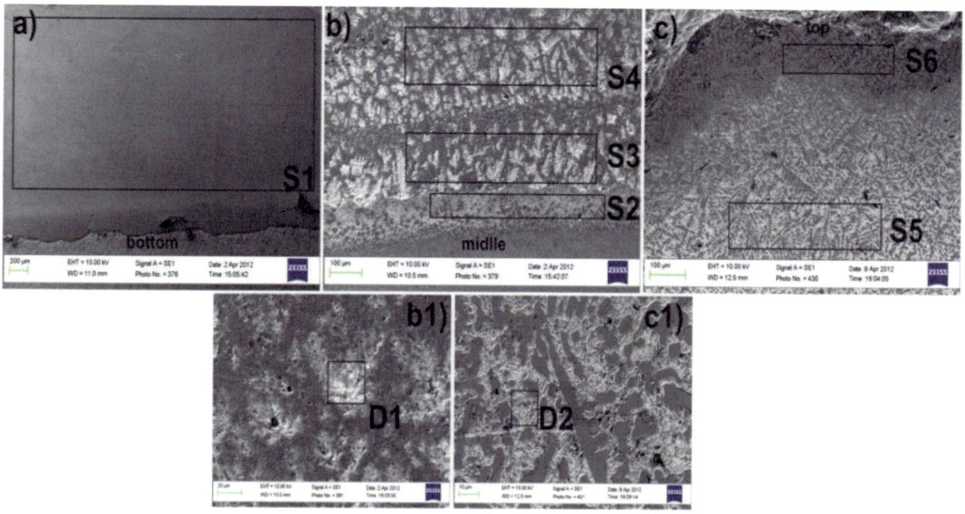

Figure 6. SEM micrographs showing typical solidification microstructures of the bottom (a), middle (b), and top (c) deposited layers with places of the EDS analysis of NiCr–Al multilayer gradient (see Table 1).

D1 area also). Finally, a dendritic structure in Fig. 6c (S5 and S6) clearly shows the intermetallide nature of the Ni_3Al phase. Fig. 6c1 corresponds to Fig. 3c1. We can assert that in the upper layers, the synthesis of the Ni_3Al intermetallide phases (S5–S6 and D2) occurs, as well as a precipitation of the $AlCr_2$ intermetallide phase on the grain boundaries. With sufficient carbon (up to 5 wt.%) and oxygen (up to 3 wt.%) content, we connect with the possibility of their hit from the environment, although the XRD pattern does not fix these elements. [46] informed that the high-temperature oxidation was a great problem after the 3D LC too.

Spectrum	C K	O K	Al K	Si K	Cr L	Fe L	Ni L
S1	1.88	1.47	-	1.03	38.99	1.28	55.35
S2	3.38	1.91	3.37	1.07	51.61	-	37.89
S3	5.66	2.82	2.02	2.94	71.61	1.24	15.71
S4	2.97	2.99	5.02	1.11	11.37	65.32	22.6
S5	3.38	1.41	28.94	0.41	14.04	-	51.35
S6	5.79	1.26	30.48	0.51	18.61	-	43.24
D1	2.89	-	12.24	0.78	50.94	-	33.15
D2	2.24	1.46	34.75	-	6.29	-	55.26

Table 1. EDS data by Fig. 6.

Nevertheless, the main question is Where is aluminum after the layerwise deposition? According to the flow data, in the middle layers, we had to contribute up to 50% of Al, while in upper ones—up to 70% of Al. Meantime, given EDS and XRD data, there was about 10–15% of Al in the middle layers and no more than 30% in the top. Future studies must be conducted to answer this question.

Thus, under the layerwise LDMD in the NiCr–Al system, the formation of the Ni_3Al intermetallic compounds was observed. The applicability of the LDMD for creating a functional gradient and building of Ni_xAl_y intermetallic structures into the nickel superalloy matrix has been experimentally studied. The microhardness values from 500 to 750 $HV_{0.1}$ were achieved, which precisely connected with the intermetallide phase's presence in the NiCr matrix.

4. SLM of the FG-layered nickel aluminides

Optimal regimes for the SLM process in the Ni–Al = 1:1 system were as follows: 100, 120, and 140 mm/s for 100 W of the LI and 160, 200, and 240 mm/s for 150 W (Fig. 2a). Optimal regimes for SLM in the Ni–Al = 3:1 system were 160 and 200 mm/s for 150 W. At last, optimal parameters of SLM of prealloyed intermetallic phase (Ni85Al15 powder) were proved to be 120 and 160 mm/s under 150 W.

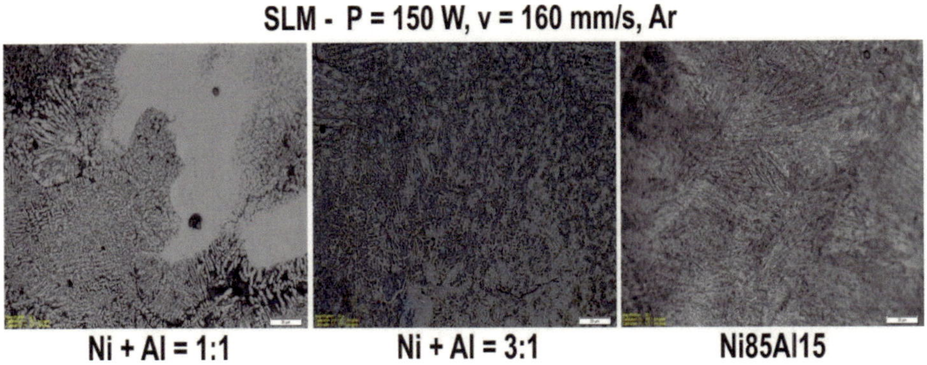

Figure 7. Comparative optical metallography after the SLM in Ni + Al = 1:1 (a); Ni + Al = 3:1 (b), and prealloyed Ni85Al15 (c). Magnification is ×500.

The most acceptable from methodical position SLM regime was chosen for all mixtures (P = 150 W, v = 160 mm/s) to carry out a more careful analysis, with subsequent comparison and experimental studying of FGS possibilities via the SLM process in the Ni–Al system with the variable contents of elements. From the analysis in Fig. 7, it is clear that crack's generation and porosity are observed in all the cases. A recommendation for future studies is to conduct the SLM process in a camera with the temperature increased (see the SLM results of the NiTi at [39, 40]. Dendrite structure obtained after melt cooling is clearly visible.

SEM images (Fig. 8) show fine dendrite structure with different grain orientations. Also, we can mention cracks and pores (Figs. 8a, b). Areas S1, S2, and S3 are the places where EDS analysis (Table 2) was carried out. Microelement analysis ensured that we really handle with

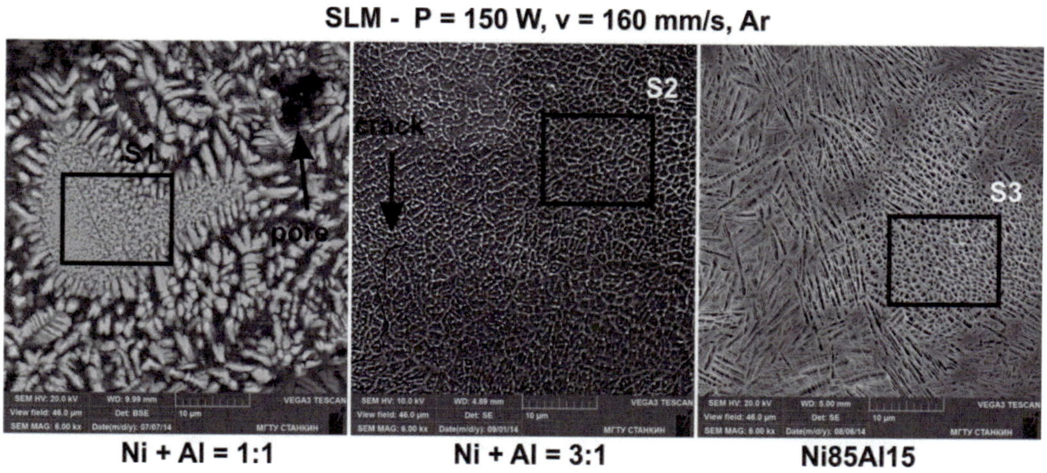

Figure 8. Comparative SEM after SLM in Ni + Al = 1:1 (a); Ni + Al = 3:1 (b); and prealloyed Ni85Al15 (c). Magnification is ×6000.

intermetallic Ni_3Al phase (Fig. 8c and Table 2—S3), intermetallic Ni_3AlSi phase (Fig. 8b and Table 2—S2), and intermetallic NiAlSi phase (Fig. 8a and Table 2—S1). The recent results obtained for kinetics of coarsening of γ precipitates in Ni superalloys have shown that Si additive varies both magnitude and sign of coherency strains between the precipitate and the matrix.

	S1	S2	S3
Element			
Al	53.55	21.94	12.20
Ni	38.07	74.83	87.80
Si	8.38	3.23	-
Total:	100.00	100.00	100.00

Table 2. EDS data by Fig. 8.

X-ray diffraction patterns showed (Fig. 9) clear reflections from AlNi phase, characteristic for JCPDS card no. 20-0019, from metastable intermetallic Ni_3Al_2 phase (JCPDS, card no. 14-0648), pure aluminum (JCPDS, card no. 01-1180), and $Al_{3.21}Si_{0.42}$ (JCPDS, card no. 41-1222).

X-ray phase analysis was supported by the EDS data. Near the high-intensity Al lines, we noticed intermetallic AlNi and Al_3Ni_2 phases too. They were synthesized during the laser-controlled SHS process. It is known that metastable Al_3Ni_2 phase has matching with Ni_3Al in whole intensity lines, so the question about Ni_3Al phase presence is open. Great solubility of Si in NiAl and Ni_2Al_3 phases can provoke improvement of strength in these intermetallic phases.

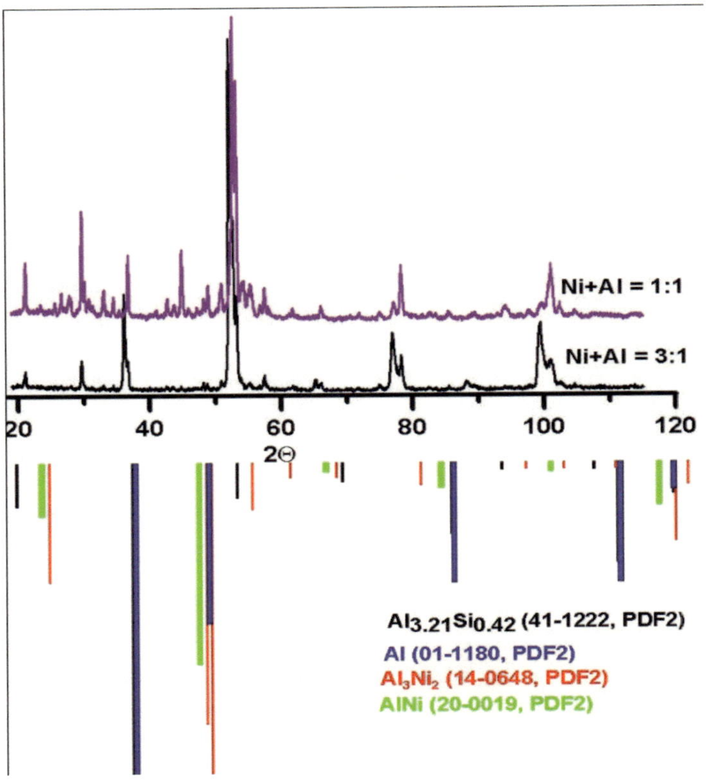

Figure 9. Comparative XRD pattern after SLM in the Ni + Al = 1:1 - (8 layers, violet) and Ni + Al = 3:1 systems (10 layers, black color line). SLM regimen is according to Fig. 7.

Hence, we have shown that layerwise SLM in Ni–Al system with alteration of Ni and Al content allows to fabricate Ni_3Al intermetallic compounds too. Applicability of SLM for creating FGS and building Ni_xAl_y intermetallic structures in nickel superalloy matrix has been experimentally studied.

5. Laser DMD and SLM of prealloyed nickel aluminide

LDMD and SLM of nickel superalloy (Ni85Al15) were successfully realized (Figs. 2c and 7c–8c) [15, 25]. The measured microhardness (Fig. 10) of the laser deposited 3D part (cube) grows from the substrate to the top irregularly. We believe this to be connected with a local hardness increase in the intermetallic phase locations. We connect certain microhardness dip ($HV_{0.1}$ 350) to the indentor hip on the layer boundary. As a whole, the microhardness values correspond to the similar measurements on the nickel aluminides after the DMD and LENS processes, but some are lower than the microhardness of the laser cladded NiAl phase [6, 52].

The qualitative X-ray analysis results are presented in Fig. 11. After the multilayer laser deposition, we have stronger lines, which are located at the angles of ~2θ of 51.4° and 52.1°,

http://dx.doi.org/10.5772/61121

which directly correspond to (111) γ-Ni_3Al and metastable (110) Ni_2Al intermetallic phases. We can propose that probably iron substrate with an interplanar spacing (110) is visible also. The crystal-lattice orientation of the γ-Ni_3Al has preferred direction [111]. On taking into account all the peaks mentioned above, it is reasonable to conclude that this XRD pattern (Fig. 11) best of all coincides with the set of the lines for the Ni_3Al intermetallide. It means that during the LDMD process, this phase remains in a stable state, which corresponds with the aim of our study.

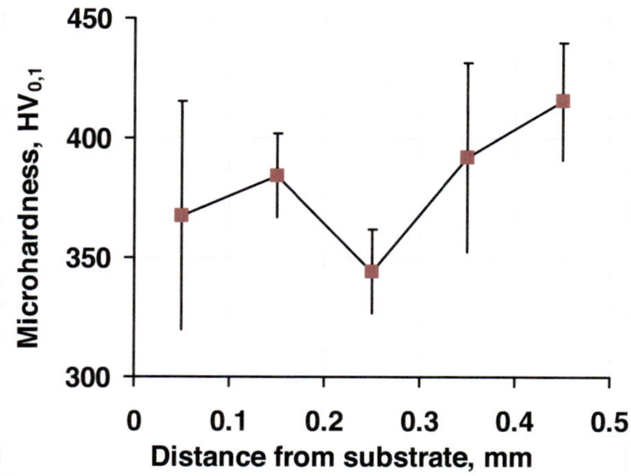

Figure 10. Microhardness distributions of the Ni85Al15 after the multilayer DMD.

SEM images with EDX analysis of the micro- and substructures are shown in Fig. 12. Microstructure study showed that the length of the main dendrite arms was about 15–20 μm, but the grain size was 40–50 μm (Fig. 12b). The secondary dendrite arms varied between 5 and 8 μm. Such structure refinement is connected with a high-speed crystallization from the melt. The dendrite orientation is mainly coplanar with the image; nearby, it is perpendicular to the image plane. Cracks (Fig. 12a) between the second and the first layers were formed on the cooling stage. This broken type indicates the shift nature of plastic deformation under melt cooling.

It should be pointed out that the microstructure mostly consists of columnar-type dendrites, which grew epitaxially from the substrate. Moreover, the growth direction of the columnar dendrites was tied to the laser scanning direction. From Fig. 12b, it is seen that the primary dendrites have almost the same orientation throughout the sample. In contrast, Fig. 12b shows that for the top and the bottom layers, the growth direction of the columnar dendrites changes by 90° in every layer. Hence, during the solidification of the melt pool, cooling mostly occurs via the substrate and the deposit. This leads to the directional growth of the grain counter to the heat dissipation and subsequently the formation of the columnar grains.

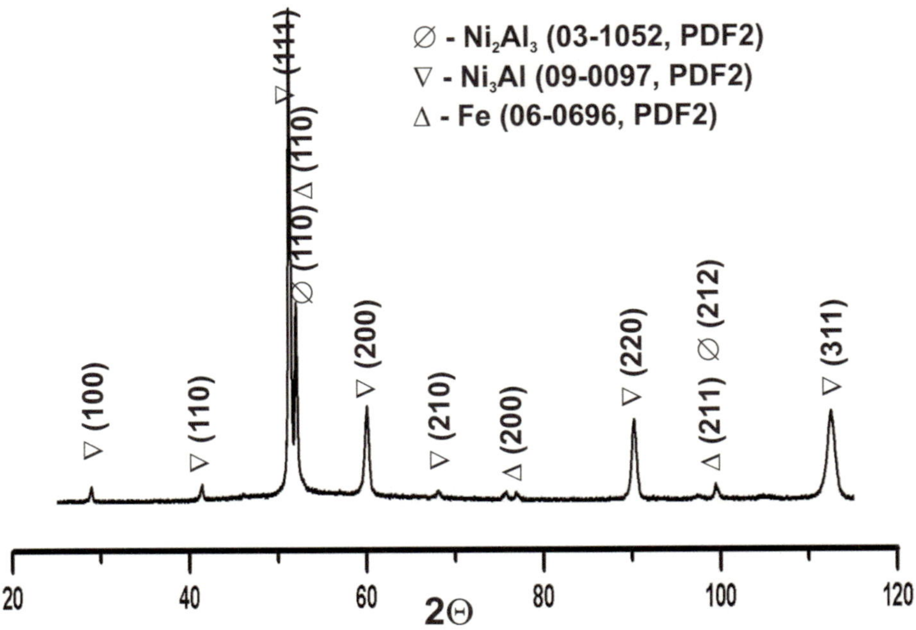

Figure 11. X-ray diffraction pattern of the Ni85Al15 after the multilayer DMD.

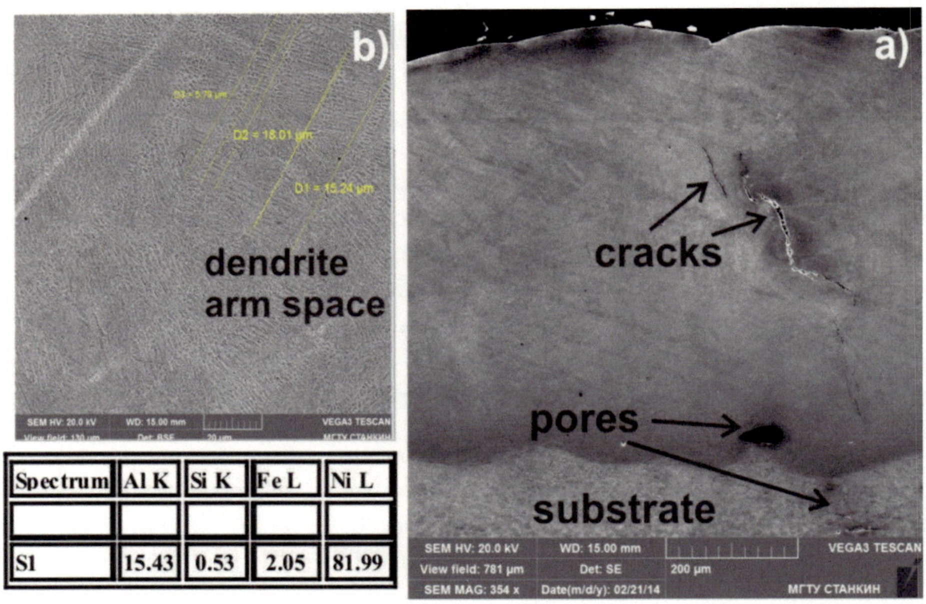

Spectrum	Al K	Si K	Fe L	Ni L
[illegible]	[illegible]	[illegible]	[illegible]	[illegible]
S1	15.43	0.53	2.05	81.99

Figure 12. SEM micrographs showing typical solidification microstructures of the laser deposited layers with the laser scanning speed of 600 mm/min, 250 W, and 9 g/min. Step between the layers was 200 μm, and EDS analysis date (table) from whole image square.

The EDS data of the composition distribution from the cladding layer to the substrate are shown in Fig. 12 into inserted table. It can be found that the matrix element Fe enters into the first cladded layer. The dilution of the substrate elements was notable with the increasing of the laser power density. Evidently, the reason is the accelerating diffusion processes for iron and the excessive overheating of the substrate at the high-power density.

6. Conclusion

The aim of this chapter was to emphasize capabilities and performances of the SLM and LDMD processes for FG part fabrication. Comparison of LDMD and SLM processes ensures that a more precise localization of laser beam is the preferred alternative for the fabrication of precise parts in the SLM process compared with the LDMD process. However, DMD warms up a powder volume more deeply, so an FGS with more homogeneous microstructure is being formed. On the other hand, the absence of initial powder around the manufactured objects leads to steep slopes on their edges after the DMD. This decreases the part's accuracy. In both cases, additional thermal heating of the synthesis zone is preferred.

The LDMD and the SLM of single layers and 3D objects of the Ni_3Al intermetallic were successfully prepared. Good metallographic characteristics and interface bonding were obtained. The fabricated microstructure consisted of γ-Ni_3Al. At the nickel and aluminum boundary, the NiAl and the Ni_3Al intermetallic phase layers are obtained, which possess high-strength characteristics. It is discovered that the concentration of NiAl and Ni_3Al phases is heterogeneous. Mechanisms of nickel aluminide intermetallic compound formation in the zone of the LDMD and SLM are studied. Thermal time parameters of the laser treatment required for the formation of these intermetallide compounds, structurally uniform and of maximum width, are revealed in the work.

7. Outlook

In controlling a multilayer structure, hardness can provide more scalabilities and customizations to apply the 3D FGM in aerospace and nuclear industries by changing the composition of the powder and by using proper CAD modeling. The studies on controlling the residual stresses and revealing LAM conditions for large and/or high-precision samples are provoking further interest.

The results of these studies can be the basis for the development and manufacture of a new class of construction materials—laminar intermetallide composites. From this aspect, nickel–aluminum alloys have a special interest due to their use for the production of units and components for aviation and space equipment, such as fuselage coverings of aircrafts, fuel injectors, screws components, components of the rocket engine nozzle, etc.

Thus, the study by means of the gradient LDMD and SLM in the Ni–Al system under different treatment regimens allowed us to discover that within the wide concentration interval of the

Al_xNi_y phase existence, the synthesis of several intermetallic compounds into nickel matrix is observed. These intermetallic compounds have an interface and differ by their element composition.

Acknowledgements

The authors are grateful to Mr. Missemer F. (DIPI, ENISE, France) for the 3D laser cladding processing, Kuznetsov S. (LPI), PhD, for the X-ray analysis data. Nazarov A., PhD, and Kotoban D. would like to thank the support of the Russian Science Foundation (grant no. 14-19-00992). Ms. Kakovkina N. is separately grateful for the financial support of the Russian Foundation of Basis Researches (grant no. 14-29-10193 ofi-m).

Author details

Igor V. Shishkovsky[1,2*], Aleksey P. Nazarov[2], Dmitry V. Kotoban[2] and Nina G. Kakovkina[1]

*Address all correspondence to: shiv@fian.smr.ru

1 Lebedev Physical Institute (LPI) of Russian Academy of Sciences, Samara Branch, Samara, Russia

2 Moscow State Technological University, STANKIN, LIAT, Moscow, Russia

References

[1] Acharya R., Bansal R., Gambone J. J., Das S. (2012) A coupled thermal, fluid flow, and solidification model for the processing of single-crystal alloy CMSX-4 through scanning laser epitaxy for turbine engine hot-section component repair (Part I and Part II). *Metallurgical and Materials Transactions B.* doi: 10.1007/s11663-014-0117-9.

[2] Amato K.N., Gaytan S.M., Murr L.E., Martinez E., Shindo P.W., Hernandez J., Collins S., Medina F. (2012). Microstructures and mechanical behavior of Inconel 718 fabricated by selective laser melting. *Acta Materialia* Vol. 60, pp. 2229–2239.

[3] Canh R.W., Haasen P (1983). *Physical Metallurgy.* 3rd revised and enlarged edition. North-Holland physics publishing, Amsterdam, Oxford, New York, Tokyo, Part 1–2 and 3, 2050 p.

[4] Chen J., Xue L. (2010) Process-induced microstructural characteristics of laser consolidated IN-738 Superalloy. *Materials Science and Engineering A.* Vol. 527, pp. 7318–7328.

[5] Chiang M.F., Chen C. (2009). Induction-assisted laser welding of IN-738 nickel–base superalloy. *Materials Chemistry and Physics*, Vol. 114, pp. 415–419.

[6] Dinda G.P., Dasgupta A.K., Mazumder J. (2009) Laser aided direct metal deposition of Inconel 625 superalloy: microstructural evolution and thermal stability, *Materials Science Engineering A*, Vol. 509, pp. 98–104.

[7] Ganesh P., Kaul R., Paul C.P., Tiwari P., Rai S.K., Prasad R.C., Kukreja L.M. (2010) Fatigue and fracture toughness characteristics of laser rapid manufactured Inconel 625 structures. *Materials Science and Engineering A*, Vol. 527, pp. 7490–7497.

[8] Gill A, Telang A., Mannava S.R., Qian D., Pyoun Y.-S., Soyama H., Vasudevan V.K. (2013). Comparison of mechanisms of advanced mechanical surface treatments in nickel-based superalloy. *Materials Science and Engineering A*, Vol. 576, pp. 346–355.

[9] Gureev D.M., Petrov A.L., Shishkovsky I.V. (1999) Formation of intermetallics phases under laser sintering of powdered SHS compositions, *Proceedings of SPIE*. Vol. 3688-36, Jan. pp. 237–242. doi:10.1117/12.337516.

[10] Hussein N.I.S., Segal J., McCartney D.G., Pashby I.R. (2008) Microstructure formation in Waspaloy multilayer builds following direct metal deposition with laser and wire. *Materials Science and Engineering A*, Vol. 497, pp. 260–269.

[11] Jia Q., Gu D. (2014a) Selective laser melting additive manufactured Inconel 718 superalloy parts: high-temperature oxidation property and its mechanisms. *Optics and Laser Technology*, Vol. 62, pp. 161–171.

[12] Jia Q., Gu D. (2014b) Selective laser melting additive manufacturing of Inconel 718 superalloy parts: densification, microstructure and properties. *Journal of Alloys and Compounds*, Vol. 585, pp. 713–721.

[13] Kamashev A.V., Panin A.S., Petrov A.L., Shishkovskii I.V. (2001). Laser-controlled synthesis of nickel-aluminum intermetallic compounds. *Technical Physics Letters*. Vol. 27, Number 6, pp. 498–499. doi: 10.1134/1.1383837.

[14] Keshavarz S., Ghosh S. (2013) Multi-scale crystal plasticity finite element model approach to modeling nickel-based superalloys. *Acta Materialia*, Vol. 61, pp. 6549–6561.

[15] Kotoban D., Grigoriev S., Shishkovsky I. (2014). Study of 3D laser cladding for Ni85Al15 superalloy. *Physics Procedia*. Vol. 56, pp. 263–269.

[16] Li J., Wang H.M. (2010). Microstructure and mechanical properties of rapid directionally solidified Ni-base superalloy Rene_41 by laser melting deposition manufacturing. *Materials Science and Engineering A*, Vol. 527, pp. 4823–4829.

[17] Li J., Wang H.M., Tang H.B. (2013) Effect of heat treatment on microstructure and mechanical properties of laser melting deposited Ni-base superalloy Rene_41. *Materials Science and Engineering A*, Vol. 550, pp. 97– 102.

[18] Liu X.-B., Yu G., Guo J., Gu Y.-J., Pang M., Zheng C.-Y., Wang H.-H. (2008). Research on laser welding of cast Ni-based superalloy K418 turbo disk and alloy steel 42CrMo shaft. *Journal of Alloys and Compounds,* Vol. 453, pp. 371–378.

[19] Liu F., Lin X., Leng H., Cao J., Liu Q., Huang C., Huang W. (2013). Microstructural changes in a laser solid forming Inconel 718 superalloy thin wall in the deposition direction. *Optics and Laser Technology,* Vol. 45, pp. 330–335.

[20] Lemmen H.J.K., Sudmeijer J., Richardson I.M., Zwaag S. (2007). Laser beam welding of an oxide dispersion strengthened super alloy. *Mater Sci.,* Vol.42, pp. 5286–5295.

[21] Montazeri M., Ghaini F.M. (2012). The liquation cracking behavior of IN738LC superalloy during low power Nd:YAG pulsed laser welding. *Materials Characterization,* Vol.67, pp. 65–73.

[22] Mumtaz K.A., Erasenthiran P., Hopkinson N. (2008). High density selective laser melting of Waspaloy®. *Journal of Materials Processing Technology,* Vol. 195 (2008), pp. 77–87.

[23] Mumtaz K., Hopkinson N. (2009). Top surface and side roughness of Inconel 625 parts processed using selective laser melting. *Rapid Prototyping Journal,* Vol. 15/2, pp. 96–103.

[24] Murr L.E., Martinez E., Amato K.N., Gaytan S.M., Hernandez J., Ramirez D.A., Shindo P.W., Medina F., Wicker R.B. (2012). Metal fabrication by additive manufacturing using laser and electron beam melting technologies. *Journal of Materials Science and Technology,* Vol. 28(1), pp. 1–14.

[25] Nazarov A., Safronov V., Khmyrov R., Shishkovsky I. (2015). Fabrication of gradient structures in the Ni–Al system via SLM process. *IUTAM Symposium on Growing Solids,* June 23–27, Moscow, Russia.

[26] Nie P., Ojo O.A., Li Z. (2014). Numerical modeling of microstructure evolution during laser additive manufacturing of a nickel-based superalloy. *Acta Materialia,* Vol. 77, pp. 85–95.

[27] Ocylok S., Weisheit A., Kelbassa I. (2010) Functionally graded multi-layers by laser cladding for increased wear and corrosion protection, *Physics Procedia* Vol. 5, pp. 359–367.

[28] Oh J.H., Kirihara S., Miyamoto Y., Matsuura K., Kudoh M. (2002). Process control of reactive rapid prototyping for nickel aluminides. *Materials Science and Engineering A,* Vo. 334, pp. 120–126.

[29] Pang M., Yu G., Wang H.-H., Zheng C.-Y. (2008). Microstructure study of laser welding cast nickel-based superalloy K418. *Journal of Materials Processing Technology,* Vol. 207, pp. 271–275.

[30] Paul C.P., Mishra S.K., Premsingh C.H., Bhargava P., Tiwari P., Kukreja L. M. (2012). Studies on laser rapid manufacturing of cross-thin-walled porous structures of In-

conel 625. *International Journal of Advanced Manufacturing Technology*, Vol. 61, pp.757–770.

[31] Qin L., Hu J., Cui C., Wang H., Guo Z. (2008). Effect of al content on reaction laser sintering of Ni–Al powder. *Science of Sintering*, Vol. 40, pp. 295–301.

[32] Qin L., Hu J., Cui C., Wang H., Guo Z. (2009). Effect of Al content on reaction laser sintering of Ni–Al powder. *Journal of Alloys and Compounds*, Vol. 473, pp. 227–230.

[33] Ren X.D., Zhan Q.B., Yuan S.Q., Zhou J.Z., Wang Y., Ren N.F., Sun G.F., Zheng L.M., Dai F.Z., Yang H.M., Dai W.J. (2014). A finite element analysis of thermal relaxation of residual stress in laser shock processing Ni-based alloy GH4169. *Materials and Design*, Vol. 54, pp. 708–711.

[34] Safdar S., Pinkerton A.J., Li L., Sheikh M.A., Withers P.J. (2013). An anisotropic enhanced thermal conductivity approach for modeling laser melt pools for Ni-base super alloys. *Applied Mathematical Modelling*, Vol. 37, pp. 1187–1195.

[35] Shishkovsky I.V., Makarenko A.G., Petrov A.L. (1999). Conditions for SHS of intermetallic compounds with selective laser sintering of powdered compositions. *Combustion, Explosion, and Shock Waves*, Vol. 35 (2), pp. 166–170. doi: 10.1007/BF02674431.

[36] Shishkovsky I.V. (2001) Synthesis of functional gradient parts via RP methods. *Rapid Prototyping Journal*, Vol. 7, N 4, pp. 207–211. doi: 10.1108/13552540110402908.

[37] Shishkovsky I.V., Kuznetsov M.V., Morozov Yu.G. and Parkin I.P. (2004) Laser-induced combustion synthesis of 3D functional materials: computer-aided design. *Journal of Materials Chemistry*. Vol. 14, pp. 3444–3448. doi: 10.1039/b406732k.

[38] Shishkovsky I., Missemer F., Smurov I. (2012). Direct metal deposition of functional graded structures in Ti-Al system. *Physics Procedia*. Vol. 39, pp. 382–391. doi: 10.1016/j.phpro.2012.10.052

[39] Shishkovsky I., Missemer F., Kakovkina N., Smurov I. (2013a). Intermetallics synthesis in the Fe–Al system via layer-by layer 3D laser cladding. *Crystals*. Vol. 3, pp. 517–529. doi:10.3390/cryst3040517.

[40] Shishkovsky I., Yadroitsev I., Smurov I. (2013b) Fabrication of 3D parts from titanium nickel via layerwise selective laser melting. *Technical Physics Letters*. Vol. 39(12), pp. 1081–1084. doi: 10.1134/S1063785013120250

[41] Shishkovsky I.V. (2015) Laser controlled intermetallics synthesis during surface cladding. In: J. Lawrence et al. (Eds.), *Laser Surface Engineering. Processes and Applications,"* pp. 237–286. Woodhead Publishing Series in Electronic and Optical Materials, Elsevier Science and Technology, on-line ISBN 978-1-78242-074-3.

[42] Sims C.T., Stoloff N.S., Hagel W.C. (1995). *Superalloys*. Part I–II. John Willey & Sons Publ. 384 p.

[43] Song B., Dong S., Coddet P., Liao H., Coddet C. (2014) Fabrication of NiCr alloy parts by selective laser melting: columnar microstructure and anisotropic mechanical behavior. *Materials and Design,* Vol. 53, pp. 1–7.

[44] Theriault, Xue L., Dryden J.R. (2009). Fatigue behavior of laser consolidated IN-625 at room and elevated temperatures. *Materials Science and Engineering A,* Vol. 516, pp. 217–225.

[45] Verdi D., Garrido M.A., Munez C.J., Poza P. (2015). Cr3C2 incorporation into an Inconel 625 laser cladded coating: effects on matrix microstructure, mechanical properties and local scratch resistance. *Materials and Design,* Vol. 67, pp. 20–27.

[46] Vilar R., Santos E.C., Ferreira P.N. (2009). Structure of NiCrAlY coatings deposited on single-crystal alloy turbine blade material by laser cladding. *Acta Materialia,* Vol. 57(18), pp. 5292–5302.

[47] Wang GH., Zuo D., Sun Y., Xu F., Zhang D. (2009). Microstructure of nanometer Al2O3 dispersion strengthened Ni-based high-temperature protective coatings by laser cladding. *Transactions of Nonferrous Metals Society of China,* Vol. 19, pp. 586–591.

[48] Wang H., Zuo D., Li X., Chen K., Huang M. (2010a). Effects of CeO2 nanoparticles on microstructure and properties of laser cladded NiCoCrAlY coatings. *Journal of Rare Earths,* Vol. 28(2), Apr., p. 246.

[49] Wang H., Zuo D., Chen G., Sun G., Li X., Cheng X. (2010b). Hot corrosion behaviour of low Al NiCoCrAlY cladded coatings reinforced by nano-particles on a Ni-base super alloy. *Corrosion Science,* Vol. 52, pp. 3561–3567.

[50] Wang H., Zuo D., Wang M., Sun G., Miao H., Sun Y. (2011). High temperature frictional wear behaviors of nano-particle reinforced NiCoCrAlY cladded coatings. *Transactions of Nonferrous Metals Society of China,* Vol. 21, pp. 1322–1328.

[51] Yuan Y., Gu Y.F., Zhong Z.H., Shi Z., Osada T., Yokokawa T., Harada H. (2013). γ precipitation in the primary γ' of a new Ni–Co-base disc superalloy. *Materials Science and Engineering A,* Vol. 579, pp. 35–40.

[52] Yu, Y., Zhou, J., Chen J. (2012). Preparation, microstructure and tribological behavior of laser cladding NiAl intermetallic compound coatings, *Wear,* Vol. 274– 275, pp. 298–305.

[53] Zakiev S.E., Kholpanov L.P., Parkin I.P., Kuznetsov M.V., Morozov Yu. G., Shishkovskii I.V. (2006) Modeling of the thermal processes that occur in the laser sintering of reacting powder compositions. *Applied Physics A,* Vol. 84(1–2), July, pp. 123–129. doi: 10.1007/s00339-006-3586-0.

[54] Zhang G.Q., Yao J., Mazumder J. (2011). Laser direct metal deposition technology and microstructure and composition segregation of Inconel 718 Superalloy. *Journal of Iron and Steel Research, International,* Vol.18(4), pp. 73–78.

http://dx.doi.org/10.5772/61121

[55] Zhang C., Li L., Deceuster A. (2011). Thermomechanical analysis of multi-bead pulsed laser powder deposition of a nickel-based superalloy. *Journal of Materials Processing Technology,* Vol. 211, pp. 1478–1487.

[56] Zhang Y.N., Cao X., Wanjara P., Medraj M. (2013). Oxide films in laser additive manufactured Inconel 718. *Acta Materialia,* Vol. 61, pp. 6562–6576.

[57] Zhao J., Cao W., Ge C., Tan Y., Zhang Y., Fei Q. (2009a). Research on laser engineered net shaping of thick-wall nickel-based alloy parts. *Rapid Prototyping Journal,* Vol. 15/1, pp. 24–28.

[58] Zhao X., Lin X., Chen J., Xue L., Huang W. (2009b). The effect of hot isostatic pressing on crack healing, microstructure, mechanical properties of Rene88DT superalloy prepared by laser solid forming. *Materials Science and Engineering A,* Vol. 504, pp. 129–134.

[59] Zhong M., Sun H., Liu W., Zhu X., He J. (2005). Boundary liquation and interface cracking characterization in laser deposition of Inconel 738 on directionally solidified Ni-based superalloy. *Scripta Materialia,* Vol. 53, pp. 159–164.

[60] Zhou S., Dai X., Zheng H. (2011). Analytical modeling and experimental investigation of laser induction hybrid rapid cladding for Ni-based WC composite coatings. *Optics and Laser Technology,* Vol. 43, pp. 613–621.

[61] Zhou S., Xiong Z., Dai X., Liu J., Zhang T., Wang C. (2014). Microstructure and oxidation resistance of cryomilled NiCrAlY coating by laser induction hybrid rapid cladding. Surface and Coatings Technology, Vol. 258, pp. 943–949.

Chapter 11

Characterization of Intermetallic Precipitates in Ni-Base Alloys by Non-destructive Techniques

V. Acharya, S. Ramesh and G.V.S. Murthy

Additional information is available at the end of the chapter

http://dx.doi.org/10.5772/61119

Abstract

The present industrial scenario requires all engineering structure to be designed considering stability of several parameters at the operating conditions (e.g. Temperature, pressure, resistance to mechanical and surface degradation). Choice of materials for any engineering component should be such that it operates safely for reliable function, without failure during in-service, giving optimum component life. Due to scarcity of various resources and cost of manufacturing, regular maintenance and evaluation of structural integrity at every stage of production is necessary. Non-destructive techniques (NDT), along with modern computational facility help in non-intrusive investigation of the component at regular intervals of the operating stages for many critical applications. This will result in increment of designed component life and also help in maximizing utilization of natural resources.

For long, Ultrasonic has been associated with defect detection, but with the recent advances in electronics in combination with computational capabilities Ultrasonic velocity measurements have also been attempted for characterization of solutionising and precipitation behavior in various alloy systems such as aluminium alloys, ferritic steel, maraging steel, nickel base alloys and titanium alloys. As the speed of sound in a homogeneous medium is directly related to both elastic modulus and density, any changes in elastic property with varying degree of inhomogeneities will affect in pulse transit time through a sample of given thickness. Due to variation in elastic modulus of the matrix in the alloy resulting from the various precipitates; it has been attributed towards the change in ultrasonic velocity of the alloy and thereby resulting in popularization of Ultrasonic testing for online monitoring of the component. The precipitation hardening has been believed to arise from the formation of very small solute clusters which uses significant scattering of the conduction electron cause during the rearrangement, while the formation of the precipitates; resulting in variation of electrical resistivity of the aged alloy.

In the present chapter, a few non-destructive techniques used to characterize different microstructural features and the precipitation behavior, evolved through various heat treatments using Direct Current Potential Drop (DCPD) technique for measuring electrical resistivity and Ultrasonic Testing for measurements of ultrasonic parameters are presented. Further, validation of the observed results on microstructural features is also presented through hardness and microscopy studies.Thus, this study in effect can be used for non-destructive evaluation of the microstructures.

Keywords: Inconel 718, Nimonic 263, Ageing, Intermetallic Precipitation, DC Electrical Resistivity, Ultrasonic Velocity, Hardness

1. Introduction

High-temperature materials are materials that possess a remarkable ability to maintain their properties at elevated temperatures. Special steels, lighter alloys, ceramics and composites can be used as high-temperature material, but due to lack of one or the other property, their application over a wide range of temperature is restricted. To achieve this step change in operating parameters, new high-temperature materials must be selected due to inherent limitations in steels. Thus, need for stronger and corrosion-resistant materials for high-temperature engineering resulted in development of "Superalloys" of "Stainless Steel" varieties. Superalloys are unique high-temperature materials that have an ability to retain most of their strength even after long exposure times above 650°C (923 K) along with good low-temperature ductility and excellent resistance to mechanical and chemical degradation in severe environments. High-temperature strength in Superalloys is due to the presence of a stable matrix having an austenitic face-centered cubic crystal (FCC) structure helping in the strengthening mechanism. The primary application of such alloys is in severe operating conditions, for example, space shuttle engine and gas turbines, which are subjected to both high temperature and pressures, alongwith corrosive environments [1,2].

The major constituents of Superalloys [1-4] are mostly group VIIIB elements from the transition metal and they consist of various combinations such as Fe, Ni, Co and Cr, as well as small amounts of W, Mo, Ta, Nb, Ti and Al. Depending on the base metal, these can be Nickel-based, Cobalt-based and Nickel-Iron-based Superalloys. Nickel-based alloys are the most complex, most widely used for the hottest parts and most preferred [5] Superalloy. Nickel-based Superalloys, used in vast applications, are mainly subjected to high temperature due to the principle characteristic of Ni as an alloy base having high phase stability of FCC nickel matrix along with strength retention upto 0.7 Tm (Tm – melting temperature). Diffusion rate of Ni is very low which leads to microstructural stability in the alloy at elevated temperatures. Co-based alloys have higher melting points than Ni and Ni-Fe based alloys, which gives them the ability to absorb stress at a higher absolute temperature. Higher chromium content in Co alloys gives superior hot corrosion resistance to gas turbine atmospheres and also shows superior thermal fatigue resistance and weldability over Ni-alloys [1,2]. Superalloys containing substantial quantities of both Ni and Fe form a distinct class of Superalloys known as Nickel–Iron-based Superalloy. The austenitic FCC matrix of Superalloys has extended solubility for some alloying elements, excellent ductility and favorable characteristics for precipitation of effective strengthening phases. Superalloy density [2] is influenced by alloying additions: aluminum, titanium and chromium reduce density, whereas tungsten, rhenium and tantalum increase it. The corrosion resistance of Superalloys depends primarily on the alloying elements added, particularly chromium and aluminum, and the environment experienced.

Superalloys were thus developed to improve high-temperature strength along with other mechanical properties. Strengthening is a phenomenon by which hardness, yield and tensile strength are increased. Superalloys utilize three basic mechanisms [6] for strengthening: intermetallic precipitation, solid solution and carbide precipitation. Intermetallic and carbide precipitation are employed in a metastable condition and thus the basic structure and distribution of the phases alter on normal thermal exposures. Thermal exposures also increase the possibility [3,4] of phase alterations such as the formation of delta, eta, mu, sigma, Laves, etc., resulting in variation of alloy properties.

1.1. Precipitation Hardening [7-10]

Precipitation hardening [7] is produced by solution treating and quenching an alloy in which a second phase is in solid solution at the elevated temperature but precipitates on quenching and ageing at a lower temperature. Thus, an element can participate in formation of intermetallic precipitates during precipitation strengthening only if it is partially soluble in the matrix. Among all other theories, Coherent Lattice theory is one of the most useful theories for the understanding of precipitation hardening. In precipitation hardening, the alloy is first solutionized by heating into a single phase region and soaking for sufficient time in order to allow the intermetallic and grain boundary precipitates to dissolve in the matrix and also to permit required diffusion. After solution treatment, the alloy is rapid quenched to get a supersaturated solid solution and to prevent formation of equilibrium precipitates due to natural ageing. In supersaturated solid solution (SSS), the alloy is in disordered condition as the solute atoms are at-random distributed in the lattice structure. The alloy during reheating to slightly higher temperature undergoes artificial ageing. Artificial ageing results in limited mobility of soluble atoms and the atoms can move over only a few interatomic distances leading to formation of a fine scale transition structure. Due to slow cooling, rearrangement of the atoms takes place, where the solute atoms move into definite positions in the lattice, forming an ordered solid solution. The ordered transitional structure thus formed will have definite lattice parameters different from the solute atoms will result in coherency of the atom. There will be considerable distortion of the matrix due to fine precipitates thus formed within the matrix and this distortion extends over a large volume only if the excess transition phase is in the form of fine discrete particles well distributed in the matrix. These fine precipitates will restrict the dislocation movement, resulting in rapid increase of strength and hardness of the alloy during ageing. Eventually, equilibrium-stable phase is formed from the transition phase, whose particles have common grain boundaries of the matrix and there is further growth of certain larger particles at the expense of neighboring smaller particles. This causes loss of coherency with the matrix and stress relief takes place in the lattice, which results in considerable decrease in strength and increase in ductility of the alloy. At this stage, the alloy is said to be overaged. The following steps [10] are associated with the process of precipitation hardening:

1. Rearrangement of atoms within the crystal structure during which micro-strains in the lattice are developed and mechanical properties are improved.
2. Optimum strengthening of the alloy by forming intermediate phase.
3. And finally, formation of stable phase from the transition phase, resulting in decrease of hardness and strength.

It is necessary to understand the above-mentioned three main steps of precipitation hardening, in order to age the alloy to optimum level for considerable increase in strength and hardness. Thus, the degree of strengthening resulting from the secondary phase particles depends on distribution of the particles in matrix.

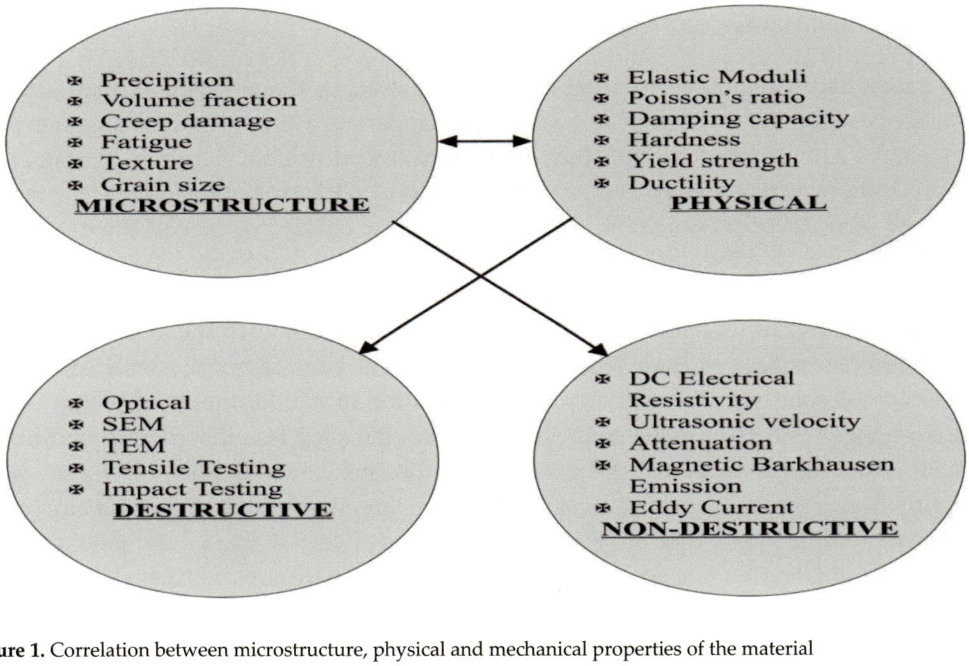

Figure 1. Correlation between microstructure, physical and mechanical properties of the material

1.2. Non-destructive techniques

Non-destructive Testing (NDT) techniques [11-14] are most commonly used for detection and characterization of flaws in the component, thus certifying whether the component is fit for the intended service or not. As crystallographic texture controls the plastic and elastic properties of the component, component material property is equally important to assess the structural integrity of the component along with the flaw characteristics. Improved instrumentation and a quantitative understanding of the materials behavior has driven the upgrowth of Quantitative Non-destructive Evaluation (NDE) to detect and characterize microstructural properties as well as flaws in materials. Two quantitative NDE approaches are now in use – 'off-line' periodic inspection during scheduled outages and 'on-line' monitoring [15, 16] during service as well as during materials processing – in order to predict future performance and reliability of components by controlling materials behavior (especially the mechanical properties) with the help of physical mechanisms. Thus, there is correlation between the microstructure of the material with its physical and mechanical properties (Fig. 1) and as these properties can be easily measured by NDE techniques, NDE can be used as a non-destructive tool at low and affordable cost for non-intrusive investigation of the component at regular intervals of the process stages by evaluating the microstructures for many critical applications. During the development of NDE, it was realized that a wide range of techniques are necessary to achieve some flexibility in non-destructive characterization and evaluation of structural materials and components with complex shapes. The present study is mainly devoted to the application of DC electrical resistivity and Ultrasonic NDE techniques to materials characterization.

1.3. Non-destructive evaluation

Non-destructive testing (NDT), Non-destructive evaluation [17-32] (NDE) and Non-destructive inspection (NDI) are the techniques that are based [33-43] on the application of determining the characteristics of materials or components and for assessing life of the component by evaluating the detected inhomogeneities and harmful defects present in the component without impairing its usefulness. During service, many factors [17] like unanticipated stresses (residual and system), operation outside designed limits (excessive temperature and load cycling), operation and environmental effects, degradation of material properties in service, etc., which are difficult to predict, affect the designed life of the component. It may not be possible to take a sample from the component and determine its health for continued service. Moreover, efforts to increase the lifetime of intricate and expensive structures like aircraft, space shuttle engine and gas turbine require on-line monitoring of the microstructure in order to improve the safety and maintainability by analyzing the current state of the component. Thus, NDE techniques used for characterization [17, 24, 25] of the material properties are of great significance during service life of the component as they help in controlling process parameters by providing timely feedback.

1.3.1. Advantages of NDT over destructive testing

- In-service testing is possible.
- Tests are made directly on the component itself and thus special coupon of the component is not required.

- Very little sample preparation of the test specimen is required.
- Less time requirement for testing.
- Repeated checks over a period of time are possible.
- A single NDT can measure many or sometimes all properties of the alloy.
- Some NDT instruments are portable and can be easily carried to the workplace.

1.3.2. Limitations of NDT

- Measurements are indirect and hence reliability has to be verified.
- Some of the NDT techniques are expensive.
- Skilled judgement and experience are required to interpret NDT results.

1.3.3. Examples of case studies carried out for assessment of components with the help of NDT

1. Characterization of microstructural and mechanical properties:
   - Measurement of grain size [17, 44-47] with the help of ultrasonic testing.
   - Microstructural control in [18] Metal Matrix Composites (MMC).
   - Estimation of non-metallic inclusions [17] in steel. There is an ASTM standard E 588 titled 'Detection of large inclusions in bearing quality steel by ultrasonic method'. Also porosity [18] can be characterized.
   - Measurement of degree of recrystallization [48].
   - Determination of elastic modulus [49] especially for brittle materials since other methods like tensile test, do not produce optimum results.
   - Estimation of strength [50].
   - Ultrasonic hardness [17] testers are also used especially for in situ measurements.
   - For monitoring [51] ductile to brittle transition temperature (DBTT).
   - Fracture toughness [52] can be estimated using ultrasonic testing as Kc (fracture toughness) is dependent on the value of E (Young's Modulus) by the equation ($Kc = \sqrt{E * Gc}$). Kc depends upon the ultrasonic attenuation while E can be calculated using ultrasonic velocity.
2. For qualification of processing techniques like the following:
   1. Nodularity of cast iron, as the increase in degree of nodularity (morphology of graphite) increases the strength, which in turn increases ultrasonic velocity [53,54].
   2. Qualification of heat treatment of precipitation hardenable 17-4 PH stainless steel [55] using ultrasonic testing.

3. Accurate measurement of case depth [56] in order to have knowledge about component's hardness.
4. Measuring degree of diffusion bonding [57-59].
5. Sheet metal formability [60].

3. Characterization of material degradation during in-service:
   - Detection of fatigue [61, 62] and creep [63] failure in turbine blade made of Superalloy.
   - Detection of hydrogen attack in low alloy steel [64, 65] and in Zircaloy-2.
   - Intergranular corrosion attack in austenitic stainless steel [66].
   - Thermal embrittlement [67] of Duplex stainless steel.
   - Ageing degradation in Ni-based Superalloy 625 [49,68].
   - Assessment of degradation of a heavy water plant [12].
   - Failure of S.S dished ends during storage with the help of in situ metallography at the inner surface of the failed dished end [12].

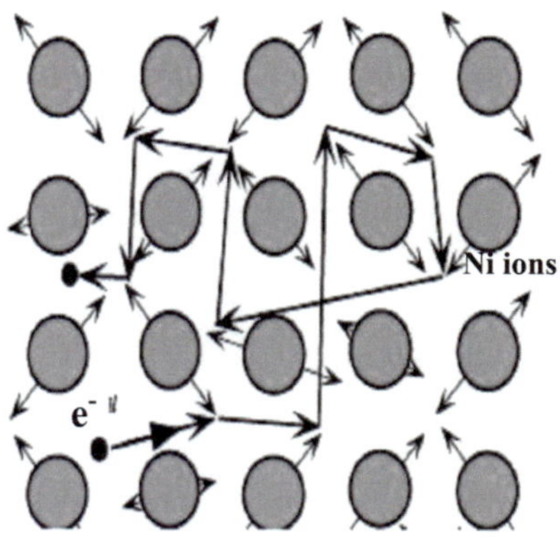

Figure 2. Random motion of electron due to thermal vibrations[71]

1.4. DC electrical resistivity

DC electrical resistivity is one of the oldest non-destructive electromagnetic techniques. The electrical resisitivity of a material depends on its physical state such as temperature and stress along with its composition and microstructure. The congential dependence of the resistivity on the microstructure can be helpful to detect various transformations in the material due to thermal exposure. Measuring material's resistivity non-destructively helps in identifying various metals and alloys and monitoring the heat treatment of the alloy; along with the detection of damage that gives rise to a change in resistivity [11,69]. The precipitation hardening has been believed [70] to arise from the formation of very small solute clusters which uses significant scattering of the conduction electron (Fig. 2) [71] and hence the electrical resistivity of the aged alloy increases. In the absence of an applied field, there is no net drift in any direction and conduction electron in the metal moves about randomly scattered (with a mean speed) by thermal vibrations of the atoms. Resistivity, in short, varies as a result of combined effects [72,73] from solution depletion, intermetallic precipitates, high-temperature solute enrichment due to the dissolution of the clusters formed during ageing and also refining of the intermetallic precipitates. Four-point direct current potential drop (DCPD) technique [69] is well suited for accurate non-destructive measurement of material resistivity. The potential drop method of evaluation is based on the principle that the electrical resistance of an alloy changes due to the presence of structural inhomogeneities. Thus, measuring bulk resistivity using DCPD one can have knowledge about performance and reliability of the component. Direct current potential drop (DCPD) technique is independent of the magnetic permeability of the material and is also useful for testing ferrous as well as low-conductivity materials. Moreover, DCPD equipment is simpler and requires less controlled parameters and there is possibility of full automation of the monitoring.

1.5. Ultrasonic testing

Ultrasonic Testing is the most preferred NDE technique for material property characterization, as UT parameters are significantly affected by changes in microstructure or mechanical properties of the material. Microstructural properties [17] like grain size measurement [44-47]; estimating presence of inclusions; measuring degree of recrystallization [48]; mechanical properties like elastic modulus, hardness [13], fracture toughness [52], and estimating strength [50] of the structure; monitoring of Ductile Brittle Transition Temperature [51] (DBTT) are correlated with ultrasonic testing parameters which include variation in the frequency and velocity [74], attenuation [52], backscatter amplitude [44], spectral analysis and acoustic microscopy. Ultrasonic material characterization has also been used to qualify various processing treatments like precipitation hardening, case hardening along with the assess of damage due to various degradation mechanisms [61-68] like fatigue, creep, corrosion, hydrogen damage, thermal embrittlement of steel, ageing degradation, etc. Elastic properties of the component can be easily recognized by the measurements of the Young's Modulus with the help of ultrasonic wave speed and attenuation measurements. Elastic moduli of pure metal are very low which will further increase with the addition of the alloying elements and during precipitation hardening. Ultrasonic velocity will be influenced [18] by the elastic moduli of the material which will be influenced by the precipitate forming elements. Information about microstructural induced changes in the elastic moduli can be deduced from the ultrasonic velocity by Equation 1:

$$v = \sqrt{c/\rho} \tag{1}$$

where v= ultrasonic velocity, c= elastic stiffness and ρ= density.

Ultrasonic velocity (v) measurement involves determination of the distance travelled by the ultrasound and dividing it by time of travel between the first and second back surface echo, as mentioned in Equation 2:

$$v = 2h/t \tag{2}$$

where, v = ultrasonic velocity, h = sample thickness and t = time of flight.

The accuracy of these measurements depends upon the accuracy with which time of flight and thickness of the component are measured. With advancement [75-78] and use of software (e.g. LabVIEW, MATLAB), one can calculate ultrasonic velocity with accuracy of 500 μsec, making it a very reliable parameter for material property characterization.

2. Research undertaken

A study is carried out to characterize different microstructural features and the precipitation behaviour evolved through various heat treatments using Direct Current Potential Drop (DCPD) technique for measuring electrical resistivity and Ultrasonic Testing for measurements of ultrasonic parameters and validating the same with hardness and microstructural features. Table 1 and Table 2 show the chemical composition of Ni-263 and IN- 718 superalloys under study.

Element	C	Co	Cr	Mo	Ti	Al	Cu	O	N	Ni
wt%	0.06	19.60	20.1	6.00	2.00	0.4	0.001	0.0013	0.006	Bal

Table 1. Chemical composition of Nimonic 263

Element	Ni	Fe	Cr	Ti	Al	Nb	Mo	Co	C	Mn
wt%	53.30	17.66	19.21	1.03	0.4	5.06	2.92	0.01	0.04	<0.10

Table 2. Chemical composition of Inconel 718

2.1. Heat treatment cycle

Temperature	Time (h)	Expected Phase
SA + 650(C + 620(C (8 h)	10, 25, 50, 75, 100	γ
SA + 750(C + 620(C (8 h)	10, 25, 50, 75	γ′ + γ″
SA + 800(C + 620(C (8 h)	25, 50, 75	γ′+ γ″ + δ
SA + 900(C	75, 100	δ

Table 3. Test matrix for IN-718 specimens

A set of Ni-263 specimens of dimensions 20 mm×20 mm×10 mm was solution annealed (SA) at 1150ºC for 1 h followed by water quenching. Referring to the TTT diagram of Ni-263 [79], the SA specimens were thermally aged for 1, 2, 4, 6 and 8 h at 650ºC and 1, 2, 4, 6, 25, 50 and 75 h at 800ºC, followed by water quenching. Similarly, a set of IN-718 specimens of dimensions 25mm x 29mm x 10mm was subjected to solution annealing at 980ºC for 1 hr and then water quenched to room temperature. Further solution annealed (SA) IN-718 specimens were aged by giving two-step ageing treatment, thereby forming new intermetallic precipitates and various phases depending on the time and temperature of ageing. Referring to the TTT diagram of IN-718 [2,80], test matrix was designed for the specimens in such a way that all the specimens have varying microstructures with difference in size and percentage of intermetallic precipitates. Figure 4 shows the experimental temperature for IN-718 in TTT diagram and Table 3 shows the test matrix for IN-718 with various expected phases.

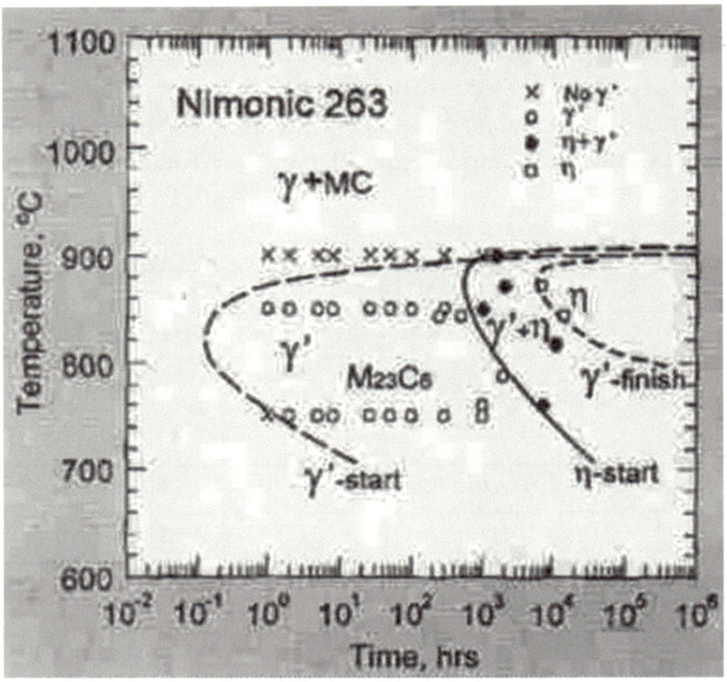

Figure 3. TTT diagram of Ni-263

2.2. Resistivity measurement

For resistivity measurement, four-probe Van der Pauw method has been used. In this method, electrical DC currents are injected into a conducting specimen through a set of two probes; while a second set of two probes is used for measuring the voltage drop across the area of contact as shown in Fig. 5. After an accurate measurement of the distance between the voltage probes, the resistivity (ϱ) of the sample is measured using Equation 3:

http://dx.doi.org/10.5772/61119

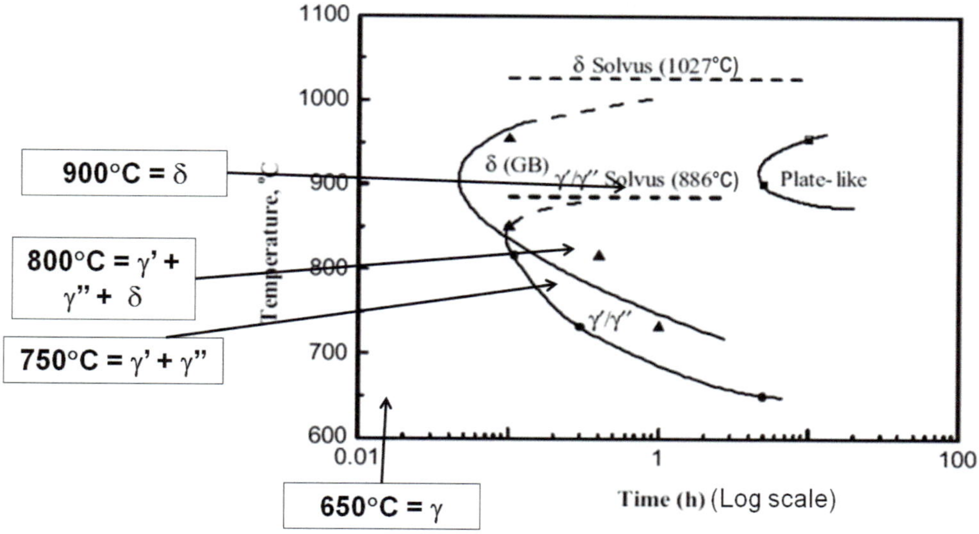

Figure 4. TTT diagram of IN-718 with experimental temperature

$$\rho = RA/L \quad (3)$$

where R = resistance or voltage drop when a constant current is passed.

A = cross-sectional area of contact of the sample.

L = distance between the voltage probes.

For better accuracy and speed of measurement, a sample holder has been used; in which four contacts are made by pressure contacts with gold tips. Keithley 2400 Sourcemeter is used as constant current source; while Keithley 2182A Voltmeter is used for measurement of the voltage drop. Resistivity for all the specimens was calculated with 0.03% error. Error percentage was determined using the maximum variation in the average resistivity value for each specimen.

2.3. Ultrasonic velocity measurements

The experimental set-up used for Ultrasonic Velocity (UV) measurement is shown in Fig. 6. A broadband pulser-receiver and a 500 MHz digital oscilloscope were used for carrying out the ultrasonic measurements. For UV measurements, the RF signals were digitized and stored for further processing. Ultrasonic velocities were measured using 5 MHz longitudinal wave transducer. For better accuracy and speed of measurement, a software is used for calculating ultrasonic velocity and attenuation of the sample by reading the digitized RF signals and the gated back wall echo from the oscilloscope stored in the computer. The accuracy obtained in the time of flight measurement is improved by 500 μsec, which led to maximum scattering of ±3m/s for ultrasonic longitudinal velocity.

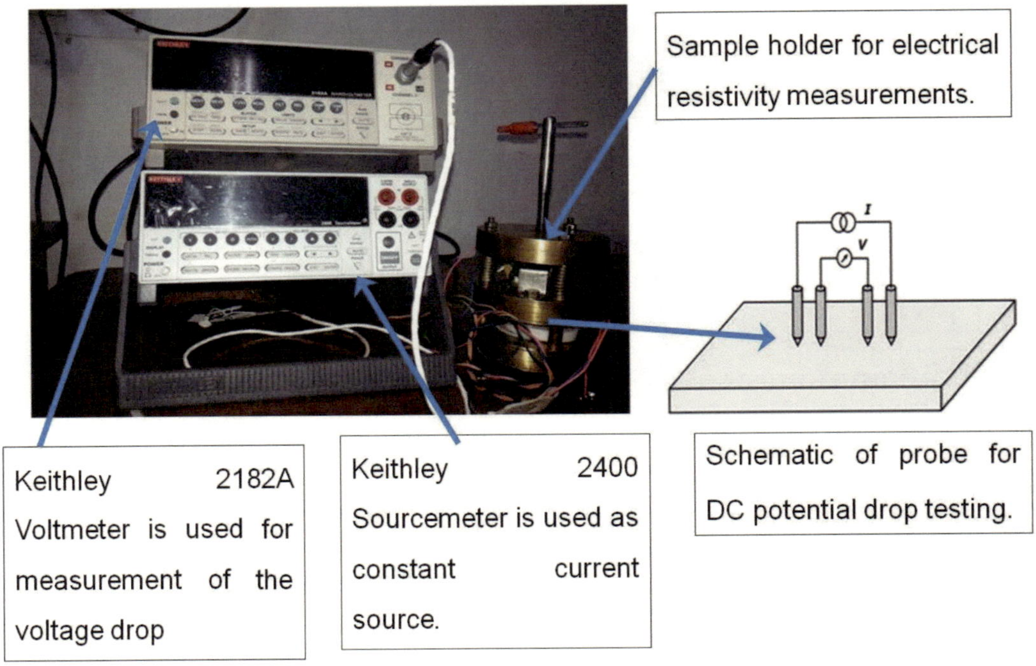

Figure 5. DC Electrical resistivity measurement set-up

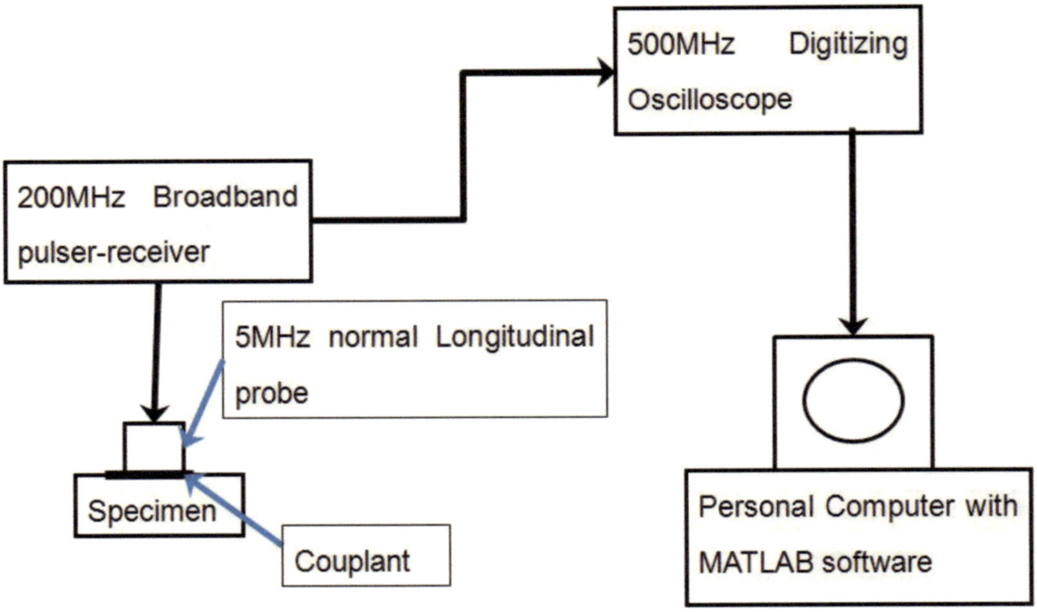

Figure 6. Ultrasonic velocity measurement set-up

http://dx.doi.org/10.5772/61119

2.4. Hardness measurements

Hardness measurements were carried out using Indentation method by applying a load of 700 N. A maximum change of ±5 BHN is obtained in hardness measurements in any specimen. Image analysis and Scanning Electron Microscopy were carried out to study the precipitation behaviour, and validate the results.

3. Observations

3.1. Resistivity observations

Table 4 and Table 5 show the variation in the resistivity with different ageing cycles. An average of 8 readings is taken as final resistance value. Resistivity measures the difficulty of electrons to move freely in the alloy. Additional ageing will lead to the formation of ordered clusters and precipitates, which will result in significant scattering of conducting electrons as they vibrate around equilibrium positions. A key factor for electron scattering is the lattice vibrations. As an electron encounters a defect in form of fine precipitates or due to the presence of other inhomogeneities, it scatters from its path by losing energy and changing direction.

Based on the experimental work, it is observed that resistivity of solution annealed specimen is the highest which will be further affected by the cluster formation; which is again dependent on the ageing time and temperature. During precipitation, the initial nucleation of the fine precipitate phase results in a maximum increase of resistivity due to the decreased conduction electron mean free path associated with forming these scattering electrons within the matrix phase. As the precipitates undergo growth and coarsening on increase of ageing temperature and time, the dispersed fine precipitate phase converts to a more widely spaced phase. Conduction electrons are now more likely to collide with solute atoms than the coarse precipitates. After a maximum, the resistivity thus begins to decrease due to the growth and coarsening of the precipitates and also due to the changes to the solute content from the increasing precipitate volume fraction. Thus, resistivity is not only dependent on kinetics of precipitation but is also affected by scattering of electrons. Resistivity, in short, varies as a result of combined effects from solution depletion, intermetallic precipitates, high-temperature solute enrichment due to the dissolution of the clusters formed during ageing and also refining of the intermetallic precipitates. These changes in resistivity at higher temperature will also be there due to the presence of pre-existing intermetallic precipitates, as they affect the rate of electron scattering and also by forming coarse grain structure which will slow down the kinetics of the precipitation by reducing nucleation rate. At lower temperature, due to increase in precipitate to the equilibrium volume fraction, the resistivity shows similar peak positions in the maximum increased resistivity. Bulk resistivity decreases on increase of ordering of the material.

3.2. Ultrasonic velocity observations

It is observed from Table 4 and Table 5 that ultrasonic velocity is influenced by the volume fraction, coherency, fineness and distribution of the precipitates. Longitudinal wave velocity is found to increase with ageing time and temperature. The increase in velocity with ageing time is found to be maximum at initial stages (indicating possible faster kinetics). Elastic moduli of pure metal are very low which will further increase with the addition of the alloying elements and this will also affect ultrasonic velocity of the material. During incubation period, there will be influence of the precipitate forming elements on the elastic moduli of the alloy. Formation of intermediate coherent transition γ''- phase in alloy will drastically affect the elastic moduli of the alloy, resulting in increase of ultrasonic velocity of the alloy. The intermetallic precipitates so formed lead to an increase in the elastic moduli of the alloy and consequently increases the ultrasonic velocities in Superalloys [49 and references therein]. However, the ageing of the alloy if carried out for shorter durations at any temperature leads to decrease in the ultrasonic velocity due to the dissolution of the intermetallic precipitates. Further, the formation of stable orthorhombic δ-phase by coarsening of γ''- precipitates increases the ultrasonic velocity. The increase in the moduli upon thermal ageing is attributed to the formation of γ' and γ''-precipitates leading to the depletion of the precipitate forming elements like Nb, Al and Ti from the solution. Again it is observed that ultrasonic velocity is only affected by the intermetallic precipitates and not by the grain-boundary carbides. Thus, ultrasonic velocity is sensitive to initial nucleation of the precipitates along with their growth and coarsening. The depletion of the precipitate-forming elements from the matrix during coarsening leads to an increase in the modulus of the matrix.

3.3. Hardness

The variations in the hardness, resistivity and ultrasonic velocity with different ageing cycles is shown in Table 4 and Table 5. For Ni-263, the solution annealed specimen exhibited lowest hardness (180 BHN) and it increases with an increase in ageing temperature from 650ºC to 800ºC. On ageing at 650°C, hardness is found to continuously increase with increase in time of exposure (2 h, 4 h, 6 h, 8 h, 25 h, 50 h, 75 h and 100 h). The strength of the specimen will decrease due to overageing of the specimen at higher temperature resulting in coarsening of grains. While on ageing at 800°C hardness initially increases with time of exposure but decreases for longer ageing time. This is in agreement with the expected phase diagram. The increase in strength by ageing is due to the presence of fine and uniformly distributed metastable coherent γ' precipitates in the matrix. While at higher temperature above 800°C there is decrease in the strength due to overageing, resulting in rapid coarsening of fine coherent γ' phase and its conversion to accicular η-phase. η-phase degrades the strength of the alloy by consuming the solution strengthening elements and this is one of the reasons for decrease in hardness at 800°C for longer exposure time. The results obtained are in line with the time temperature transformation (TTT) diagram reported in literature for the precipitation of γ' intermetallic phase in this alloy and further confirmed by microscopy studies.

Similarly for IN-718, the SA specimen exhibited lowest hardness (165 BHN) and it increases with an increase in ageing temperature from 600°C to 750°C. But there is a decrease in hardness on further increase of temperature from 750°C to 900°C. On ageing at 650°C, hardness is found to continuously increase with increase in time of exposure (10 h, 25 h, 50 h, 75 h and 100 h). The strength of the specimen will decrease due to overageing of the specimen at higher

http://dx.doi.org/10.5772/61119

temperature, resulting in coarsening of grains. While on ageing at 750°C and 800°C, hardness initially increases with time of exposure but decreases for longer ageing time. Again on ageing at 750°C, hardness is initially found to increase with increase in time of exposure and further exposure decreases the hardness values. On the other hand, for ageing at 900°C it decreases continuously. This is in agreement with the expected phase diagram. The increase in strength by ageing is due to the presence of fine and uniformly distributed metastable phase γ'' in the matrix. While at higher temperature above 750°C there is decrease in the strength due to overageing, resulting in rapid coarsening of γ''. At 800°C, it is expected from the TTT diagram that for longer time of exposure, nucleation of stable orthorhombic δ-phase will start from γ''. δ-phase degrades the strength of the alloy by consuming the precipitation strengthening elements. And this is one of the reasons for decrease in hardness at 800°C and 900°C for longer exposure time. During overageing hardness will reduce to lower values. Along with the formation of delta (δ) phase at the higher temperature, there will also be the formation of MC-type grain boundary carbides, which will also affect the hardness of the alloy. The variation in hardness with ageing time and temperature is also confirmed by electron microscopy studies.

Ageing Cycle	Ultrasonic Velocity (m/s)	Resistivity	Hardness [BHN]
1150ºC for 1 h (SA)	5960	131.60	180.40
SA+650ºC / 2 h	5980	137.70	216.10
SA+650ºC / 4 h	5992	142.97	228.43
SA+650ºC / 6 h	5993	147.97	231.20
SA+650ºC / 8 h	5995	143.77	244.83
SA+650ºC / 25 h		140.72	247.60
SA+650ºC / 50 h		140.46	265.43
SA+650ºC / 75 h		138.78	276.70
SA+650ºC / 100 h		136.43	281.90
SA+800ºC / 2 h	5982	139.21	246.60
SA+800ºC / 4 h	5987	139.37	270.70
SA+800ºC / 6 h	5988	136.87	278.53
SA+800ºC / 8 h		141.40	280.17
SA+800ºC / 25 h	5987	136.10	293.77
SA+800ºC / 50 h	5986	137.33	307.50
SA+800ºC / 75 h	5985	139.00	310.53
SA+800ºC / 100 h		136.37	303.57

Table 4. Ultrasonic Velocity, Resistivity and Hardness of Nimonic 263 in different heat treatment conditions

Ageing Cycle	Ultrasonic Velocity (m/s)	Resistivity [Ωm]	Hardness [BHN]
980°C for 1 h (SA)	5786	$1.2467*10^{-6}$	165
SA + 650°C 10 h + 620°C 8 h	5800	$1.2539*10^{-6}$	260
SA + 650°C 25 h + 620°C 8 h	5795	$1.1476*10^{-6}$	298
SA + 650°C 50 h + 620°C 8 h	5817	$1.1236*10^{-6}$	328
SA + 650°C 75 h + 620°C 8 h	5815	$1.0634*10^{-6}$	360
SA + 650°C 100 h + 620°C 8 h	5827	$1.0764*10^{-6}$	377
SA + 750°C 10 h + 620°C 8 h	5828	$1.0462*10^{-6}$	385
SA + 750°C 25 h + 620°C 8 h	5813	$0.9913*10^{-6}$	378
SA + 750°C 50 h + 620°C 8 h	5824	$0.9672*10^{-6}$	358
SA + 750°C 75 h + 620°C 8 h	5817	$1.0131 *10^{-6}$	349
SA + 800°C 25 h + 620°C 8 h	5836	$1.0007*10^{-6}$	328
SA + 800°C 50 h + 620°C 8 h	5842	$1.1207*10^{-6}$	325
SA + 800°C 75 h + 620°C 8 h	5830	$1.1139*10^{-6}$	309
SA + 900°C 75 h	5843	$1.1706*10^{-6}$	208
SA + 900°C 100 h	5795	$1.1824*10^{-6}$	206

Table 5. Ultrasonic Velocity, Resistivity and Hardness of IN-718 in different ageing cycles

4. Graphs

Figure 7 to Figure 17 are the graphs showing the variations in the hardness, resistivity and ultrasonic velocity with different ageing cycles.

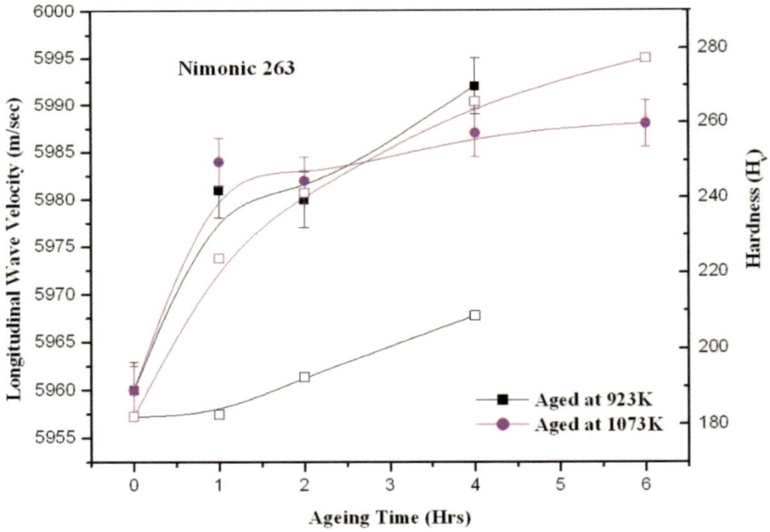

Figure 7. Ultrasonic Velocity and Hardness vs. Time for Solution Annealed (SA) specimens thermally aged at 923K (650°C) and 1023K (800°C) and studied with respect to the readings of SA specimen

http://dx.doi.org/10.5772/61119

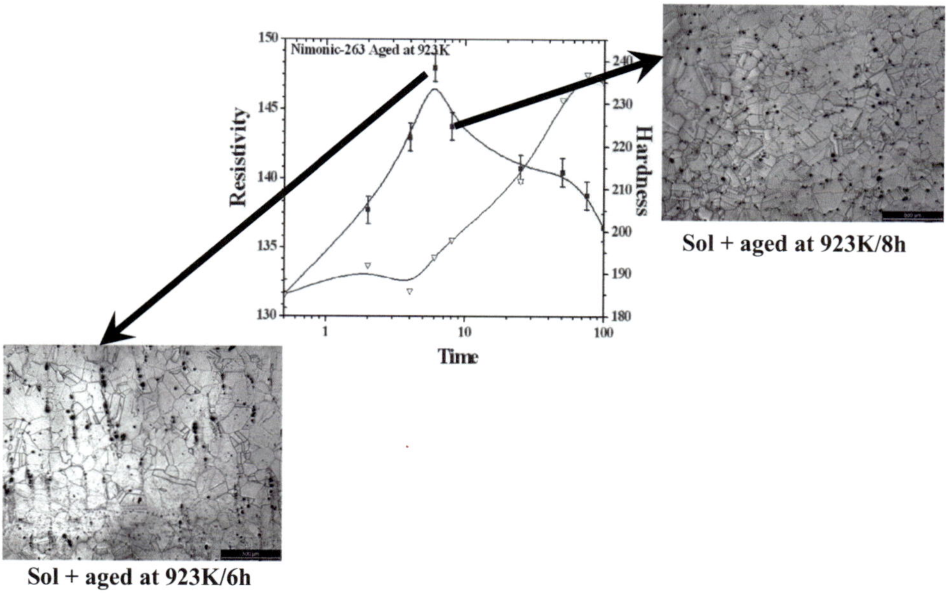

Figure 8. Resistivity and Hardness vs. Time for Solution Annealed (SA) specimens thermally aged at 923K (650ºC) and studied with respect to the readings of SA specimen

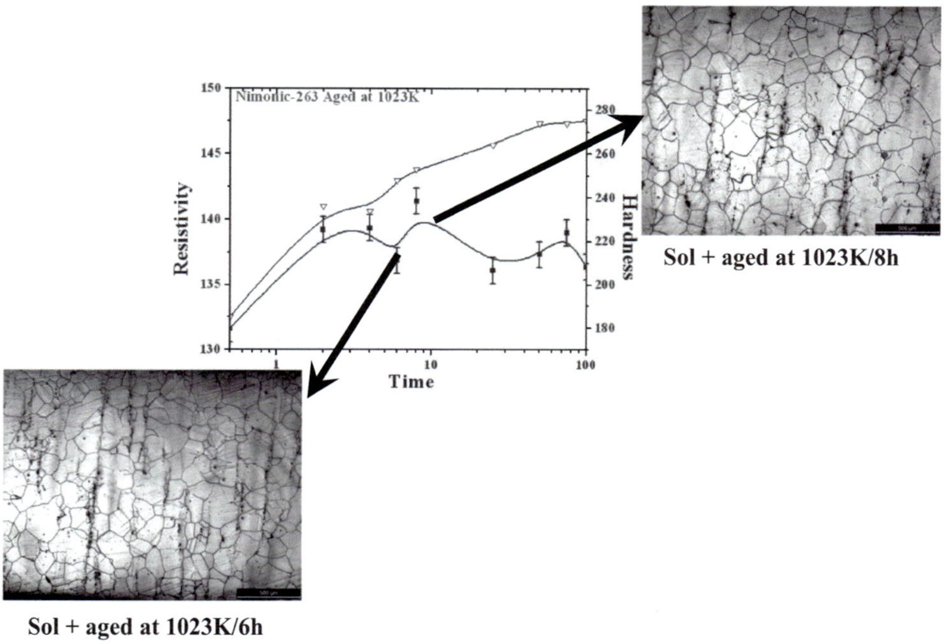

Figure 9. Resistivity and Hardness vs. Time for Solution Annealed (SA) specimens thermally aged at 1023K (800ºC) and studied with respect to the readings of SA specimen

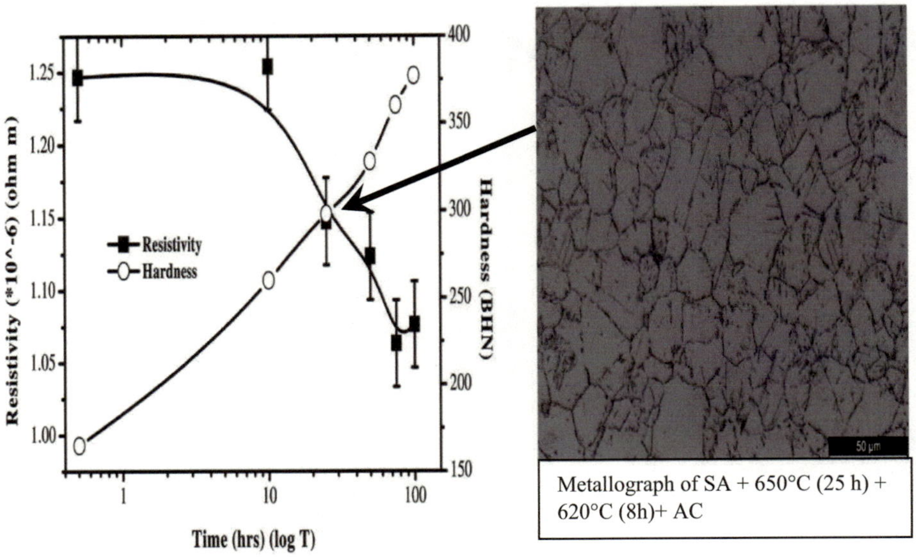

Figure 10. Resistivity and Hardness vs. Time for Solution Annealed (SA) specimen thermally aged at 650°C (10 h,25 h, 50 h,75 h,100 h) followed by ageing at 620ºC (8 h) and studied with respect to the readings of SA specimen

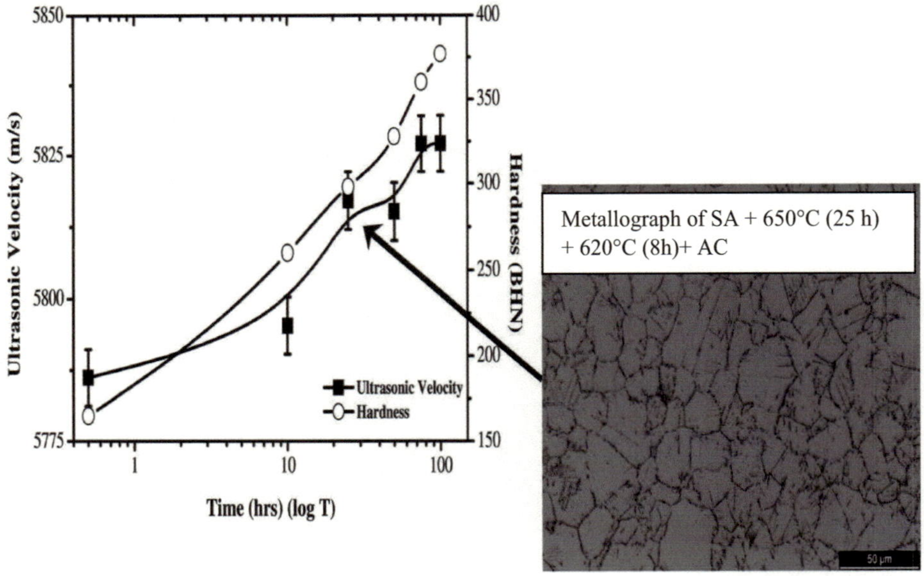

Figure 11. Ultrasonic Velocity and Hardness vs. Time for Solution Annealed (SA) specimen thermally aged at 650°C (10 h,25 h,50 h,75 h,100 h) followed by ageing at 620ºC (8 h) and studied with respect to the readings of SA specimen

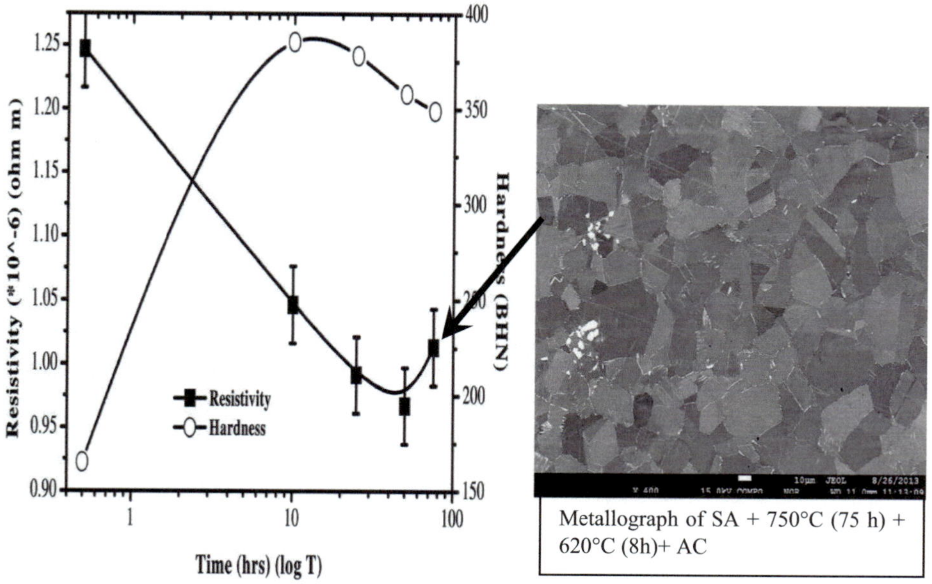

Figure 12. Resistivity and Hardness vs. Time for Solution Annealed (SA) specimen thermally aged at 750°C (10 h,25 h, 50 h,75 h) followed by ageing at 620ºC (8 h) and studied with respect to the readings of SA specimen

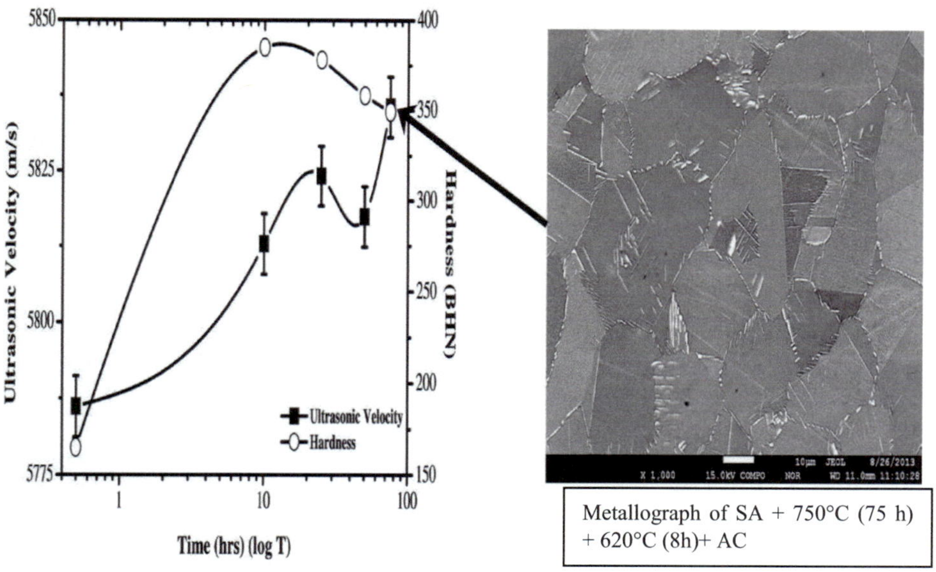

Figure 13. Ultrasonic Velocity and Hardness vs. Time for Solution Annealed (SA) specimen thermally aged at 750°C (10 h,25 h,50 h,75 h) followed by ageing at 620ºC (8 h) and studied with respect to the readings of SA specimen

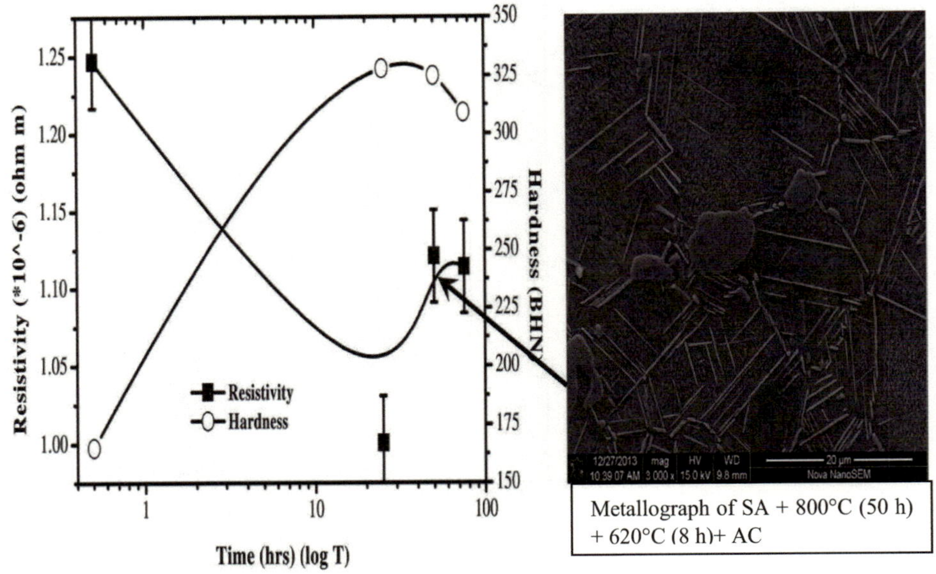

Figure 14. Resistivity and Hardness vs. Time for Solution Annealed (SA) specimen thermally aged at 800°C (25 h,50 h, 75 h) followed by ageing at 620ºC (8 h) and studied with respect to the readings of SA specimen

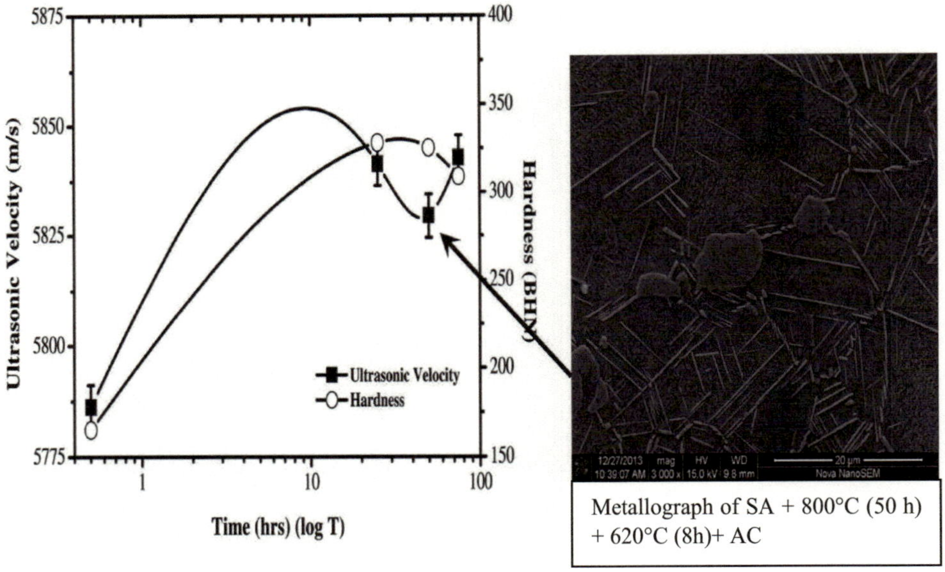

Figure 15. Ultrasonic Velocity and Hardness vs. Time for Solution Annealed (SA) specimen thermally aged at 800°C (25 h,50 h,75 h) followed by ageing at 620ºC (8 h) and studied with respect to the readings of SA specimen

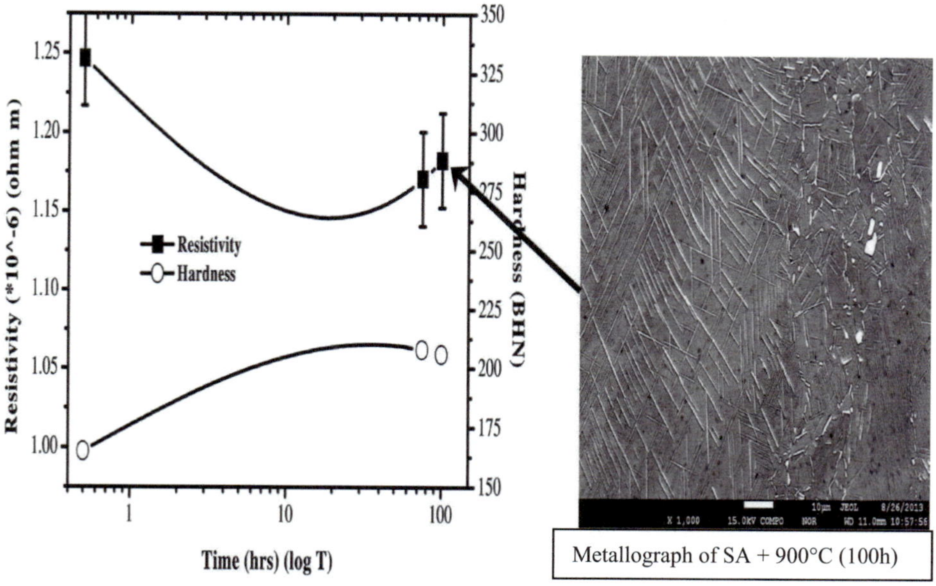

Figure 16. Resistivity and Hardness vs. Time for Solution Annealed (SA) specimen thermally aged at 900°C (75 h, 100 h) and studied with respect to the readings of SA specimen

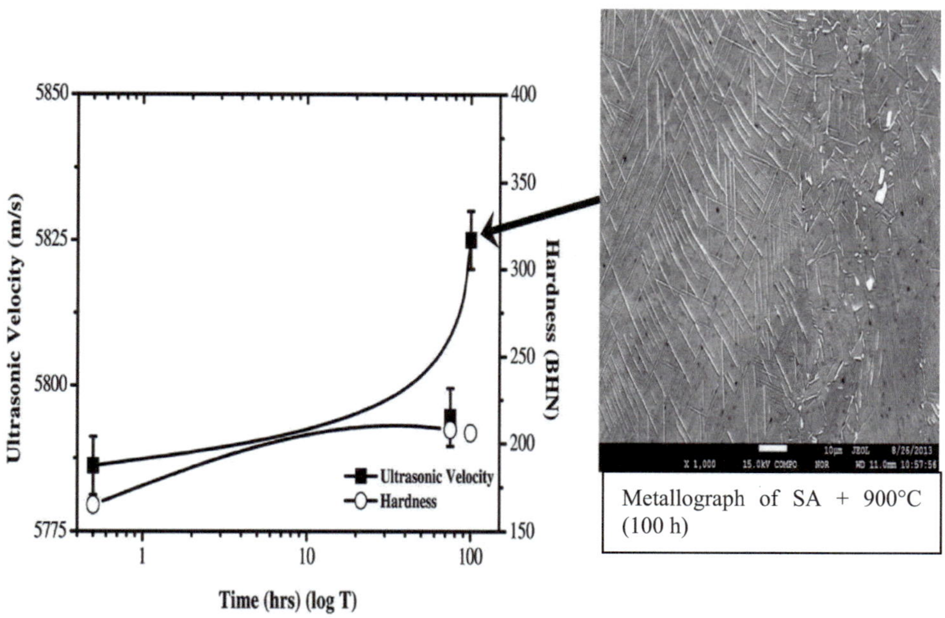

Figure 17. Ultrasonic Velocity and Hardness vs. Time for Solution Annealed (SA) specimen thermally aged at 900°C (75 h, 100 h) and studied with respect to the readings of SA specimen

5. Conclusions

From the experimental study it is found that resistivity is sensitive to the structural variations that occur on ageing and ultrasonic velocity is more sensitive towards initial formation of fine precipitates and during their nucleation and growth, while hardness is only affected after the formation of the precipitates to a critical size which can hinder the dislocation movement. There will be considerable distortion of the matrix due to the presence of the coherent phase having lattice parameters different from those of the solvent and this distortion will extend over a distance more than the size of precipitate. It is this distortion that interferes with the movement of dislocations and accounts for the increase in hardness and strength during ageing. These observations are consistent with electron microscopy studies.

Acknowledgements

This work was carried out at CSIR-National Metallurgical Laboratory (NML), Jamshedpur. The authors thank the Director, CSIR-National Metallurgical Laboratory (NML) for his kind permission to carry out and publish this work.

Author details

V. Acharya, S. Ramesh and G.V.S. Murthy*

*Address all correspondence to: gvs@nmlindia.org

Materials Science and Technology Division, CSIR- National Metallurgical Laboratory, Jamshedpur, India

References

[1] Roger C. Reed. *The Superalloys: Fundamentals and Applications*. Cambridge University press; 2006, pp. 1-101.

[2] Matthew J. Donachie and Stephen J. Donachie. *Superalloys: A Technical Guide*. 2nd edn., ASM International; 2002.

[3] Sims C.T. and Hagel W.C. *The Superalloys*. Wiley, New York; 1972.

[4] Sims C.T., Stoloff N.S. and Hagel W.C. *Superalloys II*. Wiley, New York; 1987.

[5] Åsa Martinsson. *Ageing Influence on Nickel-Based Superalloys at Intermediate Temperatures (400ºC-600ºC)*. Master Thesis. Luleå University of Technology; 2006.

http://dx.doi.org/10.5772/61119

[6] Collins H.E. and Quigg R.J. 'Carbide and Intermetallic Instability in Advanced Nickel-Base Superalloys', *ASM Trans. Quart.*, March 1968, L6 1, pp. 139.

[7] Dieter G.E. *Mechanical Metallurgy*. McGraw-Hill Book Co., London; 1988.

[8] Martin J.W. *Precipitation Hardening*. Pergamon Press, New York; 1968.

[9] Sidney H. Avner. *Introduction to Physical Metallurgy*, 2nd edn., Tata McGraw-Hill Education; December 1997.

[10] Rajan T.V., Sharma C.P. and Ashok Sharma. *Heat Treatment: Principles and Techniques*. 2nd edn. PHI learning Pvt. Ltd.; 2011.

[11] Stanley R.K., Moore P.O. and McIntire P. (eds.). *Nondestructive Testing Handbook*, 2nd edn., vol. 9, Special Nondestructive Testing Methods, American Society for Nondestructive Testing, Columbus, OH,1995.

[12] Baldev Raj, Jayakumar T. and Thavasimuthu M., *Practical Non-Destructive testing*, 3rd edn., Narosa Publishing House Pvt. Ltd., 2011.

[13] Metals Handbook: *Mechanical Testing*, 9th edn., vol. 8.

[14] Metals Handbook: *Non-Destructive Evaluation and Quality Control*, 9th edn., vol. 17.

[15] Victor Giurgiutiu and Adrian Cuc, 'Embedded Non-destructive Evaluation for Structural Health Monitoring, Damage Detection, and Failure Prevention', Sage Publications, *The Shock and Vibration Digest*, March 2005, vol. 37, no. 2, pp. 83-105.

[16] Farrar C.R. and Worden K. 'An Introduction to Structural Health Monitoring', *Proc Trans Royal Soc A*, 2007, 365, pp. 303-315.

[17] Nanekar P.P. and. Shah B.K. 'Characterization of Material Properties by Ultrasonic", *BARC Newsletter*, Issue No. 249.

[18] Thompson R.B. 'Theory and Application of Ultrasonic Microstructural Characterization', *J Minerals Metals Mater Soc* (*TMS*) (*JOM*), October 1992.

[19] Eds. Buck O. and Wolf. S.M., *Nondestructive Evaluation: Application & Materials Processing*, Materials Park, OH: ASM, 1984.

[20] Wadley H.N.G. (ed.). *NDE of Microstructure for Process Control*, Materials Park, OH: ASM, 1985.

[21] Bobo S.P. *Review of Progress in Quantitative Nondestructive Evaluation*, vol. 9B, Thompson D.O. and Chimenti D.E. (eds.), New York: Plenum Press, 1990, pp. 2097-2169.

[22] Eds. Thompson D.O. and Chimenti. D.E. *Review of Progress in Quantitative Nondestructive Evaluation*, vol. 1-11, New York: Plenum Press, pp. 1982-1992.

[23] Ruud C.O. and Green R.E. Jr. (eds.). *Nondestructive Methods for Materials Property Determination*, New York: Plenum Press, 1984.

[24] Bussiere J.F. (ed.). *Nondestructive Characterization of Materials II*, New York: Plenum Press, 1987.

[25] Holler P. (eds.). *Nondestructive Characterization of Materials*, Berlin: Springer-Verlag, 1989.

[26] Buck O. and Wolf S.M. (eds.). *Nondestructive Evaluation: Microstructural Characterization and Reliability Strategies*, Warrendale, PA: TMS, 1981.

[27] Liaw P.K., Buck O. and Wolf S.M. (eds.). *Nondestructive Evaluation and Material Properties of Advanced Materials*, Warrendale, PA: TMS, 1991.

[28] Baikand J.M. and Thompson R.B. *J Nondestruct Eval*, 4, 1984, pp. I77-196.

[29] Margetan F.J., Thompson R.B. and Gray T.A. *J Nondestruct Eval*, 7, 1988, pp. 131-152.

[30] Nagy P.B. and Adler L. *Review of Progress in Quantitative Nondestructive Evaluation*, Thompson D.O. and Chimenti D.E. (eds.), New York: Plenum Press, 1991, vol. 10A, pp. 177-184.

[31] Yalda-Mooshabad I., et al. *J Nondestruct Eval.*

[32] Giurgiutiu V. and Cuc A., 'Embedded Non-destructive Evaluation for Structural Health Monitoring, Damage Detection, and Failure Prevention', Sage Publications, *The Shock and Vibration Digest*, March 2005, vol. 37, no. 2, pp. 83-105.

[33] *Analytical Ultrasonic in Material Research and Testing*, NASA CP 2383, 1984.

[34] Banerjee S. and Shah B.K. 'Characterization of Industrial Materials', In G. Sridhar, S. Ghosh Chowdhary and N.G. Goswami (eds.), *Material Characterization Techniques – Principles and Applications*, 1999, pp. 1-15.

[35] Dobmann G., et al. 'Non-Destructive Characterization of Materials: A Growing Demand for Describing Damage and Service Life Relevant Ageing Process in Plant Components', *Nucl Engin Design*, 1997, vol. 171, pp. 95-112.

[36] Thomson R.B. 'Laboratory Non-Destructive Evaluation Technology Material Characterization', *J Non-Destruct Eval*, 1996, vol. 15, no. 3 & 4.

[37] Raj B., et.al. 'NDE Methodologies for Characterization of Defects, Stresses and Microstructure in Pressure Vessels and Pipes', *Int J Pressure Vessel Piping*, September 1997, vol. 73, issue 2, pp. 133-146.

[38] Shah D.N. 'Testing and Characterization of Materials', Workshop, IIM, Mumbai Chapter, 1990.

[39] Shah B.K. 'Advances in NDT Techniques for Flaw and Materials Characterization', Seminar, IIM, Mumbai Chapter, 1992.

[40] Mitra A. (ed.) 'Recent Trends in Nondestructive Evaluation of Materials', Workshop, 1997.

[41] Bhattacharya D.K. 'Characterization of Microstructure in Steels by Magnetic Techniques an Overview', *ISNT J Non-Destructive Eval*, 1997, vol. 17, no. 1.

[42] Dobmann G. et.al. 'Ageing Material Evaluation and Studies by Non-Destructive Techniques (AMES-NDT), A European Network Project', *J Nucl Engin Design*, 2001, vol. 206, pp. 363-374.

[43] Theiner W.A., et.al. 'Non-Destructive Analysis of the Structure of Pressure Vessel Steels by Micro-magnetic Testing Techniques', *J Nucl Engin Design*, December 1983, vol. 76, issue 3, pp 251-260.

[44] Willems H. and Gobbeles K. 'Characterization of Microstructure by Backscatter Ultrasonic Waves', *Mat Sci*, November- December 1981, vol. 15, pp. 549-553.

[45] Mak D.K. 'Determination of Grain Size, Hysteresis Constant and Scattering Factor of Polycrystalline Material Using Ultrasonic Attenuation', *Canadian Metall Quart*, vol. 25, no.3, pp 253-255.

[46] Diamand R.D. 'Development of Instrument for On-Line Measurement of Grain Size in Copper Alloys and Stainless Steels', 21st Annual British Conference on Non-Destructive Testing NDT 0 86, pp. 225-242.

[47] Badidi Bouda A., et.al. 'Grain Size Influence on Ultrasonic Velocities and Attenuation', *Proc of NDT & E International*, January 2003, vol. 36, 1, pp. 1-5.

[48] Generazio E.R. 'Ultrasonic Attenuation Measurements to Determine Onset, Degree and Completion of Recrystallization', *Mater Eval*, August 1998, vol. 46, pp. 1198-1203.

[49] Murthy G.V.S., Sridhara G., Anish Kumar and Jayakumar T. 'Characterization of Intermetallic Precipitates in a Nimonic Alloy by Ultrasonic Velocity Measurements', *Mater Character*, March 2003, vol. 60, no. 3, pp 234-239.

[50] Nanekar P.P., et.al. 'Non-Destructive Characterization of Ceramics and Concrete Structure', Testing and Quality Control, conducted by ASM India Section, May 2001, Mumbai.

[51] Smith R.L., et.al. "Ultrasonic Attenuation, Microstructure and Ductile to Brittle Transition Temperature in Fe-C Alloys", *Mater Eval*, February 1983, vol. 41, pp. 219-222.

[52] Vary A. 'Concepts for Interrelating Ultrasonic Attenuation, Microstructure and Fracture Toughness in Polycrystalline Solids', *Mater Eval*, April 1988, vol. 46, pp. 642-649.

[53] Jean F. Bussiere, et.al. 'Analysis of the Effect of Graphite Morphology on the Elastic Properties of Cast Iron', Proc. of 3rd Symposium on Non-Destructive Characterization of Materials, October 1988, Saarbrucken, FRG, pp. 353-660.

[54] Shah B.K., et.al. 'Qualification of Beta Heat Treatment of Uranium Fuel Rods by Ultrasonic', *Insight, J British Instit Non-destruct Testing*, November 1999, vol. 41, no. 11, pp. 707-709.

[55] Nanekar P.P., et al. 'Ultrasonic Characterization of Precipitation Hardenable 17-4 PH Stainless Steel', Proc. of 47th Annual technical meeting of Indian Institute of Metals, Hyderabad, November 1993.

[56] Weston-Bartholomew W. 'Use of Ultrasonic Goniometer to Measure Depth of Case Hardening', *Int Adv Nondestruct Testing* 1979, vol. 6, pp. 111-123.

[57] Y. Lu, Parthasarathi S. and Wadley H.N.G. 'Advanced Sensing, Modeling And Control Of Materials Processing', In Matthys E.F. and Kushner B. (eds.), Warrendale, PA: TMS, 1992, pp. 204-221.

[58] Margetan F.J. and Thompson R.B. *Review of Progress in Quantitative Nondestructive Evaluation 1992*, D.O. Thompson and D.E. Chimenti (eds.), vol. 11B, New York: Plenum Press, pp. 1717-1724.

[59] Thompson R.B. et al. *Review of Progress in Quantitative Nondestructive Evaluation 1992*, D.O. Thompson and D.E. Chimenti (eds.), vol. 11B, New York: Plenum Press, pp. 1685-1691.

[60] 'On-Line Monitoring of Sheet Metal Texture', *J of Nondestruct Eval.*

[61] Green R.E., et.al. 'Ultrasonic and Acoustic Emission Detection of Fatigue Damage', *Int Adv Nondestruct Testing,* 1979, vol. 6, pp. 125-177.

[62] Joshi N.R. and Green R.E. 'Ultrasonic Detection of Fatigue Damage', *Fract Mech* 4, 1972, pp. 577-583.

[63] Gerd Dobmann, et.al. 'Nondestructive Characterization of Materials (Ultrasonic And Magnetic Techniques) For Strength And Toughness Prediction And The Detection Of Early Creep Damage', *Nucl Engin Design,* 1992, 157, pp. 137-158.

[64] Birring A.S., et al. 'Ultrasonic Detection of Hydrogen Attack in Steels', *Corrosion,* vol. 45, no. 3, pp. 259-263.

[65] Kruger S.E., et al. 'Hydrogen Damage Detection by Ultrasonic Spectral Analysis', Proc. of NDT & E International, 1999, vol. 32, pp. 275-281.

[66] Shah B.K. 'Monitoring of Intergranular Corrosion in Austenitic Stainless Steels AISI 304 by Non-Destructive Testing Methods', M.Tech Thesis, IIT Bombay, 1984.

[67] Ikuta E., et al. 'Ultrasonic Evaluation of Thermal Embrittlement', Proc. of the 13th International Conference in the Nuclear and Pressure Vessel Industries, Kyoto, Japan, 1985, pp. 285-289.

[68] Shah B.K, et.al. 'Ultrasonic Characterization of Ageing Degradation During Long Term High Temperature Exposure in Nickel Base Alloy 625', Proc. of 14th World Conference on NDT, New Delhi, December 1996, pp. 2235-2238.

[69] Bowler N. 'Theory of Four-Point Direct Current Potential Drop Measurements on a Metal Plate', *Proc Trans. Royal Soc A,* 2007, 463, pp. 817-836.

[70] Panseri C. and Federighi T. *J Instit Metals*, 1966, vol. 94, pp. 99-107.

[71] Kasap S. O. *Principles of Electronic Materials and Devices*, 2nd edn., Tata McGraw-Hill, Boston 2002.

[72] Dubost B., Bouvaist J. and Reboul M. 'Aluminum Alloys: Their Physical and Mechanical properties', *Engin Mater Advis Serv Ltd.*, Chalottesville, 1986, vol. 2, pp. 1109-1123.

[73] Blitz. J. 'Electrical and Magnetic Methods of Non-Destructive Testing', Springer, 1997, vol. 3.

[74] Papadakis E.P. 'Ultrasonic Velocity and Attenuation Measurement Methods with Scientific and Industrial Applications', *Physic Acoust*, 1976, vol. 12, pp. 277-374.

[75] Holbrook J. and Bussiere J.F (eds.). *Nondestructive Monitoring of Materials Properties*, Pittsburgh, PA: MRS, 1989.

[76] Wadley H.N.G., et al. (eds.). *Intelligent Processing of Materials and Advanced Sensors*, Warrendale, PA: TMS, 1987.

[77] Wadley H.N.G. and Eckhart W.E., Jr. (eds.). *Intelligent Processing of Materials*, Warrendale, PA: TMS, 1990.

[78] Matthys E.F. and Kushner B. (eds.). *Advanced Sensing, Modeling and Control of Materials Processing*, Warrendale, PA: TMS, 1992.

[79] Zhao J.C. and Henry Michael F. 'The Thermodynamic Prediction of Phase Stability in Multi Component Superalloys. *JOM*, 2002; pp. 37-40.

[80] Chandrasekar R., Chester C.H. Lo, Frishman A.M., Larson B.F. and Nakagawa N. 'Quantification of Precipitates and Their Effects on the Response of Nickel-base Superalloy to Shot Peening', AIP Conf. Proc. 1430, 2012, pp. 1437-1444.

Section 2

Coatings and Films

Chapter 12

Coatings for Superalloys

Mathias C. Galetz

Additional information is available at the end of the chapter

http://dx.doi.org/10.5772/61141

Abstract

High-temperature coatings for superalloys can be divided into three categories: Two of them, diffusion and overlay coatings, are both used to protect a system from oxidation and corrosion. The third type, thermal barrier coatings, protects the substrate from thermal degradation.

All types of coatings are designed to avoid direct contact between a hot gas or detrimental components from the environment and the base metal alloy. The aim is to enhance the lifetime of high temperature applications, which is otherwise often limited by corrosion or thermal degradation. In this chapter, the different coatings and their durability are discussed, which ideally matches the aging process of the substrate underneath. Besides the corrosive attack the coatings must also often be able to withstand certain mechanical and thermal mechanical loads as well as erosion. Nevertheless, all coatings have a limited thickness and get consumed in service. Therefore it is essential to understand and choose coatings that show slow degradation mechanisms.

Keywords: Diffusion coatings, MCrAlYs, TBC

1. Introduction

Superalloys are operated in industrial environments containing corrosive species such as sulfur, chlorine- or carbon-containing compounds, water vapor, alkali and alkaline-earth metal salts, or ashes such as vanadates [1,2]. At elevated temperatures, such compounds cause a wide range of attack types on most metallic alloys such as oxidation, carburization, sulfidation, hot corrosion, or a combination of different mechanisms [3,4]. But even in air the oxidation resistance of nickel- and cobalt-based superalloys is not sufficient for continuous operation above 1000°C [5]. Oxide scales grow too fast and the subsurface zone of the alloys is changed and loses its mechanical strength [6-8]. Despite these limitations, advances and improvements of industrial technology have led to more efficient processes and more powerful engines with increased operation temperatures, and the alloys used for their construction are pushed toward the applicable limit even as far as mechanical properties are concerned, such as fatigue, tensile strength, and creep resistance [9]. Therefore higher additions of alloying elements for corrosion protection such as Cr, Al, or Si cannot be used as they either lead to embrittlement or lower the melting point and therefore the creep strength at the high target temperatures [10,11]. The only way is to apply high-temperature coatings to face aggressive corrosive high-temperature atmospheres and make processes possible which could not be operated efficiently and reliably without such coatings.

Diffusion- and metal-based overlay coatings are both used to protect a system from oxidation and corrosion [12]. Such coatings are designed to avoid direct contact between the base metal alloy and a hot gas carrying detrimental species by growing a thin, self-healing oxide scale in situ.

Besides oxidation and corrosion protection, the other major factor affecting the overall life of substrates at high temperatures are heat-flux effects. Nickel- and cobalt-based alloys are employed in gas temperatures up to 1400°C [13]. Without proper protection, the temperatures could easily induce softening or even reach the melting point of the alloys. The employed materials are the actual major constraints that determine the maximum gas temperatures in many processes of today's aircraft as well as energy conversion industry. Without cooling and protection by coatings, the efficiency and speed of jet engines would also be limited to a very low level. In these cases, only a combination of two coatings allows operation; a so-called bond coat (the bond coat belongs to the class of diffusion or overlay coatings, which are discussed in detail below) in combination with a ceramic layer on top to form a thermal barrier coating system. The ceramic layer also reduces the attack indirectly by lowering the metal surface temperature and thus decelerates the degradation of the metallic coating and substrate underneath.

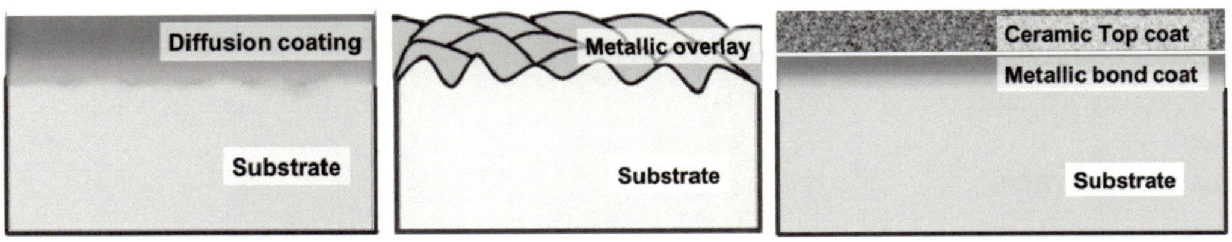

Figure 1. Schematic sketches of the general configuration of the three different types of coatings used for high-temperature applications (from left to right): diffusion coatings, metal overlay coatings, and thermal barrier coatings

Only by combining one of the different types of coatings schematically shown in Figure 1 with a load-carrying superalloy underneath, safe operation over a reasonable lifetime can be ensured. In operation the coatings must be able to sustain certain mechanical and thermo-mechanical loads as well as erosion on top of the oxidative or corrosive attack. Ideally, the durability or lifetime of protective coatings for parts such as turbine blades or vanes matches that of the aging processes of the substrate underneath. However, since all coatings are more or less thin films, the mechanical lifetime of the substrate is often longer than the period of coating degradation at high temperature. Therefore, after a certain time in service, the coatings residues are stripped and the coatings are remanufactured several times during the lifetime of the blades [14].

2. Requirements for protective coatings at high temperatures

Unlike aqueous corrosion conditions at low temperatures, where usually either an anodic protection is applied or even very thin barrier coatings can protect from degradation by separating an electrolyte from the material, high-temperature degradation processes are controlled by transport processes in or through coatings via diffusion. High temperatures are especially demanding because all metals tend to become thermodynamically stable in their oxidized form or tend to react with other gas components such as nitrogen, carbon, or sulfur. For example, even highly corrosion-resistant platinum suffers from significant oxidative degradation above 1000°C in air [15]. Which corrosion products form in an environment depends on the partial pressure or activity of different potential reaction partners in the atmosphere. The idea behind all oxidation- and corrosion-resistant coatings is to create a reservoir of scale-forming elements from which a thermodynamically stable, slow growing, adherent scale can be formed, consisting of corrosion products, usually thermally grown oxide on the surface. An overview of the requirements for coatings for high temperatures is provided in Table 1.

• Slow-growing, thermodynamically stable scale formation and high concentration of the scale-forming elements
• Thermal stability, no detrimental phase changes
• Erosion resistance
• Good adhesion
• Low interdiffusion of the substrate
• Mechanical compatibility: similar modulus and thermal expansion to avoid stresses at the interface
• No crack initiation, not too brittle
• Good processability
• Low price

Table 1. Requirements for high-temperature coatings

Even with stable protective oxide scales, the service conditions will usually create flaws such as cracks in the scale. Most important is the self-healing potential or ability for reformation, properties which for example a pure ceramic coating cannot provide. In Figure 2 a pure ceramic coating is compared with a thermally grown oxide scale that builds up in situ. In the first case, if a crack occurs, the atmosphere can attack the metal below, which cannot protect itself. Ceramics also always possess a certain porosity and cannot prevent the gas from reaching the metal below, which means they offer no effective long-term corrosion protection. Ceramic coatings are therefore almost exclusively used to lower substrate temperatures and in combination with one additional coating of the diffusion or metallic overlay type which can form slow-growing oxides. Diffusion or overlay coatings possess a sufficient reservoir of protective elements underneath the scale to allow the reformation of the protective scale many times before the reservoir of the protective elements is depleted.

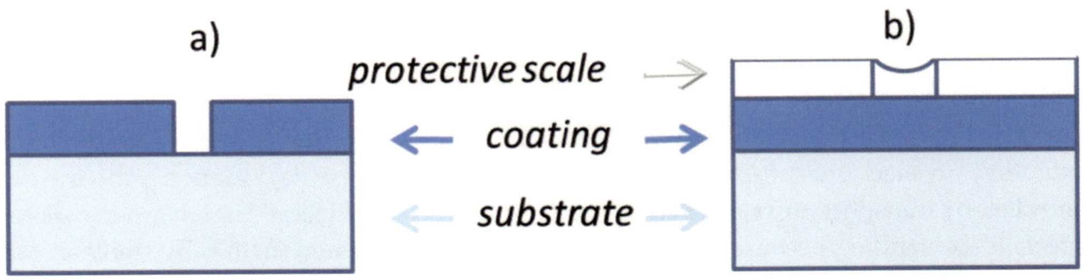

Figure 2. Comparison between a ceramic coating without self-healing properties (a) and a reservoir coating that allows the reestablishing of the protective scale in case of local failure (b)

The potentially protective elements, which qualify for scale formation, must form thermodynamically more stable corrosion products than the main elements in the superalloy such as nickel or cobalt. The graph in Figure 3 is a simplified Ellingham-Richardson Diagram that shows the standard free energy of oxide formation as a function of the temperature. Generally, for scale formation on nickel- or cobalt-based alloys, elements with a very high affinity to oxygen such as aluminum, chromium, or silicon are important, but also titanium or tantalum would theoretically qualify. The drawback for example in the case of titanium is that it allows fast oxygen diffusion through its oxide scales, which have high growth rates even at moderate temperatures. Additionally, they often tend to spall off easily.

So not only the criterion of thermodynamically preferred reaction has to be fulfilled, the scales also have to be slow-growing. Only scales that form a dense and stable crystal lattice effectively limit diffusion and guarantee a slow growth and weight gain as shown in Figure 4. From the curve, the growth rate constant k_p of the thermally growing oxide can be determined according to the function:

$$(\Delta\mathrm{m})^n = k_p t \tag{1}$$

where Δm is the weight gain per surface area of the metal [g/cm^2], t is the exposure time [s], and n is the rate exponent. If the scale is protective, stable, and adherent, the value of n is usually close to 2 and the weight gain shows a parabolic diffusion-controlled behavior. The diffusion is generated by the anion and cation concentration gradient in the scale. The speed

http://dx.doi.org/10.5772/61141

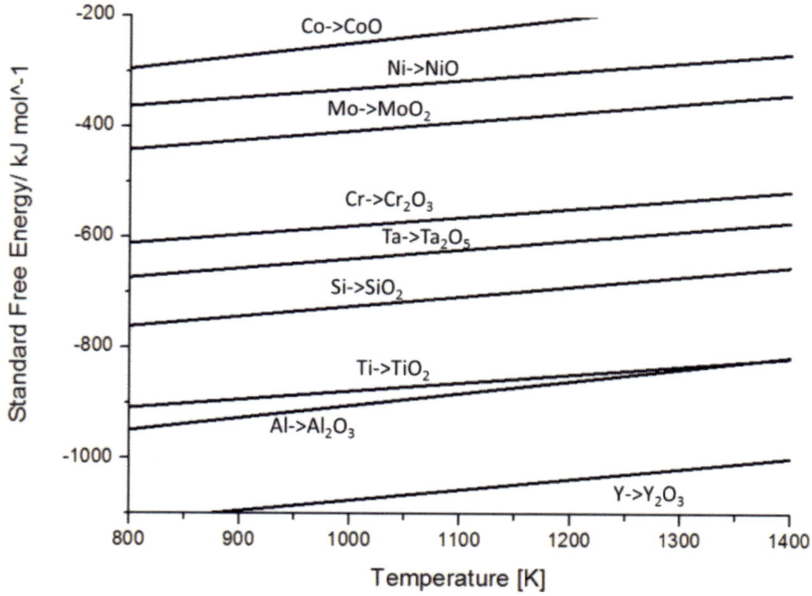

Figure 3. Simplified Ellingham-Richardson Diagram

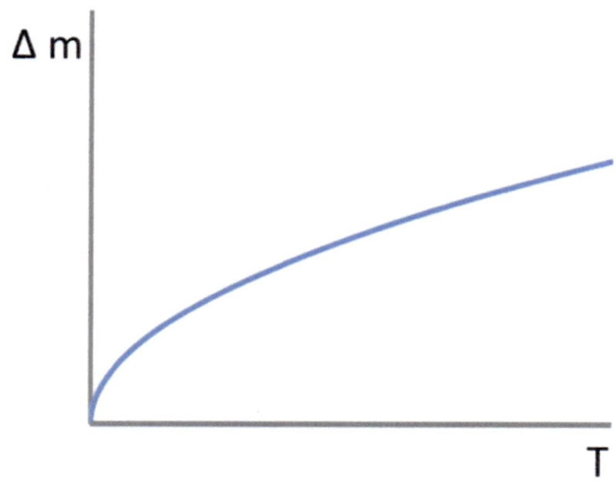

Figure 4. Parabolic weight gain of ideal scale growth

of diffusion is determined by the crystallographic structure of the oxide, the defect structure, and the microstructure, for example grain size and grain boundary distribution. In most cases, the faster diffusing species is a metallic cation, but anion diffusion is also possible [16], for example it is predominant in silica formers in a wide range of temperatures. Figure 5 shows the parabolic rate constants at 1000°C normalized to the behavior of slow-growing alumina for nickel, chromium, and silicon. It can be seen that Cr, Al, and Si offer much slower growth rates than the base metal nickel. Cobalt is not included in the diagram since its oxidation rate is even ten times higher than that of nickel.

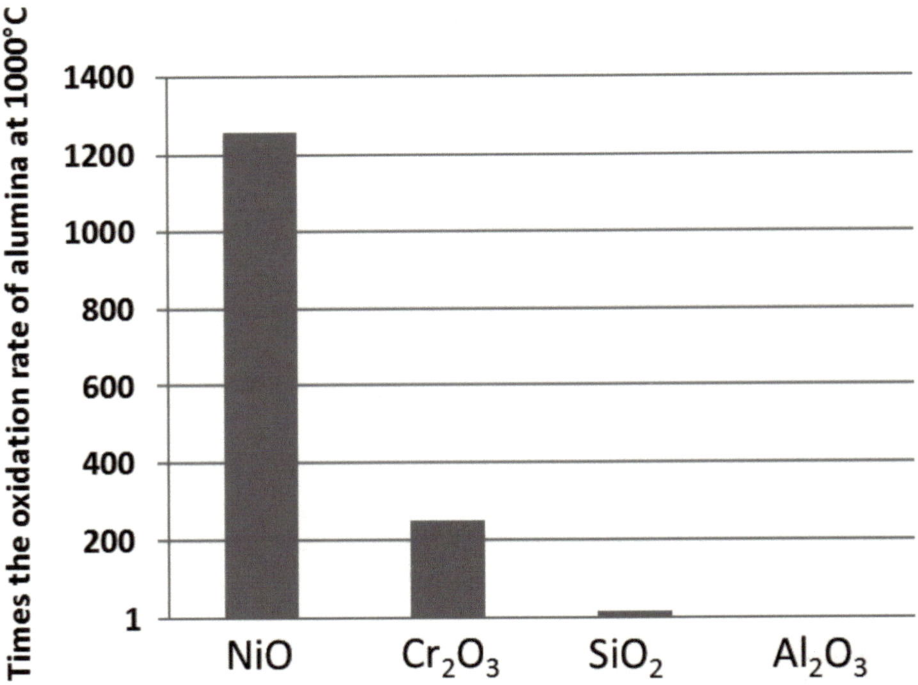

Figure 5. Parabolic rate constants of Fe, Co, Ni, Cr, Si, and Al oxide formers at 1000°C; data from [17] as k_p ($g^2cm^{-4}s^{-2}$)

Further requirements derive from the mechanical interaction of the scale with the coating underneath. During formation growth stresses occur due to the volume change, when a scale is formed by including elements from the environment such as oxygen. As a first hint on the stresses, the classical approach of Pilling and Bedworth from 1923 can be used, according to which the Pilling-Bedworth Ratio is defined as:

$$PBR_{metal} = \frac{Volume\ of\ oxide}{Volume\ of\ metal} \tag{2}$$

It must be higher than 1 in order to guarantee a continuous scale formation, as ceramic scales are highly prone to failure by tensile stresses. On the other hand, it must not be too high either, because when high compressive stresses occur they also induce failure by cracking and buckling of the scales. In Table 2 the values of different oxides on pure metals are shown.

Oxide	MgO	Al_2O_3	ZrO_2	NiO	TiO_2	CoO	Cr_2O_3	Ta_2O_5	Nb_2O_5
PBR	0.81	1.28	1.56	1.65	1.75	1.86	2.07	2.5	2.68

Table 2. Pilling-Bedworth Ratio of different oxides on pure metals [18]

http://dx.doi.org/10.5772/61141

It should be noted that the PBRs of alloys are not the same as those of pure metals, so for example the PBR of an alumina scale growing on intermetallics from the Ni-Al system is significantly different to the PBR of pure Al (Table 3). So the alloy system on which an oxide grows at high temperature must be considered as well.

α-Al2O3 on	Ni_3Al	NiAl	$NiAl_3$
PBR	1.71-1.88	1.64-1.78	1.48-1.57

Table 3. Pilling-Bedworth Ratio of α-Al_2O_3 on different intermetallic aluminides [19]

After formation of the scales, the occurring stresses are dominated by the mismatch in the thermal expansion coefficients between the ceramic scale, the coating, and the alloy beneath.

When the mismatch is too high, for example with silica scales on nickel- or cobalt-based alloys, extensive cracking occurs during cyclic exposure, providing less protection than alumina and chromia, oxides which have a higher thermal expansion coefficient closer to that of the alloys underneath [20,21]. Still, even alumina or chromia scales suffer some cracking in service, but coatings for high temperatures are designed to allow self-healing of the scales.

As already mentioned, such high-temperature coating systems can be classified as diffusion coatings or overlay coatings. Ceramic thermal barrier coatings also fall into the category of overlay coatings, but these will be covered separately.

3. Diffusion coatings

Diffusion coatings are based on the enrichment of protective metallic elements close to the surface by diffusion from a reservoir outside. As for all diffusion processes, the parameters time, temperature, and phase composition are critical. The process conditions have to be optimized for each material because the phase composition and the microstructural evolution of diffusion coatings depend highly on the substrate which is coated. Figure 6 illustrates the general idea of diffusion coatings. A reservoir with a very high concentration of the element that should be enriched is created at the surface either in the gas phase or in a solid/liquid state. Due to the concentration gradient at high temperatures, elements start to diffuse into the substrate, and elements of the alloy have a tendency to diffuse outward. Depending on which process is faster, the inward or outward diffusion process is more pronounced. In both cases these processes eventually achieve the aim of changing the composition close to the surface by enrichment with the elements from the diffusion reservoir.

Several different manufacturing methods have been developed for diffusion coatings. Figure 7 gives an overview of the most important procedures. These include pack cementation, which can be done in-pack or out-of-pack (also called above the pack, which resembles the chemical vapor deposition (CVD) processes). Other processing techniques include slurry-reaction coatings, galvanic and metal foil coatings, or a hot-dipping process in a metal melt. They all

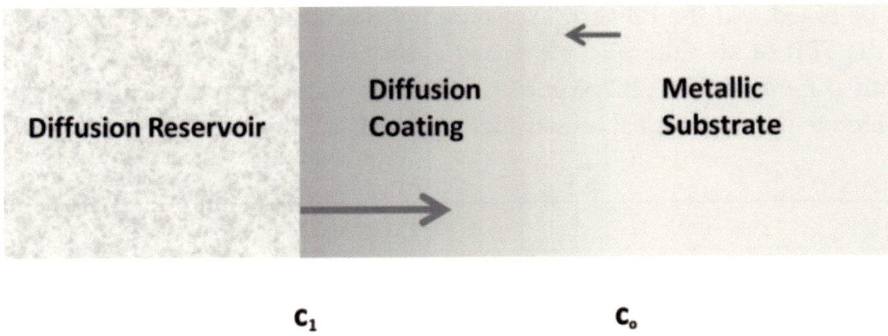

Figure 6. Principle of diffusion coatings manufacturing

have in common that a reservoir of the diffusing metal is first applied to the surface and later diffused via heat treatment.

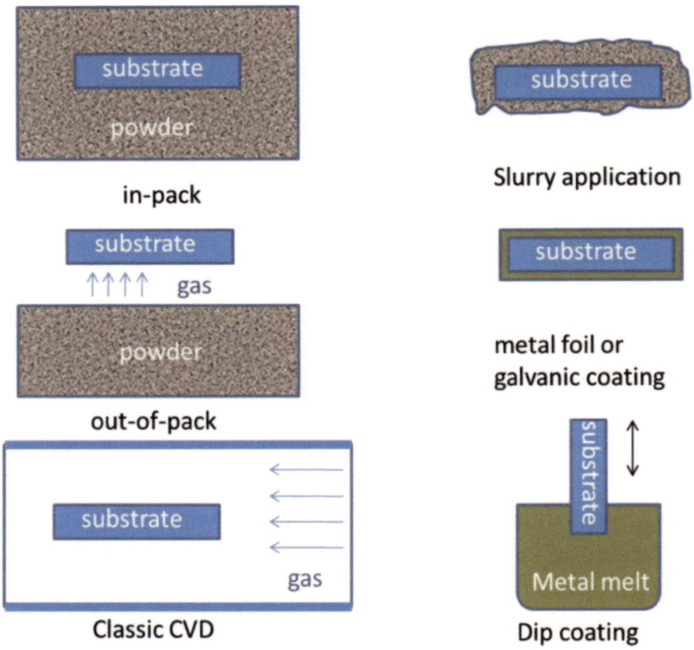

Figure 7. Different methods to produce diffusion coatings

3.1. Pack cementation processes

As early as in 1914 the first "pack cementation" process was reported by Allison and Hawkins [22]. They diffused aluminum into iron and steel, but the process did not receive much attention up to the 1950s. Since then, pack cementation with chromium has been widely used to enhance the corrosion behavior of low-alloyed steels [23]. In the 1970s, the aluminizing process for nickel-based alloys became popular, especially in the aircraft industry. Today more than 80% of stage 1 airfoils are coated with aluminum diffusion coatings [12], most of them by the pack cementation process. It falls into the category of chemical vapor processes and is carried out at high temperatures (600-1200°C) in an inert gas atmosphere, in vacuum, or under reducing conditions. The the powder pack is composed of an inert filler (usually Al_2O_3), a halogen-carrying compound as an activator, and a metallic powder of the elements to be enriched. In the 1990s, with a better theoretical understanding of the processes, co-deposition of more than one element at a time became possible and was first developed for Al in combination with Cr [24-28] to improve the corrosion behavior when sulfur is part of the hot gas. Recently, investigations have been focusing on the modification of aluminide scales by co-diffusion with reactive elements such as Y, Ce, or Hf [29,30]. Small amounts of these elements are known to improve scale adhesion [31]. Platinum modification can also be beneficial under certain conditions since it facilitates the formation of stable alumina scales. Pt has to be galvanically applied and interdiffused before aluminum is enriched via pack cementation [32]. Silicon can also be added to the aluminide coating. Similar to chromium, it enhances the resistance against sulfur-carrying gases [33]. However, its application is usually limited to small amounts because silicon bears the risk of forming low-melting phases in nickel- or cobalt-based systems [34].

In industrial applications, the reaction is conducted in tightly sealed containers in which the components are placed. The parts can either be embedded in the reactive powder mixture ("in pack") or suspended "out-of-pack" (also called "above the pack"). In both cases either an inert or a reducing atmosphere is used. The entire container is heated to 100-200 °C to remove any residual moisture and oxygen. In the following step it is heated to the actual process temperature, which depends on the diffusing elements and on the substrates. For superalloys the temperature typically lies in the range of 900-1100°C. The holding time is dependent on the particular system and usually varies between 1 and 10 h. The optimized parameters must be evaluated for each system, usually with the help of thermodynamic calculation programs to investigate the temperature and activity of the different metal halogens which carry diffusion metals to the surface [35,36]. If the activities of the diffusing metal halides at the surface are too high, brittle phases occur frequently; if their activities are too low, too little diffusion takes place [37].

Figure 8 summarizes the major steps of the classical pack cementation process. Metal halides are formed and transported via the gas phase to the surface of the component to be coated, where the halide interacts with the sample surface and dissociates into the metal and the halogen anion. The metal atom diffuses into the substrate surface zone. The formation of the coating resembles that of diffusion couples, and the driving force for the interdiffusion is the activity or concentration gradient between the environment (which contains the diffusing elements such as Al) and the surface of the components [29]. After releasing the metal at the surface, the halogen reacts with new metal atoms from the powder to keep the process continuously alive. In any case, halogens (most common are chlorine and fluorine, but bromine or iodine can also be used) serve as an activator. The contents of activator and metallic powder determine the activity of metallic halogens and therefore the amount of metal which can react

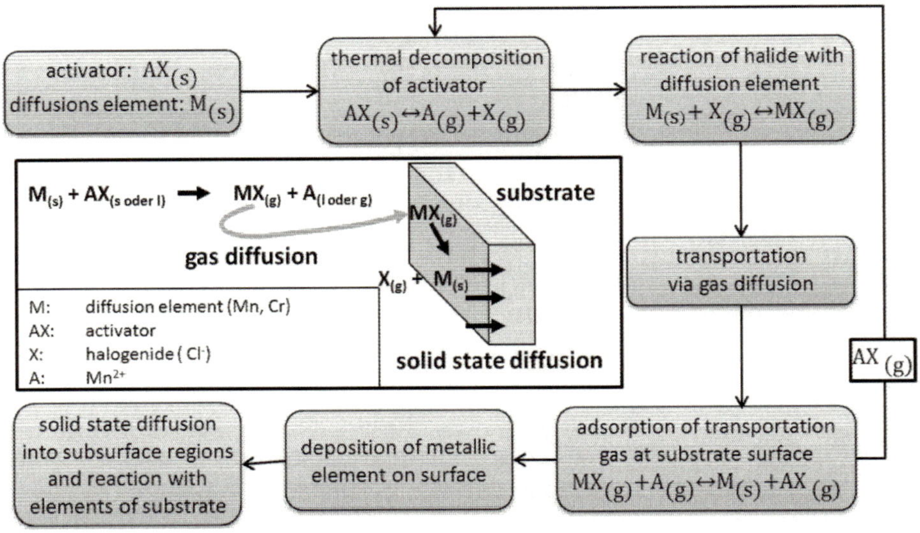

Figure 8. Major steps during the pack cementation process [38]

at the surface. The activator is added in the form of compounds such as NH_3Cl, AlF_3, or as HCl gas. The metallic powder, for example Al, can be used as pure element powder but also in an alloy form with nickel or chromium, for example CrAl powder, to lower its activity.

In any case, at the surface several intermetallic phases form according to their thermodynamic stability and local phase composition. Typical intermetallic phases that can form within the substrate metals by aluminization are $CoAl_3$, Co_2Al_5, or CoAl in cobalt-based alloys and $NiAl_3$, Ni_2Al_3, NiAl, or Ni_3Al in nickel-based alloys. As mentioned earlier, the formation of the phases at the surface is controlled by the activity (the powder composition and activator), the temperature, and the duration of the process. Goward and Boone [39] classified diffusion coatings into low- and high-activity coatings, based on observations on nickel-based alloys. A "high activity" pack structure is usually observed in pack cementation processes at lower temperatures in combination with a high aluminum activity at the surface. In this case, Ni2Al3 is formed by inward diffusion of aluminum as the dominant mechanism and the diffusion of Ni is rather low. Additionally, in aluminum-rich phases the diffusion of aluminum is favored, even in aluminum-rich NiAl. With lower Al content, its mobility decreases as well. After a high-activity process, Ni2Al3 is present at the surface, while closer to the coating–substrate interface aluminides less rich in aluminum can be found. Ni2Al3 coatings usually require a second heat treatment to transform the rather brittle phase into the desired NiAl phase. Instead, if a lower activity is used in the pack in combination with a higher temperature, Ni diffusion is faster and NiAl forms directly. As a result, these coatings grow by outward diffusion and elements of the substrate that have little solubility in the NiAl phase such as refractory metals are enriched in the substrate close to the interface as part of the so-called interdiffusion zone. In Figure 9, a high-activity coating is compared with a low-activity coating on a nickel-based superalloy.

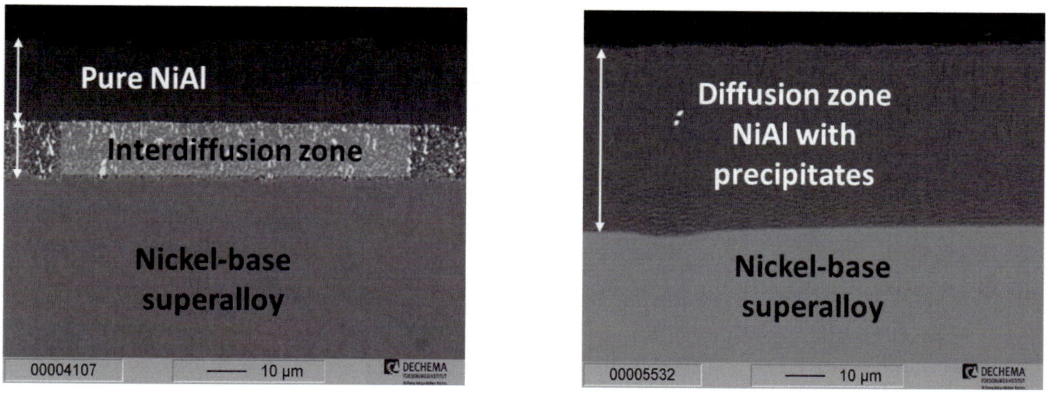

Figure 9. Comparison of a high-activity coating (left) with a low-activity coating (right) on a nickel-based superalloy substrate (PWA 1484)

Such coatings can be applied even inside of pipes. In this case, the tube is filled with the powder mixture and then sealed at the ends. The tube represents both the container and the part to be coated. This is an industrially established technique for huge components such as 15-m-long furnace tubes [40]. After the coating process, in case of "in pack," sometimes inert particles from the ceramic filler remain on the surface and must be removed. If the parts have small holes or channels, these can even be blocked by residues of the inert filler. To avoid such drawbacks, turbine blades and vanes are usually coated by using the "out of pack" technique. In all cases the huge advantage is that, since it is a gas phase process, complete coverage of the surface can be ensured.

3.2. Slurry coatings

Compared with the pack cementation process, slurry coatings were developed much later in the 1970s and 1980s [39,41]. Although the processing and formation mechanisms are totally different, they offer similar microstructure and features as coatings that were applied by pack cementation. In Figure 10, the diffusion layers obtained with these two different coating methods are shown for a nickel-based alloy (CM247). The coating on the left was produced by the pack cementation described in detail above. The other coating was achieved via the slurry aluminizing route. After 1000 h of exposure in air at 1050°C, diffusion layers of both coatings look very similar and are still protective.

Thus, the slurry route represents an interesting alternative for many systems or components, especially since slurry systems can be applied on components via common immersion, painting, or spraying methods, which is a big advantage for large parts. The slurries usually contain a metal powder, sometimes an activator, and either an organic, water-based binder or chromium-phosphate acidic binder [42,43]. Water-based activator-free systems are preferred nowadays, as the use of chromium-phosphates or halogen activators is hazardous and toxic. When the coating is applied, it also has to be heat-treated. In a first step, organic binders are burnt-out in the temperature range of 300-450°C and subsequently the metal powder reacts and diffuses into the substrate at higher temperatures between 600 and 1100°C [44]. In contrast

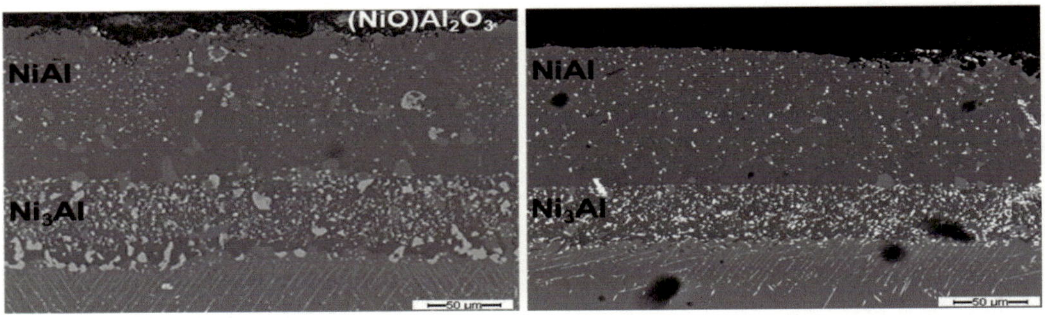

Figure 10. Diffusion coatings on a CM 247 nickel-based alloy after exposure in air at 1000°C for 1050 h. The left coating was applied by the slurry process (slurry composition: 40% Al, polyvinyl alcohol, water), the coating on the right via pack cementation (powder mixture: 5% Al, 1% NH4Cl, rest Al2O3, 1000°C, 1 h in Ar/5% H2)

to the pack cementation process no gas phase is present. Instead, above the melting point metallic melt from the metal powder wets the substrate and reacts via a combustion synthesis (exothermic formation of intermetallics) mechanism, which is extremely fast [44]. In the beginning of the process aluminum-rich phases are formed that usually require further heat-treatment so that the layers can be converted into the desired NiAl or CoAl phases (similar as for pack cementation low-activity coatings). Modification of coatings applied by this slurry technique with other elements such as chromium and silicon is also possible by alloying with aluminum particles used for the slurry or by mixing two metallic powders. Most recently, even low-activity coatings have been manufactured by the slurry technique in one step, as described in detail elsewhere [43].

Figure 7 shows an overview of application methods for diffusion coatings, including aluminum foil, galvanic or dip coatings, whose formation mechanisms can be compared to slurry coatings where a certain amount of aluminum is directly deposited on the surface and heat-treated. However, since it is much more difficult to apply homogeneous layers, such techniques are hardly used compared with the slurry technique.

4. Overlay coatings

4.1. MCrAlY-coatings

In contrast to diffusion coatings, no elements of the substrate are incorporated in overlay coatings. Therefore, such coatings offer the possibility to apply totally different compositions compared to the base materials, and their properties can be perfectly optimized to fulfill the requirements listed in Table 1, such as being corrosion-resistant as well as mechanically and thermally compatible with the substrate. Although this flexibility allows a wide range of compositions, almost all systems for superalloys are based on a general MCrAlY composition and contain usually more than four elements with M = Co or Ni or a mixture of them plus aluminum, chromium, and an element from the group of the reactive elements such as Yttrium (MCrAlY is the common term for such types of coatings and Y stands for Yttrium) [12]. One requirement is an Al content of 10-12 wt%, which is less than in the diffusion coatings. However, a higher chromium content favors alumina formation and makes such systems reliable alumina formers. The still rather high amount of aluminum in MCrAlY coatings for superalloys forms intermetallic phases, but in this case only the β-NiAl or CoAl phase is present in the coatings. In contrast to the diffusion coatings, such phases are surrounded by a metallic nickel- or cobalt-based gamma solid solution matrix. The resulting two-phase microstructure β+γ (see Figure 11, left) increases the ductility of the coating over purely intermetallic coatings and thereby gives higher thermal fatigue resistance.

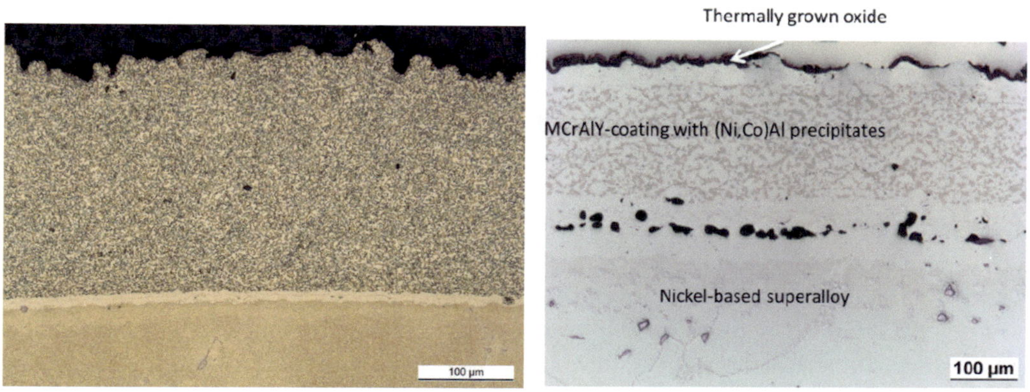

Figure 11. NiCoCrAlY coating on nickel-based superalloy; on the left a light microscopy picture after manufacturing after application [45], on the right a MCrAlY oxidized for 500 h at 1050°C in air is shown. Close to the surface the β-precipitates are dissolved and at the interface interdiffusion and interdiffusionpores (Kirkendall-pores) can be seen

One other advantage over diffusion coatings is the significant amount of Cr in the MCrAlY coatings, which enhances the corrosion and oxidation resistance. Regarding only oxidation resistance, NiCrAlY shows the best protection, but when sulfur is present in the gas, Co-based MCrAlY systems have a higher resistance. Additionally, NiCrAlY shows a phase transition above 1000°C. Therefore, usually well-balanced NiCoCrAlY or CoNiCrAlY compositions are used in industrial applications, also depending on the type of superalloy. Besides the major elements, these systems are enhanced by the addition (<1%) of reactive elements (RE) such as yttrium, cerium, lanthanum, hafnium, or zirconium. The addition of rhenium was also shown to be beneficial [46,47].

When employed at high temperatures, the coatings degrade mainly by Al depletion due to both oxide formation and interdiffusion with the substrate. The beginning of the β dissolution is already visible close to the surface and at the interface to the superalloy after 500 h at 1050°C as shown in Figure 11, right. Similar to diffusion coatings, this aluminum depletion alters the microstructure, as visible close to the surface and at the interface to the substrate. When the β-phase is totally dissolved, the coating quickly loses its protective properties. Compared to diffusion coatings, the composition and performance is optimized, but the manufacturing of overlay coatings is much more expensive. Overlay coatings are produced by one of the following processes: Physical vapor deposition (PVD) process, thermal spraying process, or overlay welding (also called cladding). For MCrAlY overlay coatings, mainly thermal spray processes or rarely PVD processes are used.

Thermal spraying processes are a group of surface coating processes in which a spray material is partly or fully melted inside of a spray gun (electrical arc discharge is usually used as the source of energy) and accelerated toward the surface of the component to be coated in the form of micrometer-size particles. The resulting coatings are formed by the accumulation of numerous sprayed particles. Adhesion occurs primarily due to a mechanical interlocking with the surface roughness of the substrate, while metallurgical processes hardly take place and a subsequent heat treatment is required. When MCrAlY powders are sprayed in the molten state, aluminum and chromium are likely to oxidize "in-flight." Therefore, only the following three thermal spray methods are commonly used to coat superalloys with MCrAlYs: Low pressure plasma (LPPS), vacuum plasma (VPS), and high velocity oxyfuel (HVOF) spraying. Atmospheric plasma spraying (APS) is only used for the application of zirconia thermal barrier coatings for turbine blades, as discussed later. For plasma spray processes an anode and a cathode are incorporated into a spray gun producing an electric arc to ionize the operating gas. The dissociation and ionization produces heat that allows melting and spraying even refractory elements or ceramics. Several gases can be used, but typically one of the following is chosen: argon, hydrogen, helium, or their mixtures. When the powder is injected in the plasma jet, it is melted by the high temperature of the plasma torch and also propelled toward the substrate. Because of the required low pressure or inert atmosphere for MCrAlYs, the effort and cost for this process is high. High-speed processes such a high-velocity oxy-fuel spraying (HVOF) are much more efficient and have earned high industrial acceptance. By increasing the particle velocity during spraying, unwanted oxidation reactions can also be minimized without the requirement of operating in vacuum [48]. Furthermore, due to the high deformation during the solidification process in the HVOF process, very low porosities can be achieved (<1%). Another option is Physical Vapor Deposition (PVD) processes [12,49]. PVD processes are conducted in high vacuum to avoid reactions between the metal atoms and the process gas. In order to transform the coating material into the gas phase, generally two methods are employed: heating until high vapor pressures occur; or bombardment with high energetic electrons or ions to release atoms from the target. For the components of high temperature coatings, which usually have high evaporation temperatures, the second method is preferred. Various modifications of this method exist, but the electron beam heating is the most well-established one. Concentrated electron beam rays induce a high energy density on the surface of the target so that locally extremely high temperatures occur and even refractory alloys or ceramic material can be evaporated. In order to apply alloys such as MCrAlY coatings, it is important to note that the partial pressures of the different elements are very different. The composition of the melt bath or of the master alloy has to be adapted with respect to the difference in the vapor pressures during the coating process. It can be calculated, but it must be considered that they change even during the process. Especially in the beginning of the coating process, the parameters are far from equilibrium and therefore cannot produce the desired composition. For metallic alloys, different elements have to be added during the process to regulate and adapt the master alloys in a way that the manufactured coating always has the right composition. Secondly, the achieved microstructure is highly textured, with

http://dx.doi.org/10.5772/61141

elongated grains oriented perpendicular to the surface of the substrate. This microstructure and process is of highest technical relevance for two-component ceramic yttrium oxide-zirconium oxide layers, serving as thermal barrier top coats for superalloys.

4.2. Thermal barrier coatings

Thermal barrier coatings (TBC) are used to reduce the heat flow into the metal below in order to reduce oxidation rates and to protect it from thermal softening and accelerated creep. The main application of these systems is in stationary gas and aircraft turbines. Modern TBC systems decrease the temperature load in the superalloys beneath by 200°C, or vice versa allow an increase in operating temperature in that range to improve the turbine efficiency. TBCs consist of a combination of a bond coat which can be either a MCrAlY overlay coating or a diffusion layer with a ceramic top coat. A schematic TBC system is given in Figure 12. It shows the temperature gradient and the typical configuration on a superalloy used, e.g., as a turbine blade.

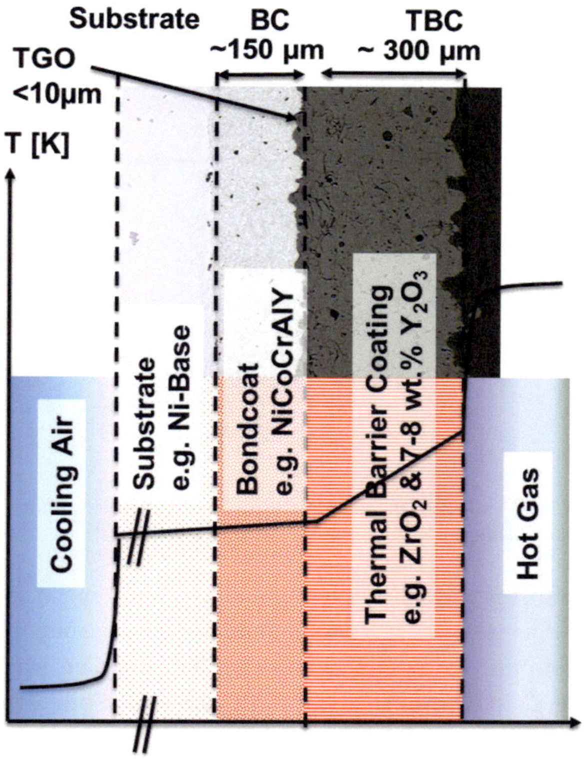

Figure 12. Different layers of a thermal barrier system and the typical temperature gradient

All classical TBC systems, which have been developed and improved throughout the last 30 years, rely on a partially stabilized zirconia as ceramic top coat, which offers a very good combination of necessary properties [46]. Zirconia was chosen because of its low thermal conductivity in combination with a rather high thermal expansion coefficient closer to metals than that of most other ceramics [50]. In addition, it is compatible with the thermally grown alumina oxides (TGO) scales on top of the bond coat. One challenge is that ZrO is allotropic and shows three phase modifications (Figure 13).

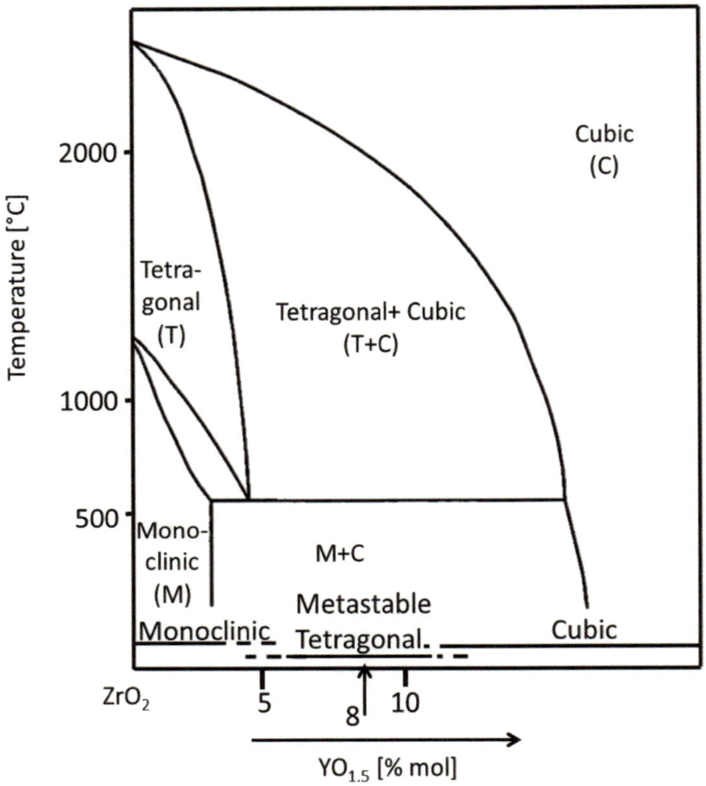

Figure 13. Phase diagram of ZrO2-Y2O3 showing the allotropy of zirconia and the technically used range in which the tetragonal lattice modification remains stable

Especially the phase transformation from tetragonal to monoclinic at about 1170°C that goes along with a volume change of about 3-9% would induce detrimental cracks. This phase transformation can be suppressed by adding oxides of Yttirum, Cer, Magnesium, Niobium, or Calcium [51]. In Figure 13, the impact of yttrium addition on the phase formation is shown. In commercial technical systems, about 8 % YO1.5 is added to receive metastable partially stabilized tetragonal zirconia and suppress detrimental phase transformations. Another modification that is technically used is MgO-stabilized zirconia, which requires about 20-25 wt.% MgO. This zirconia slowly destabilizes above about 1000°C and can only be used for components of diesel engines and not in turbines. Any phase destabilization or an increased sintering behavior at higher temperatures determines the upper temperature limit of the

http://dx.doi.org/10.5772/61141

application of zirconia, because it destroys the necessary porous structure and induces cracks. Such limitations trigger the investigation of other ceramics for even higher-temperature applications than the systems today allow. Especially several ziconates with a low thermal conductivity are looked at, of which the most promising candidate is gadolinium zirconate. Such coatings are developed as a two-layer ceramic system with classical zirconia under the novel ceramics [52]. Due to the fact that ceramics are prone to tensile strains, the microstructure has to be designed carefully in order to allow at least a certain strain tolerance. The ceramic layers today are either manufactured via atmospheric plasma spraying (APS) or electron beam physical vapor deposition (EB-PVD). In both cases, a certain strain tolerance is achieved by adjusting the microstructure [53].

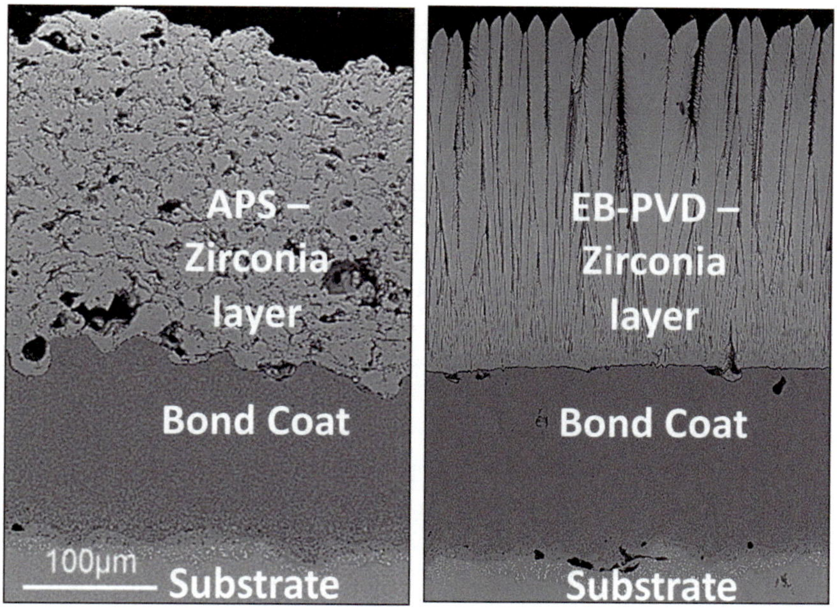

Figure 14. Comparison of TBC systems with APS-sprayed zirconia and EB-PVD zirconia top layers showing the different morphology of the ceramic coating on top of a sprayed MCrAlY bond coat on the left and a pack cementated bond coat on the right

In Figure 14, on the left the APS-sprayed microstructure is schematically shown. It possesses a pancake-like structure. The strain tolerance is created by a heavily intertwined network of fine cracks. This network separates the ceramic coating into "segmented flakes" providing a higher tolerance upon strain application through the temperature gradient or difference in thermal expansion to the material below. EB-PVD (Figure 13 on the right) coatings always exhibit the typical columnar structure. This method has the big advantage that, during the EB-PVD process, the crystals of the scales grow epitactically and perpendicular to the surface. The ceramic columns are separated from each other by small gaps, providing an exceptional tolerance to deleterious tensile strains parallel to the surface. These coatings are preferentially used for aircraft turbine blades, while for stationary turbine blades or aircraft nozzle parts often APS coatings are employed, which can be produced at lower cost. One drawback of EB-PVD zirconia top coats is that they are sensitive to calcium-magnesia-alumina silicates (CMAS) attack [54]. CMAS can derive from the use of turbines in sandy desert areas or when volcanic ash enters a turbine. The CMAS melt in the heat chamber and are deposited on the turbine blades. When they penetrate into the TBC, they cause failure and spallation. Other factors with an influence on the lifetime are hot corrosion mechanisms by sulfur, calcium, or vanadium deposits on the coatings that can reduce the lifetime [55], often by reacting with the stabilizers (e.g., yttrium), thereby destroying the resistance against phase transformation of the zirconia. When the ceramic scale remains intact, in service an alumina scale grows slowly under the zirconia ceramic. This thermally grown oxide scale (TGO) must not exceed about 3 μm in thickness to guarantee mechanical integrity and avoid spallation [56].

5. Outlook

By using diffusion coatings, overlay coatings, and ceramic top coats for high temperatures, the otherwise often very short lifetime of unprotected materials can be enhanced and makes processes and applications possible for which otherwise no material is available. At the same time, high-temperature processes have been constantly changing due to changes in fuel or operating conditions, and the thermal operation limit of superalloys has continuously increased over the years. In the future, even systems such as molybdenum-based alloys might become interesting with a thermal application potential well above the nickel- and cobalt-based superalloys of today [9]. Such developments induce also a strong driving force to further develop high-temperature coating systems in order to keep pace with such increased operating conditions, and to allow efficient and reliable operation of metallic high-temperature materials.

Author details

Mathias C. Galetz

Address all correspondence to: galetz@dechema.de

DECHEMA-Forschungsinstitut, High Temperature Materials, Frankfurt am Main, Germany

References

[1] R.C. Reed. *The Superalloys: Fundamentals and Applications,* Cambridge University Press; 2006, 283-335

[2] R. Subramanian, A. Burns, W. Stamm. Advanced Multi-Functional Coatings for Land-Based Industrial Gas Turbines. Proceedings of ASME Turbo Expo 2008: Power for Land, Sea and Air. ASME. 2008, 1-10.

[3] D.J. Young. *High Temperature Oxidation and Corrosion of Metals*, Elsevier; 2008, 139-455.

[4] P. Kofstad. *High Temperature Corrosion*, Essex UK: Elsevier Applied Science; 1988, 139-181.

[5] N. Vialas. Substrate effect on the high temperature oxidation behavior of a Pt-modified aluminide coating. Part II: long-term cyclic-oxidation tests at 1050-¦C, *Oxidat Metals*. 68 (2007) 223-242.

[6] S. Dryepondt, D. Monceau, F. Crabos, E. Andrieu. Static and dynamic aspects of coupling between creep behavior and oxidation on MC2 single crystal superalloy at 1150 -¦C, *Acta Materialia*. 53 (2005) 4199-4209.

[7] V. Seetharaman, A.D. Cetel. *Superalloys*, TMS, 2004, pp. 207-214.

[8] A. Srivastava, S. Gopagoni, A. Needleman, V. Seetharaman, A. Staroselsky, R. Banerjee. Effect of specimen thickness on the creep response of a Ni-based single-crystal superalloy, *Acta Materialia*. 60 (2012) 5697-5711.

[9] M. Heilmaier, M. Krüger, H. Saage, J. Rösler, D. Mukherji, U. Glatzel, R. Völkl, R. Hüttner, G. Eggeler, C. Somsen, T. Depka, H.J. Christ, B. Gorr, S. Burk, Metallic materials for structural applications beyond nickel-based superalloys, *J Minerals Metals Mater Soc*. 61 (2009) 61-67.

[10] H. Harada. Superalloys: Present and Future. Current Topics on High Temperature Materials, JSPS Report of the 123rd Committee on Heat Resisting Materials and Alloys. 48 (2007) 357-363.

[11] C.T. Sims. A History of Superalloy Metallurgy for Superalloy Metallurgists. In: M. Gell, C.S. Kortovich, R.H. Bricknell, W.B. Kent, J.F. Radavich (eds.). *Proceedings of Superalloys 1984*. Warrendale, Pennsylvania, USA: The Metallurgical Society of AIME; 1984.

[12] G.W. Goward. Progress in coatings for gas turbine airfoils, *Surf Coatings Technol*. 108-109 (1998) 73-79.

[13] R.C. Clarke, M. Oechsner, N.P. Padture. Thermal-barrier coatings for more efficient gas-turbine engines, *MRS Bull*. 37 (2012) 891-898.

[14] S. Poupard, J.-F. Martinez, F. Pedraza, Soft chemical stripping of aluminide coatings and oxide products on Ni superalloys, *Surf Coatings Technol*. 202 (2008) 3100-3108.

[15] H. Jehn. High temperature behaviour of platinum group metals in oxidizing atmospheres, *J Less Common Metals*. 100 (1984) 321-339.

[16] C.E. Birchenall. Diffusion in Oxides: Assessment of Existing Data and Experimental Problems. Proceedings of the Symposium "Mass Transport in Oxides," October 1967. 1967, p. 119-127.

[17] N. Birks, G.H. Meier, F.S. Pettit. High Temperature Corrosion. In: J.K. Tien, T. Caulfield (eds.). *Superalloys, Supercomposites and Superceramics*. Academic Press; 1989, p. 439-489.

[18] A.S. Khanna. *Introduction to High Temperature Oxidation and Corrosion*, ASM International; 2002, 203-206.

[19] C. Xu, W. Gao. Pilling-Bedworth ratio for oxidation of alloys, *Mater Res Innov*. 3 (2000) 231-235.

[20] M. Schütze. Mechanical aspects of high-temperature oxidation, *Corr Sci*. 35 (1993) 955-963.

[21] M. Schütze. Mechanical properties of oxide scales, *Oxidation Metals*. 44 (1995) 29-61.

[22] Allison H.B.C, L.A. Hawkins. *Gen Elect Rev*. 17 (1914) 947-951.

[23] F.C. Kelley. Chromium Impregnation, *Metals Handbook*, American Society for Metals, Cleveland, Ohio; (1948) 705-706.

[24] P.A. Choquet, E.R. Naylor, R.A. Rapp. Simultaneous chromizing and aluminizing of iron-base alloys, *Mater Sci Engin A*. 121 (1989) 413-418.

[25] F.D. Geib, R.A. Rapp. Simultaneous chromizing-aluminizing coating of low-alloy steels by a halide-activated, pack-cementation process, *Oxidation Metals*. 40 (1993) 213-228.

[26] W. Da Costa, B. Gleeson, D.J. Young. Codeposited chromium-aluminide coatings II. Kinetics and morphology of coating growth, *J Electrochem Soc*. 141 (1994) 2690-2697.

[27] W. Da Costa, B. Gleeson, D.J. Young. Codeposited chromium-aluminide coatings I. Definition of the codeposition regimes, *J Electrochem Soc*. 141 (1994) 1464-1471.

[28] B. Gleeson, W.H. Cheung, W. Da Costa, D.J. Young. Co-deposited chromium-aluminide coatings and their high-temperature corrosion resistance. Proceedings of the 8th Asian-Pacific Corrosion Control Conference, Bangkok. 2000, p. 43-48.

[29] R. Bianco, R.A. Rapp. Pack cementation aluminide coatings on superalloys: codeposition of Cr and reactive elements, *J Electrochem Soc*. 140 (1993) 1181-1190.

[30] Z.D. Xiang, P.K. Datta. Formation of Hf- and W-modified aluminide coatings on nickel-base superalloys by the pack cementation process, *Mater Sci Engin A*. 363 (2003) 185-192.

[31] B.A. Pint, J.A. Haynes, T.M. Besmann. Effect of Hf and Y alloy additions on aluminide coating performance, *Surf Coatings Technol*. 204 (2010) 3287-3293.

[32] M.C. Galetz, X. Montero, H. Murakami. Novel processing in inert atmosphere and in air to manufacture high-activity slurry aluminide coatings modified by Pt and Pt/Ir, *Mater Corrosion*. 63 (2012) 921-928.

[33] H.W. Grünling, R. Bauer. The role of silicon in corrosion-resistant high temperature coatings, *Thin Solid Films*. 95 (1982) 3-20.

[34] E.Y. Lee, D.M. Chartier, R.R. Biederman, R.D. Sisson. Modelling the microstructural evolution and degradation of M-Cr-Al-Y coatings during high temperature oxidation, *Surf Coatings Technol*. 32 (1987) 19-39.

[35] V.A. Ravi, P.A. Choquet, R.A. Rapp. Chromizing-aluminizing coating of Ni- and Fe-base alloys by the pack cementation technique. In: T. Grobstein, J.K. Doychak (eds.). *Oxidation of High-Temperature Intermetallics. Minerals, Metals Materi Soc.;* 1989, p. 127-145.

[36] R. Bianco, R.A. Rapp, N.S. Jacobson. Volatile species in halide-activated diffusion coating packs, *Oxidat Metals*. 38 (1992) 33-43.

[37] A. Naji, M.C. Galetz, M. Schütze. Improvements in the thermodynamic and kinetic considerations on the coating design for diffusion coatings formed via pack cementation, *Mater Corrosion* (2015).

[38] D. Fähsing, M.C. Galetz, M. Schütze. Eindiffundieren metallischer Elemente. In: G. Spur, H.-W. Zoch (eds.). *Handbuch Wärmebehandeln und Beschichten*. Carl Hanser Verlag GmbH & Co; 2015, in press.

[39] G.W. Goward, D.H. Boone. Mechanisms of formation of diffusion aluminide coatings on nickel-base superalloys, *Oxidation Metals*. 3 (1971) 475-494.

[40] B. Ganser, K.A. Wynns, A. Kurlekar. Operational experience with diffusion coatings on steam cracker tubes, *Mater Corrosion*. 50 (1999) 700-705.

[41] B.G. McMordie. Oxidation resistance of slurry aluminides on high temperature titanium alloys, *Surf Coatings Technol*. 49 (1991) 18-23.

[42] A.J. Rasmussen, A. Agüero, M. Gutierrez, M.J.L. Östergaard. Microstructures of thin and thick slurry aluminide coatings on Inconel 690, Surf Coatings Technol 202 (2008) 1479-1485.

[43] X. Montero, M.C. Galetz, M. Schütze. Low-activity aluminide coatings for superalloys using a slurry process free of halide activators and chromates, *Surf Coatings Technol*. 222 (2013) 9-14.

[44] M.C. Galetz, X. Montero, M. Mollard, M. Günthner, F. Pedraza, M. Schütze. The role of combustion synthesis in the formation of slurry aluminization, *Intermetallics*. 44 (2014) 8-17.

[45] W. Stamm. Siemens AG, Personal communication 2015

[46] A. Robert. Current status of thermal barrier coatings – An overview, *Surf Coatings Technol* 30 (1987) 1-11.

[47] N. Czech, F. Schmitz, W. Stamm. Improvement of MCrAlY coatings by addition of rhenium, *Surf Coatings Technol*. 68/69 (1994) 17-21.

[48] D. Toma, W. Brandl, U. Köster. The characteristics of alumina scales formed on HVOF-sprayed MCrAlY Coatings, *Oxidation Metals*. 53 (2000) 125-137.

[49] R.D. Jackson, M.P. Taylor, H.E. Evans, X.-H. Li. Oxidation study of an EB-PVD MCrAlY thermal barrier coating system, *Oxidation Metals*. 76 (2011) 259-271.

[50] X.Q. Cao, R. Vassen, D. Stoever. Ceramic materials for thermal barrier coatings, *J Eur Cera Soc*. 24 (2004) 1-10.

[51] R.L. Jones. Some aspects of the hot corrosion of thermal barrier coatings, *J Therm Spray Technol*. 6 (1997) 77-84.

[52] Y. Zhang, J. Malzbender, D.E. Mack, M.O. Jarligo, X. Cao, Q. Li, R. Vaßen, D. Stöver. Mechanical properties of zirconia composite ceramics, *Cera Int*. 39 (2013) 7595-7603.

[53] L. Singheiser, R. Steinbrech, W.J. Quadakkers, D. Clemens, B. Siebert. Thermal barrier coatings for gas turbine applications – Failure mechanisms and life prediction. In: D. Coutsouradis (ed.). *Materials for Advanced Power Engineering*: Proceedings of the 6th Liege Conference, Part III. Kluwer Academic Publishers; 1998, p. 977-996.

[54] N. Chellah, M.-H. Vidal-Sétif, C. Petitjean, P.-J. Panteix, C. Rapin, and M. Vilasi. 8th International Symposium on High-Temperature Corrosion and Protection of Materials, 2012.

[55] B. Bordenet, W. König, G. Witz, and H.-P. Bossmann. 8th International Symposium on High-Temperature Corrosion and Protection of Materials, 2012.

[56] H.E. Evans. Oxidation failure of TBC systems: An assessment of mechanisms, Surf Coatings Technol. 206 (2011) 1512-1521.

Chapter 13

Properties of the Ultrathin Multilayer Ground State of Fe/Pd

A.V. dos Santos

Additional information is available at the end of the chapter

http://dx.doi.org/10.5772/61289

Abstract

The technological evolution in the recent years allowed us to improve computers. Consequently, the numeric calculus with computational methods had a big progress. In theoretical physics that had a benefit with this technology, we can highlight the calculation of the solids' electronic structure. Using the first principles method, LMTO (Linear Muffin-Tin Orbital) with the ASA (Atomic Sphere Approximation) approximation, we will study the band structure in the magnetic multilayers. The choice of these methods was based on two aspects: (a) The computer available to perform the job; a CRAY super computer that belongs to the Supercomputer National Center. Because the LTMO method presents good vectorization, it allows us to perform calculations with many atoms in the unitary cell, which would be too difficult with a smaller computer. (b) The fact that the LMTO method already presented good results in studies like intermetallic alloys and of iron nitride substituted.

Keywords: Band structure, calculation of the first principles, Ultrathin Multilayer

1. Introduction

1.1. Calculation of band structure: A brief history

In this section, a brief history of the calculation of band structure is provided, invoking the main methods used in this way. This does not intend to supplement, but to give the reader a good basis, especially for those who are not experts in the area, to understand the concepts used in this chapter. The understanding of the distribution of electrons in solids gives an appreciation of some of their physical properties. The most interesting properties of the new material industry, such as magnetization, come from the distribution of electrons in a solid. Theoretically, we should treat a system of several electrons distributed randomly in the material. However, we cannot follow this infinite number, and we must use the Born-Oppenheimer approximation. We treated electrons and the nuclei as if they were separate from each other. So, we will calculate the energy of the ground state, depending on the position of the atomic nuclei. This implies that the basic problem is to calculate the steady states of the interacting electrons system moving in a periodic electrostatic field originated by the fixed nuclei i.e., to obtain the electronic structure of solids and consequently the structure of the same band. The identification of the electronic structure of solids involves considering an infinite number of the interacting fermions. But what does this mean? This leads us to solve the Schrödinger equation for an electron moving in an average field of other electrons over the field of the nuclei. Therefore, the field is determined by the distribution of the electronic charge, the adjustment for correlation, and the exchange effect. And it is usually calculated self-consistently as we shall see later on.

In the recent years and with the advent of powerful computers, which are faster and more accurate. Furthermore, the appearance of linear methods [01- 02] for calculating the structure of the ordered solid bands. We can see that some of the main calculation methods had a breakthrough. It comes from a theoretical but also a practical point of view because there was a technological breakthrough in providing fast and accurate computers. We can then identify the physical properties with greater ease and certainty of outcomes [05-06]. An important work of Hohenberg-Kohn-Sham [07-08] provides the theoretical base pair to the other methods because they teach us to work with the mean-field theory using the electron density function and the energy of the ground state.

Other methods have also emerged with the DVM [10] ("Discrete Variational Method"), SPR-LMTO [11] ("Spin Polarized Relativistic Linear Muffin-Tin Orbital"), and the FLMTO [12] (TUB Linear Muffin-Tin Orbital "). Currently, the first purest strains of methods are employed to study the defects in the crystal lattice [13].

The calculation of the first principles band structure is a very important tool in the study of thin films and multilayers. Such systems are investigated by different techniques, such as X-ray and Mossbauer duration or magnetization measurements [15-19].

In the modern calculation of the first principles applied to magnetic multilayers, Jarlborg and Freeman [21, 22]did some work that can be used as a primary reference. These calculations were motivated by the comments made by TJ Taler et al. [23] Ni / Cu alloys use ferromagnetic resonance. However, it should be considered that at that time, there was a difficulty in the experimental physics to build very thin layers (of the order 5-15 angstroms), which made the comparison between the calculations and the very difficult experience.

More recent studies using the first principles methods, such as performed by Blügel et al. [24] using the FLAPW ("Full Linear Augmented Plane Waves") that presented a good agreement with the experimental work carried out by Celinski et al. [19] In the case of magnetic multi-layers, in the recent years, there has been a promising advance in both the theoretical and experimental study, which can be checked in fairly comprehensive references on the subject [25]. However, it is necessary to continue the theoretical investigation of the magnetic properties and the structural ultrafine multilayer because there are still questions to be answered about the charge transfer, magnetic moments, and the "stress" in the interface region, which causes changes in the density of the states. Another problem to be studied is the multilayer stability into the very thin structures. Properties, such as the hyperfine field and isomer shift of multilayer stability, are also a point of interest. In this work, we will hold extremely thin multilayer calculations worrying about the interface region between the materials that make up the multilayer.

2. Introduction to calculate the electronic structure of ultrathin multilayers

In this chapter, we will investigate the magnetic properties of ultrathin multilayers and hyperfine Iron and Palladium (Fe / Pd). The multilayer research is done in a few different stoichiometries in order to have a better understanding of their magnetic properties and behavior in the pressure. We begin by studying the electronic structure of the ultrathin multilayers of Fe and Pd, with a composition of 50% of each element. Below, we will investigate the behavior of two different systems: the dual system ultrathin multilayers of iron and a palladium (2 Fe / Pd), and the inverse system (2 Pd / Fe). This enables a comparison between these three different systems, which will show a clear change in the band structure of the systems when we change its stoichiometric structure. The theoretical research for both Fe / Pd ultrathin multilayers, as in the case of the 2 Fe / Pd and 2 Pd / Fe systems, is made using the LMTO, method of Andersen [01-02], and approximation of the atomic spheres (ASA). To study the electronic structure of multilayer systems, in the ferromagnetic case, we performed the calculations with a spin polarization and the parameterization of von Barth and Hedin [36] for energy correlation and exchange of an electron gas. In the self-consistent calculations for the non-magnetic case, we used the Hedin-Lundqvist potential [34]. This LMTO calculation does not include the spin-orbit interaction as it is very small and put term fixes combined [01]. To solve the equation, Schrödinguer used the s atomic orbitals that present the results. We then started with the calculations of the structure of the Fe / Pd ultrathin multilayers.

3. Magnetic structure and electronics of the Fe/Pd ultrathin multilayers

Calculations were performed using a structural model in which the bilayers grow in the direction (0,0,1) with a packaging sequence ABA type with tetragonal structure. The ratio of the axes was $c / a = 1.41$ with the Fe atoms occupying the position (0,0,0) and Pd atoms at (0,1/2,1/2). We chose the muffin tin-spheres involving atoms with the same radius for both the Fe and the Pd atoms. We performed self-consistent calculations for various lattice parameters through which we obtained the amount of the theoretical equilibrium. Figure 1 shows the total energy as a function of lattice parameter for the ferromagnetic and nonmagnetic states. We noted that the stability of the ferromagnetic state in relation to the non-magnetic state is good because the energy difference in the volume balance between the ferromagnetic state and the non-magnetic is $\Delta E = -22.98$mRy by atom. The lattice parameter in the balance was calculated using a third-degree polynomial, which was the best fit of the plotted points for both the ferromagnetic state and the non-magnetic state. We obtained the lattice parameters a = 7,188u.a. for the ferromagnetic state and a = 7,044u.a. for the non-magnetic state. The value found for the ferromagnetic state is greater than the one found for the ordered alloy of 50% Fe and 50% Pd on self-consistent calculations. With the growth of the interatomic distances, Fe / Pd bilayers, and the consequent reduction of metal-metal interaction, such a system shows a slightly different magnetic behavior of the pure materials.

All tables are taken from the PhD thesis in reference 41. In table 1, we have the magnetic moment on the Fe site equal to 2,870μB in the Fe/Pd ultrathin multilayers. We noted that there was an increase in the magnetic moment on the website of Fe, compared to the pure iron that is 2,217μB [41].This is in agreement with the experimental work that measured the magnetic moments in the Fe / Pd multilayer [19].Here, we noticed a strong magnetic moment featuring a super magnetic alloy. In Table 1, it is easy to see that the greatest contribution of the electrons in the Fermi level N (E_F) is the electrons d-down the Fe site due to the loads transferred to this site.

	Fe			**Pd**		
	Up	**Down**	**Up - Down**	**Up i**	**Down**	**Up-Down**
N	5.4686	2.5986	2.8700	5.1563	4.7759	0.3804
n_s	0.3318	0.3275	0.0043	0.2824	0.2997	-0.0173
n_p	0.3718	0.3707	0.0011	0.3048	0.3326	-0.0278
N_d	4.7650	1.9004	2.8646	4.5691	4.1436	0.4255
$N(E_F)$	1.8851	13.8763		1.4616	3.5053	
$Nt(E_F)$		15.7614		4.9669		
NC(ef)				20.7283		
v				3.5924		
ΔQ		0.06744			-0.06744	
E_F				0.6532		

Table 1. Parameters calculated for the spin polarization ofFe / Pd bilayers using a self-consistent potential. Here, n is electron / spin; N in states / atom spin Ry; Nc states / Ry unit cell; y in mJ / molK2; EF Ry and ΔQ into electron.

4. Density states of the Fe/Pd ultrathin multilayers

The calculation of the density of states of d in the bilayers of the Fe and Pd sites are shown in Figures 2 and 3, respectively. We noted that there is a narrowing of the bands of the ultrathin multilayers compared to a league and the ordered FePd. Therefore, the metal-metal interaction causes a small change in the spin up and down states of the Fe and Pd metals. Figure 3 noted that the Pd site spin up and spin down states are busy unlike the one found for the DOS-d Fe site, which has many empty d spin-down states above the Fermi energy. In the Fe site, there are magnetic moments as the area below the Fermi energy is different between the up and down states, which is largely featuring a high magnetic moment in this place. This is in

http://dx.doi.org/10.5772/61289

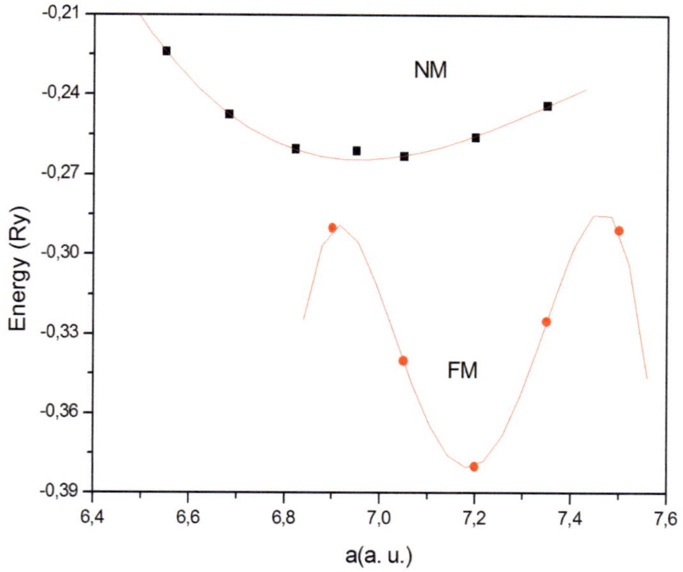

Figure 1. Total energy as a function of lattice parameters, the Fe / Pd bilayer. We did not present the error because they are smaller than the points.

agreement with the results in Table 1. The Fe-Pd interaction decreases the occupation of the d-down states in the ultrathin multilayers compared to the Fe-Pd and Fe nitrides 4PdN.

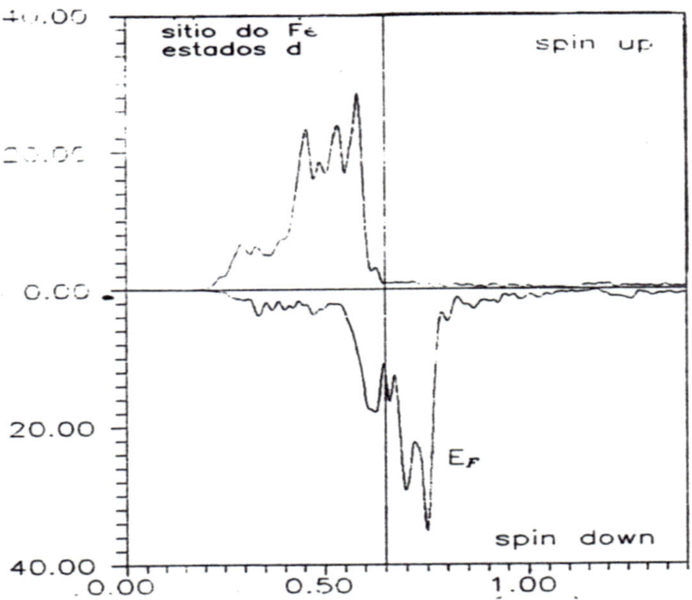

Figure 2. Density of states designed to electrons with spin up and down for d in the Fe / Pd bilayer in the Fe site.

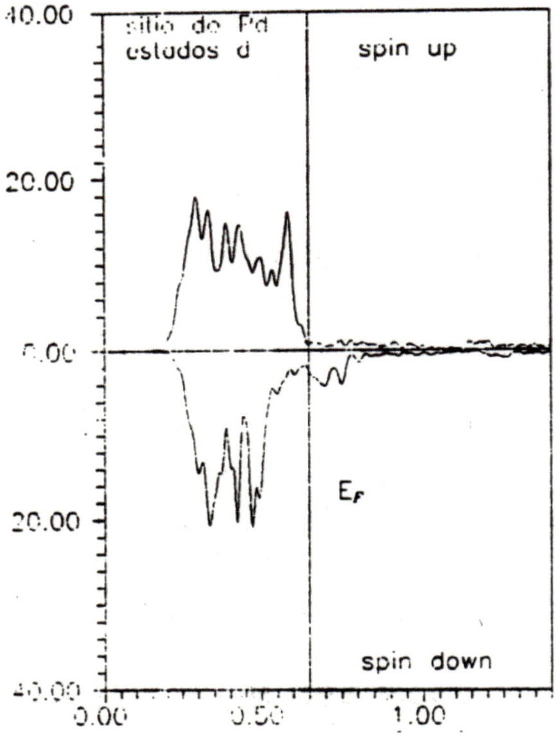

Figure 3. Density of states designed to electrons with spin up and down states for d in the Fe / Pd bilayer in the Pd site.

5. Pressure influence on the magnetic properties of the Fe/Pd ultrathin multilayers

Now, we investigate the behavior of the magnetic properties against pressure. The pressure effect is simulated by reducing the spacing of the lattice. We perform these self-consistent calculations for some lattice parameters of the Fe / Pd ultrathin multilayers, but we keep the tetragonal structure unchanged. The results of these calculations are shown in Figure 4. In this figure, we seethe variation of the magnetic moment on the site of the Fe and Pd due to the lattice parameter. In the Fe site, the magnetic moment decreases until the lattice parameter which sharply drops to zero. In the Pd site, the magnetic moment remains constant until it also falls abruptly to zero in the same lattice parameter a = 6,321ua. We can associate this lattice parameter limit value at a certain critical pressure. This type of behavior also happens in other leagues, known as Invar [56,57,60,61] alloys. Similar behavior of the magnetic moment has been obtained for iron carbides in ferromagnetic calculations [61]. We have published data from experimental studies on this fact in ultrathin multilayers of intermetallic materials systems.

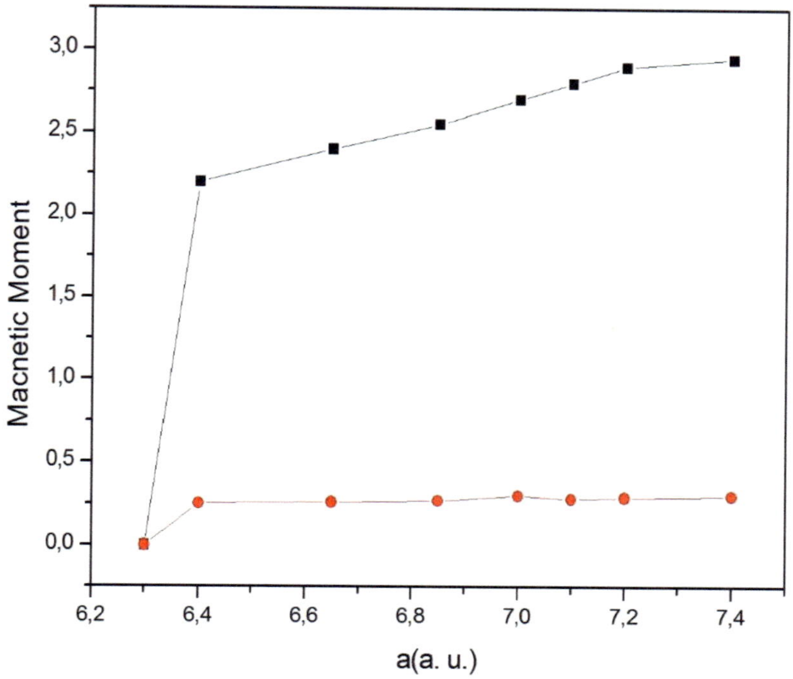

Figure 4. Magnetic moment due to the lattice parameter for the Fe / Pd, Fe, and Pd monolayers on site.

Fe				**Pd**		
	Up	**Down**	**Up-Down**	**Up i**	**Down**	**Up -Down**
n	4.1460	4.1459	0.0001	4.8537	4.8537	0.0000
n_s	0.3136	0.3136	0.0000	0.2841	0.2841	0.0000
n_p	0.4168	0.4168	0.0000	0.3672	0.3672	0.0000
N_d	3.4156	3.4155	0.0001	4.2024	4.2024	0.0000
$N(E_F)$	12.5957	12.6732		5.6340	6.6350	
$Nt(E_F)$		25.2689			11.2670	
$Nc(E_F)$			36.5359			
y			6.3320			
ΔQ		0.2921			-0.2921	
E_F			1.1399			

Table 2. Parameters calculated for spin polarization potential monolayers using self-consistent calculations. Here, n is in electron / spin; N in states / atom spin Ry; Nc states / Ry unit cell, y in mJ / molK2, EF Ry and ΔQ electrons. We obtain the critical pressure Pc = 109Kbar.

This result is of the order of the critical pressure of league ordered Fe-Pd [55], and also comparable to the theoretical and experimental results for Fe4Ni [59], Fe4PdN [57], and Fe_4SnN [60], which are made known to have the Invar behavior type and magnetic collapse with pressure feature. The issue of electronic redistribution due to this transition from the ferro-magnetic state to the non-magnetic state will be discussed. To do so, we focus our attention in table 2, which shows some parameters obtained through the self-consistent calculation of the lattice parameter of the magnetic collapse shown in Figure 4. A comparison of Tables 1 and 2 show a large increase in the charge transfer to the site Fe. This excess charge will populate the state d spin down the Fe site, and also, there is an inversion of spin up electrons in the electron spin down that will cause the magnetic collapse. The results presented in Table 2 show that for high pressures, there is a redistribution. But this is not merely an average of the spin up and down, which existed before the occupation in Table 1. This redistribution is a consequence of the strong metal-metal interactions in front of the reduction of the lattice spacing. On the other hand, Table 2 shows an increase in the specific heat coefficient y. This value corresponds to a large number of states at the Fermi energy for the non-magnetic state. This growth occurs mainly in the up states, both in the Fe and the Pd sites.

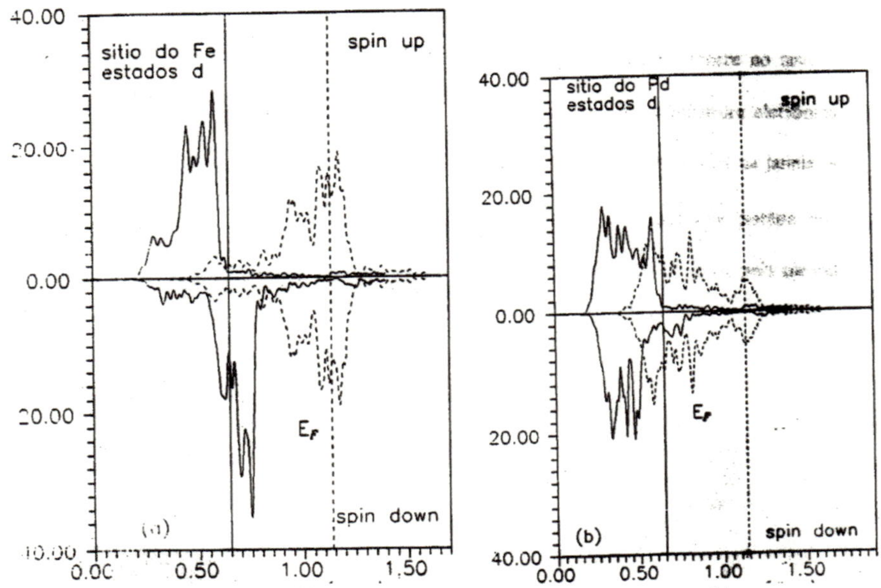

(a) (b)

Figure 5. a -The DOS in the Fe site in the non-magnetic state (dotted lines) and in the ferromagnetic state (solid lines) are presented. b–The density of states designed for electrons with spin up and down states for d in the Fe / Pd bilayers: (a) at the Fe site in balance volume (solid lines), the magnetic collapse (dotted lines) and (b) Pd site in volume balance (solid lines), the magnetic collapse (dashed lines).

We noted that there was a widening of the band - d of the non-magnetic state in relation to the ferromagnetic state. Also, the DOS moves as a whole to higher energies, and the Fermi energy increases.

In Figure 5b,we see the DOS in the non-magnetic (dashed lines) and ferromagnetic (solid line) states to the site of Pd. This site also promotes the DOS to a higher energy. Even in the magnetic collapse, some free states remain above the Fermi energy at the Fe site.

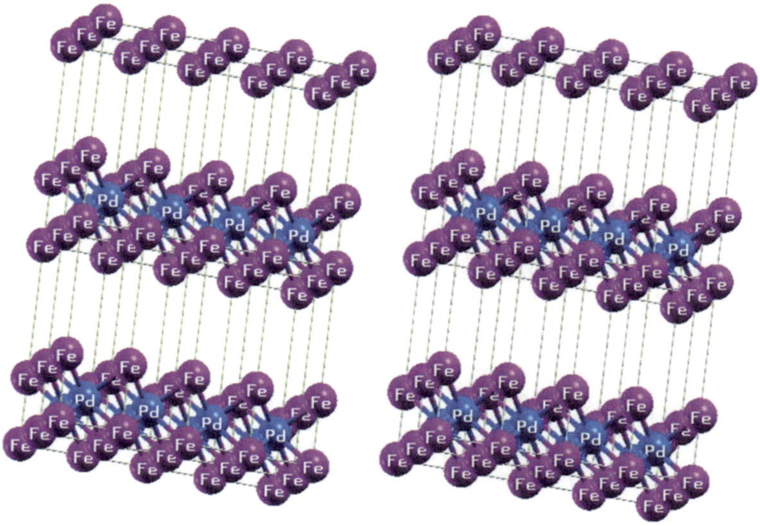

Figure 6. The unit cell of 2Fe / 2 Pd and Pd / Fe ultrathin multilayers.

6. The electronic structure of 2Fe/Pd and 2Pd/Fe ultrathin multilayers

In the rest of this chapter, we will study the 2 Fe/Pd and 2 Pd/Fe ultrathin multilayers. To accomplish the calculation of the band structure of the ultrathin multilayers of Fe-Pd, we introduced some modifications to the unit cell structure so we can have a better description of the actual physical system (1/2, 1/2, 0) for Pd and (0, 0, 0), (0, 1/2, 1/2) to the Fe atoms as bilayers in Figure 6. The structure used herein is a tetragonal structure of c / a = 2.23.

With this structure, we note that all Fe atoms have the same vicinity in the 2 Fe / Pd system. The above two systems used 1500 points in the power window and 1330 points k in the reciprocal space (we changed the number of points because we changed the crystal lattice). In the rest of this chapter, we will present the results of the calculations of the electronic structure and some comparisons between the systems. In the magnetic properties of the multilayer 2 Fe / Pd and 2 Pd / Fe, we performed the calculations of the total energy for some lattice parameters. With these calculations, we obtained the volume of theoretical equilibrium for both systems. Figure 7a shows the total energy states for the non-magnetic (NM) and ferromagnetic (FM) states of 2 Pd / Fe, depending on the lattice parameters. We used a three-degree polynomial to get the best fit curve. Figure 7b has the total energy as a function of the lattice parameter for the non-magnetic states (NM) and ferromagnetic (FM) states for the 2 Fe / Pd system.

We have noted that in both systems, there is a good stability of the ferromagnetic state in relation to the non-magnetic state. The energy of the difference in the volume balance between the ferromagnetic and non-magnetic state is ΔE = -13mRy by atom for the 2 Fe / Pd system. While in the 2 Pd / Fe system, the difference is even greater: ΔE = -20,7mRy by atom. Compared with the bilayer of Fe / Pd, we noted that they are more stable, since ΔE is even higher than the two systems: 2Fe / 2 Pd and Pd / Fe. These differences will be further explored when we discuss the influence of pressure on the magnetic properties.

The minimum total energy shows that the lattice parameter in the balance is 7,069ua for the ferromagnetic state and 6,890u.a for the non-magnetic state in the 2 Fe / Pd system. The lattice parameter for the ferromagnetic state was equal to 7,297uaand 7,209ua for the non-magnetic state. Thus, it is evident that if we put more layers of Pd, there will be an expansion of the Fe-Pd system, since the Pd atom is larger than the Fe atom. The magnetic moment calculated for the Fe site for both systems is high. We noted that the three systems studied in this chapter have magnetic moments on the site Fe, greater than the pure Fe, which is 2,217 μ_B. This is already known in the literature. Furthermore, this result agrees with the experimental [19] and theoretical studies performed by Richter et al. [29] to calculate the magnetic moment of Fe in place relative to the thickness of the multilayer. This suggests that increasing the number of Pd layers in the system increases the magnetic moment on the site Fe. Now, if we compare the magnetic moment on the Fe site with certain ordered alloys, we can note that it is larger and even greater than in nitrides [34,35,36,38,41]. In the Pd site, the magnetic moment has a different behavior from that found on the website of Fe, and the largest magnetic moment is in the 2 Fe / Pd system.

	Fe			**Pd**		
	Up	**Down**	**Up-Down**	**Up**	**Down i**	**Up-Down**
n	5.3700(5.5074)	2.6704(2.5247)	2.7096(2.9697)	5.1263(5.1523)	4.7863(4.8312)	0.3400(0.3211)
n_s	0.3268(0.3327)	0.3279(0.3327)	-0.00 11 (-.0043)	0.2834(0.2828)	0.2991(0.2949)	-O.OI57(-O.OI21)
n_p	0.3668(0.3676)	0.3756(0.3676)	-0.0088(-.0022)	0.3097(0.2995)	0.3291(0.2995)	-0.0194(-0.0091)
Nd	4.6864(4.8071)	1.9669(1.8309)	2.7195(2.9762)	4.5332(4.5700)	4.1581(1.5197)	0.3751(0.3423)
N(EF)	1.7855(1.2308)	12.6785(18.5056)		4.5332(4.9316)	1.9580(1.5197)	
N_T(EF)	14.4640(19.7364)			6.4913(6.4513)		
N_r(EF)				45.9223(27.3666)		
Y				7.9587(4.7428)		
E_F				0.6789(0.6109)		
ΔQ	0.0453(0.0351)			-0.0892(-0.0161)		

Table 3. Parameters calculated for the spin polarization in ultrathin multilayers volume balance of 2 Pd / Fe (in parentheses) and 2 Fe / Pd (outside the parentheses) using a self-consistent potential. Here, n is electron / spin; N in states / atom spin Ry; Nc states / Ry unit cell; y in mJ / molK2; EF Ry and ΔQ into electrons.

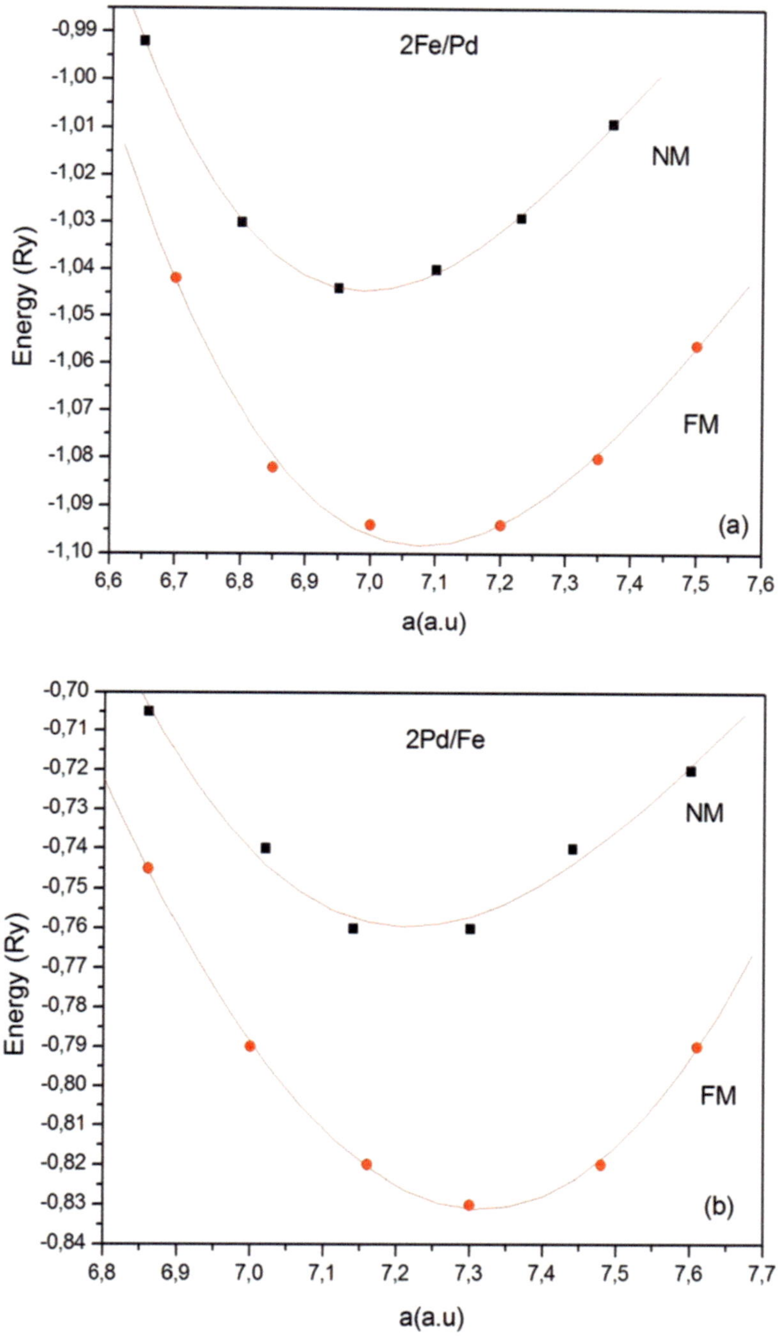

Figure 7. Total energy as a function of lattice parameters, the Fe / Pd bilayer. We did not present the error because they are smaller than the points. According to the energy parameter of the lattice ultrathin multilayers: (a) 2 Fe / Pd and (b) 2 Pd / Fe.

Table 3 -The specific heat coefficient, which is proportional to the total density of states and the Fermi level of the unit cell. We note that the 2Fe /Pd systems and 2 Pd / Fe is greater than that of the Fe / Pd bilayer where we now have a larger number of sites in the unit cell. The 2 Fe / Pd system is greater than 2 Pd / Fe, because the density of the states at the Fermi level is higher at the site of Fe in the Pd site. The Fermi energy is higher in the 2 Fe / Pd system. This shows that there is an increase in the Fermi energy when we increase the number of layers of Fe.

7. State density of the 2Fe/Pd and 2Pd/Fe ultrathin multilayers

The density of states for the electrons Fe and Pd multilayers are shown in Figures 8a, 8b, 9a, and 10b. Figures 8a and 8b show the DOS to the 2Fe / Pd system to the site of Fe (8a) and the site of Pd (8b) that spin in both directions, which is the balancing lattice parameter. In these figures, we noted initially that the Pd site states are all busy for both spin directions, but there are many empty down spin states in the Fe site. It is found in Figures 9a and 9b that DOS-d spin both directions of the equilibrium lattice parameter of the 2 Pd / Fe. In this system, we also found that for the Pd site, there are virtually no empty states above the Fermi energy, but in the Fe site, there are still several unoccupied states above the Fermi energy, as in the 2 Fe / Pd system.

Here, we find a change in the DOS form of a system to another. In the Pd site, there is a reversal peak height of up states in the 2 Pd/ Fe system with the next largest peaks at the Fermi energy. Also, a peak appears on the Pd site in the spin down states in the 2 Fe / Pd system where the area of energy is between 0,4Ry and 0,6Ry. This reflects the interaction with the states of Fe d down because if we look at the DOS on the Fe site in Figure 5.8a, we see that there is a peak in this energy range. Furthermore, this shows the influence of the two Fe sites on Pd since this interaction does not happen in the 2 Pd / Fe system.

8. Hyperfine parameters

It is evident that the Mossbauer Effect is an effective ferment when dealing with the magnetic structure of the nuclei and their interaction with the neighborhood. The hyperfine property related only to the site of the iron atom, the hyperfine field (in kg), and isomer shift (in mm / s) will be theoretically discussed. Initially, we propose that the magnetic field in the core is given by H = Hext + HFC + Horb + Hdip.

So, we understand that: Hext is the external applied magnetic field at the nucleus; and HFC is the hyperfine interaction (or Fermi contact term), which comes from an unbalanced spin density of the s-electrons can see these settings in references 30 and 31 with details.

To find the Fermi contact term we must use the following equation: gN is the nuclear gyro-magnetic ratio and y- (0) is the wave function at the nucleus for the spin-up and spin-down s-electrons.

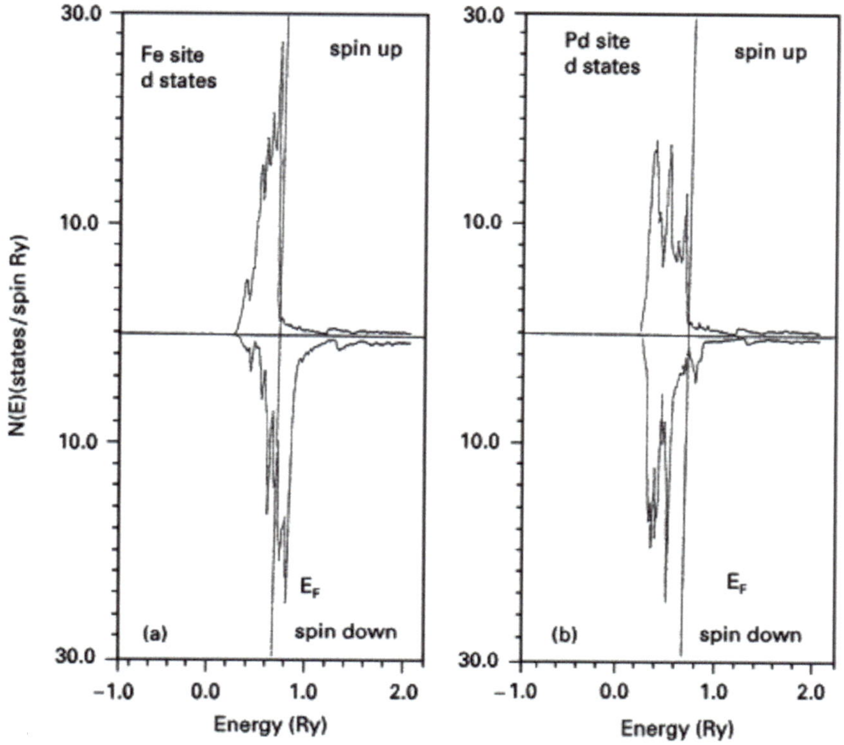

Figure 8. Density of states designed with spin up and down, to the states in d 2FePd / Pd ultrathin monolayers: (a) at the site of Fe and (b) the Pd site.

Another property to be calculated is the isomer shift of a given Beheerder atom comparing the nucleus electronic density ϱ (0) with a reference nuclei: α-Fe (BCC). In this case, Fe BCC with lattice parameters of a = 2865A °. Thus, the isomer shift is calculated by the Equation 6: where ϱa = Density of electrons in the nuclei; ϱs = Density of electrons in the reference nuclei; and α = constant of proportionality.

To understand the electron density in the nucleus, the isomer shift of a given atom is calculated comparing the nucleus electronic density ρ (0) with reference nuclei: α-Fe (BCC) in this case. The Fe BCC with lattice parameters of a=2865A°.Thus, the isomer shift is calculated by the Eq. 6: $IS = [\rho_a(0) - \rho_s(0)]\alpha$, where: ϱ_a = Density of electrons in the nuclei; ϱ_s = Density of electrons in the reference nuclei; andα = constant of proportionality.

In Table 4, we have the values of the HFC and IS systems for 2 Fe / Pd and 2 Pd / Fe. In this table, it is clear that if we change the neighborhood Fe site, there is a major change in the Fermi contact field. When we have2 Fe / Pd multilayer, we have the lowest Fermi contact term of the three systems, 210 kOe, which is less than the experimental value 330kOe of pure Fe. In the Fe bilayer / Pd multilayer and 2 Pd / Fe, we found the value higher than the experimental pure Fe. In the 2 Fe / Pd system, the site Fe receives loads of the Pd site, which will narrow the gap between the s electrons up and down causing a reduction in the Fermi contact field that is 210

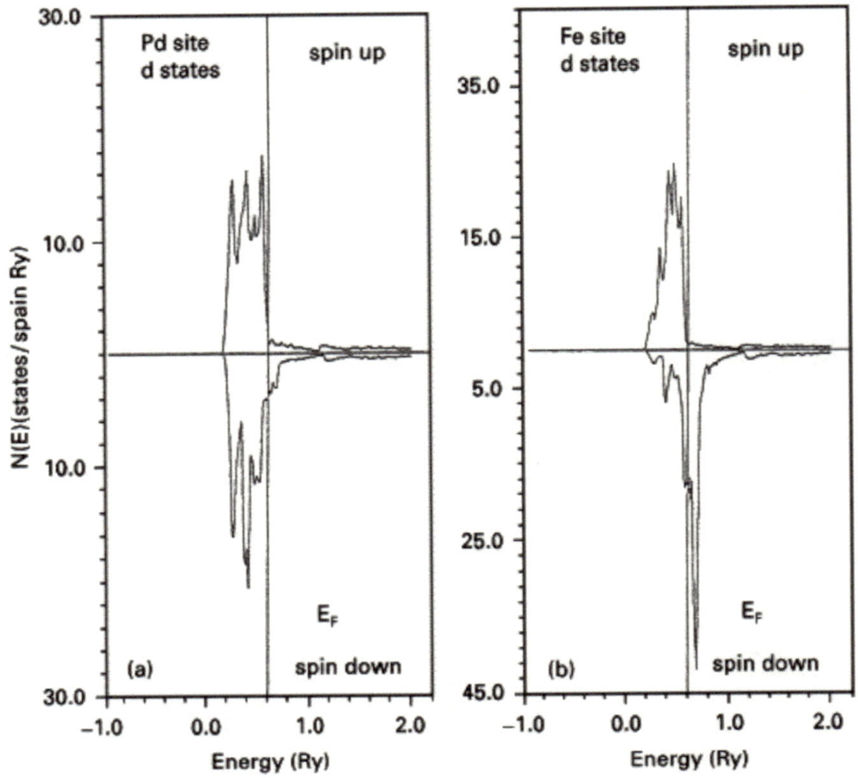

Figure 9. States density designed to electrons with spin up and down states for d in the 2 Pd / Fe multilayer: (a) the site Fe and (b) the Pd site.

kOe. When we have two layers of Pd, the charge transfer to the s states increases the difference between the electrons up and down, increasing the Fermi contact field (see table 4). This behavior of the Fermi contact field is consistent with the magnetic moment of the behavior on the site Fe, which confirms the empirical relationship of the magnetic moment that is proportional to the hyperfine field. Table 4 also shows the values calculated for the IS, which is greater for the bilayer Fe / Pd. The above results can be used for comparison with the experimental results of M. Li et al. [68,69], which show that when a Pd layer is placed in the 8Angstron, the Fermi contact field increased from 330 kOe to 335 kOe, which agrees qualitatively with our calculations. The differences are due to the calculations simulated at zero Kelvin temperature.

		H_{FC}(kOe)	IS(mm/s)
2Fe/Pd 2Pd/Fe	Fel	210	0.316
	Fe	388	0.456

Table 4. Hyperfine parameters calculated for 2 Pd / Fe and 2 Fe / Pd using spin polarization. Fermi contact term (HPC) in kOe, isomeric deviation (S) in mm / s.

9. Influence of pressure on the magnetic properties and structure of the 2Fe/ Pd and 2Pd / feelectronic ultrathin multilayers

In this section, we analyze the behavior of the magnetic properties and electronic structure of the 2 Fe / Pd and 2 Pd / Fe multilayer with pressure. We do this by varying the lattice parameter, which simulates a variation in pressure. Every self-consistent calculation submitted, the lattice parameter shows a variation of 2%. In Figures 10a and 10b, the magnetization shows as a function of the lattice parameter. Figure 10a shows the magnetization in the Fe site for both systems. We have noted that there is a different behavior between the two systems against the pressure. For the 2Fe / Pd system, there is an abrupt decrease in the magnetic moment of Fe, leading to the collapse of the magnetic moment. This transition from a ferromagnetic state to the non-magnetic state was obtained for the Fe / Pd bilayers [42].This behavior is observed experimentally in nitride [33] and in accordance with the calculations performed by bands [46]. In the case of 2 Pd / Fe, the pressure is not enough for a drop to zero in the magnetic moment. This behavior was observed in the alloys ordered Fe-Pd, depending on the Pd concentration in the league [34]. Figure 10b shows a similar behavior in the Pd site. However, in the 2 Fe / Pd system, there is a small increase of pressure in the magnetic moment in the Pd sites, but this is not very big. The 2 Pd / Fe system shows a different behavior: the magnetic moment does not drop to zero with increasing pressure and leaves a significant magnetization, as Fe-Pd alloys with the same proportions of Fe and Pd [32].

In figures 7a and 7b, we see the graph of the total energy as a function of the lattice parameter for the ferromagnetic and nonmagnetic states. From these data, we obtain the critical pressure of 92 kbar for the 2Fe / Pt system and 277 kbar for 2 Pd / Fe. Earlier in this chapter, we obtained a PC bilayer Fe / Pd with 109 kbar. It is clear that if we put more layers of Pd with respect to Fe, the PC system increases. These results are in the order of magnitude of ordered alloys Fe3Pd [34] and results to carbides Fe4C [42] and certain iron nitrides replaced [34,35,36,37,39]. In this situation, we noted that the multilayer has a similar behavior league that ordered the invar type.

Analyzing the decrease of the magnetic moment with increasing pressure, we studied the electron redistribution that occurred in the multilayer systems. Table 5 has some parameters obtained through self-consistent calculations for the spin polarization of the 2Fe / Pd systems (outside the parentheses) and 2 Pd / Fe (in the parentheses). Comparing Table 5 with Table 1, we noticed that there was an increase in the load transfer to the Fe site for both systems. This is due to the decrease of interatomic distances and the interpenetration of the electronic clouds. There was also a considerable increase in the density of states at the Fermi level for both compounds, which causes an increase in the specific heat coefficient y. The more general point is that there was an electronic redistribution in both systems. In the 2 Fe / Pd system, the transferred load Pd site to the Fe site population almost exclusively, the states d-down, together with a reversal of spin up electrons down, makes the non-magnetic system. Therefore, despite the pressure, the 2 Pd / Fe ferromagnetic state remains.

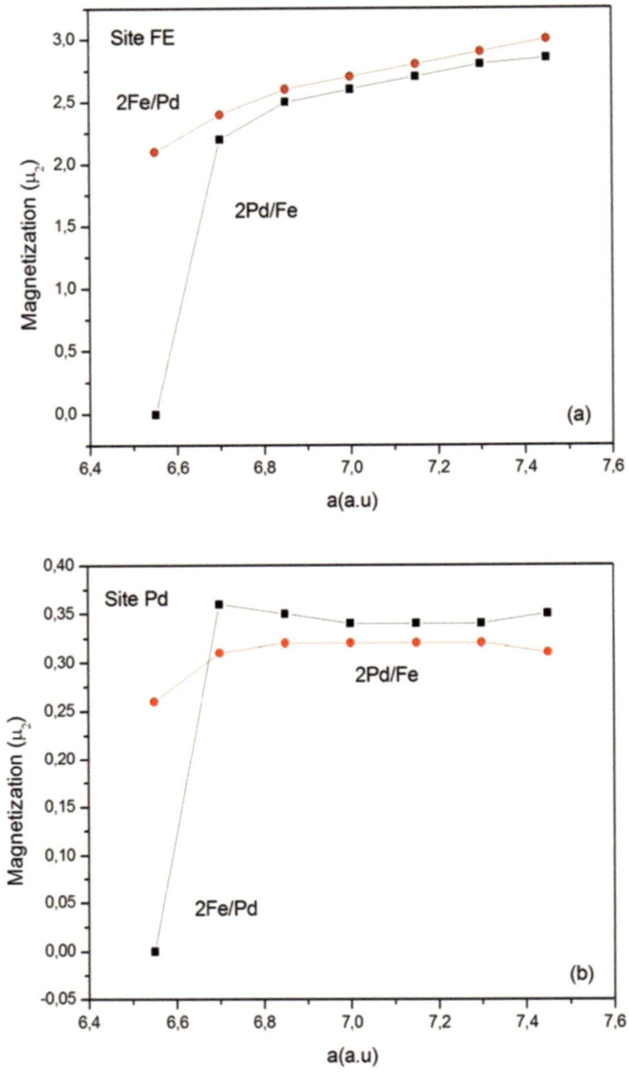

Figure 10. Magnetic moment due to the lattice parameter for ultrathin multilayers: (a) at the site of Fe and (b) Pd site.

Table 5 shows high pressure and low volume. And there is an electronics redistribution of the s electrons and this redistribution is not purely an average spin-up and down.

Figures 11a-d present the DOS for the spin-up and spin-down of the Fe site for both systems. In water, 6.5120 parameters with dotted lines for the2Fe / Pd system are shown, and full lines are seen for the 2 Pd / Fe system. We also noted that the d-DOS, as a whole, is moved to higher energies for both systems, but the 2 Pd / Fe d-DOS system is still taken to higher energies. There is also an increase in the Fermi energy of the system. It is clear that the 2 Fe / Pd system undergoes a transformation from the ferromagnetic phase to the non-magnetic phase, as in

http://dx.doi.org/10.5772/61289

this system, there is practically no more difference between the area below the Fermi energy of up situations for the down states. In the Fe site of the 2 Pd/ Fe system, there is a difference between these areas that features a magnetic moment in this site.

The results shown in Figures 12 and 13 show a strong dependence of the HFC and IS with the spacing of the lattice, which is also checked for nitrides experimentally [33]. In all cases, the absolute value of the pressure decreases with HFC. This may be associated with the reduction in the contribution of the s electrons to the spin density at the core of Fe. We noted that for the 2Fe / Pd system in low volume. In the 2 Pd / Fe system with pressure, there was a decrease to zero for the HFC, and the same happens with the magnetic moment of the Fe site in this system (Figure 10). This confirms the proportionality between the HFC and magnetic moment. The IS behavior on the Fe sites due to the reduction in lattice spacing is quite similar for both the 2 Pd / Fe as well as for the 2Fe / Pd system. These results suggest that the difference between the hyperfine parameters can be related to the expansion of the lattice due to the larger radius of the atomic Pd atoms. In fact, the reduction in HFC when a Pd atom is replaced by the Fe atoms (2 Pd / Fe and 2 Fe / Pd) can be viewed as a simulation of applying pressure in the 2 Pd / Fe system (Figure 12).

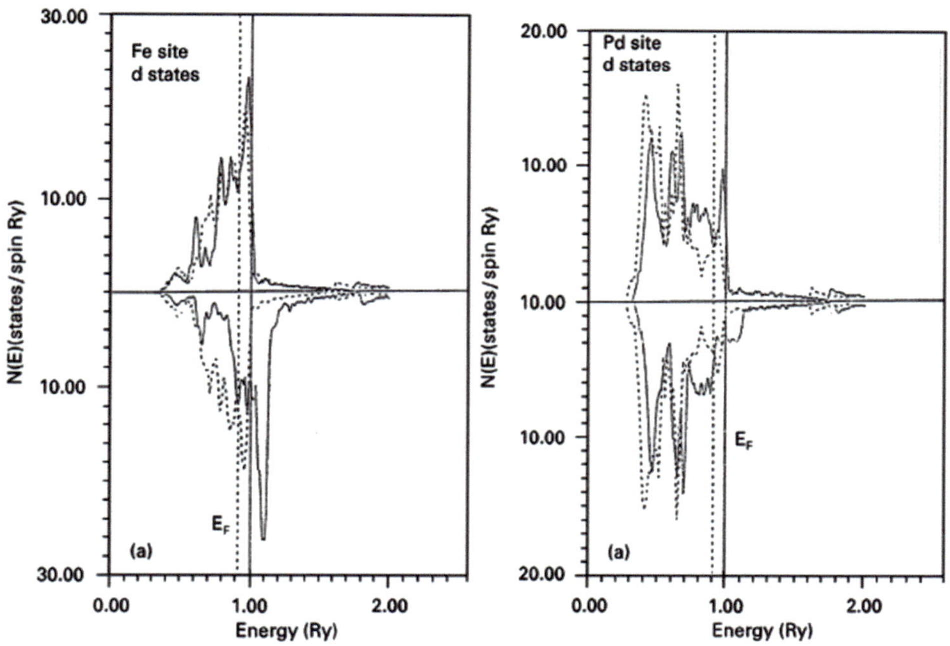

Figure 11. States that the density is designed to the electrons with spin up and down states for d, the multilayer 2 Fe / Pd, and 2 Pd / Fe. The volume of the magnetic breakdown are as follows: (a) the Fe site (solid lines 2 Pd / Fe), (dotted lines 2 Fe / Pd).

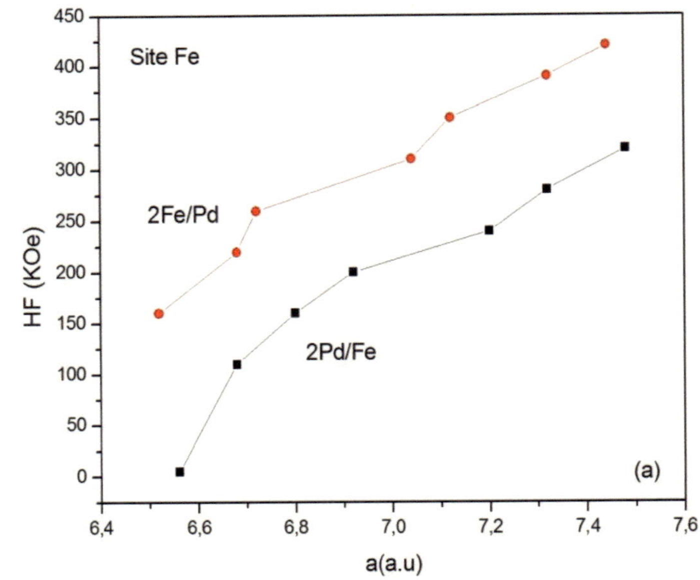

Figure 12. Hyperfine fields due to the lattice parameter.

10. Conclusion

In this paper, we used the LMTO method to the nearest ASA to investigate the electronic structure of ultrathin alloys, and the electronic structure of multilayer was analyzed.

The self-consistent calculations performed for the Fe / Pd ultrathin multilayers show a difference in the electronic structure when the stoichiometry of the multilayer systems is changed. As we increase the number of Pd layers relative to Fe, there is an increase in the magnetic moment on the website of Fe as well as an increase in the volume of the unit cell. Comparing the magnetic moment on the Fe site in the multilayer Fe / Pd alloys with certain ordered FePd and nitrides [32,33,39,40], we noticed a considerable increase. The hyperfine properties calculated for the multilayer Fe / Pd compared with the experimental results of M. Li et al. [42,43] have reasonable results from a qualitative point of view. The differences are due to the structure of the experimental multilayer, which is not ideal.

When subjected to pressure, the 2 Fe / Pd systems (2 Pd/ Fe and Fe / Pd) presented different behaviors among themselves. The 2Fe / Pd and Fe / Pd systems exhibited a magnetic collapse with an increasing pressure, while the 2 Pd / Fe magnetic moment were maintained in both the Fe and Pd sites. The hyperfine properties also changed with pressure.

Author details

A.V. dos Santos

Address all correspondence to: vandao@urisan.tche.br

Universidade Regional Integrada do Alto Uruguai e das Missões – URI, Campus Santo Ângelo, Santo Ângelo, RS, Brazil

References

[1] Andersen, K. O., *Phys. Rev B.*, 12, 3060, 1975.

[2] Andersen, K. O., Jepsen, O., *Physica B.*, 91, 317, 1977.

[3] Gyorffy, B. L., *Phys. Rev B.*, 5, 2382, 1972.

[4] Gyorffy, B. L., Stocks, G. M., Electrons in Disordered Metals and Metallic Surfaces., New York, Plenum, 1984.

[5] Andersen, K. O., Jepsen, O., Glotzel, D., Canonical Description of The Band Structures of Metals., In: International School of Physics Enrico Fermi, 1983, Verenna, Amsterdam, North-Holland, 1985. Course 89, p.59-170.

[6] Stocks, P. M., Winter, H., Electronic Structure of Complex Solids, New York, Plenum, 1984.

[7] Hohenberg, P., Kohn, W., *Phys Rev.*, B 136, 3864, 1964.

[8] Koitn, W., Sham, L. J., *Phys Rev., E* 140, 1133, 1965.

[9] Slater, J. C., *Phys Rev.*, B 81, 385 1951.

[10] Elzain, M. E., Yousif, A., *Hiperfme Interaction.*, 94, 511, 1994.

[11] Guo, G. Y., Ebert, H., *Hiperfme Interaction,* 97/98, 11, 1996.

[12] Kraft, T., Marcus, P. M., Scheffler, M., *Phys Rev.*, B 49, 511, 1994.

[13] Fu, C. L., Painter, G. S., *ActaMater.* 45 481, 1997.

[14] Perdew, J. P., Chevary, J.[a], Vosko, S. H. Jackson, K. A., Pederson, M. R., Singh., Fiolhais C., *Phys Rev.*, B 46, 6671, 1992.

[15] Garcia, P. F., Meinhaldt, A. D., SUNA, *A., Appl. Phys. Leu.*, 47, 178, 1985.

[16] Draasma, H. G., Jonge, W. J. M.,./. *Magn. Magn. Mat.*, 66, 351, 1987.

[17] Broeder, F. J. A. den., Kniper, D., Donkersloot, H. C., Horing, *W.^ppl.* Phys. Lati., 49, 507, 1989.

[18] Bayreuther, G., Hiprfme Interaction., 49, 237, 1989.

[19] Celinski, Z., Heinrich, B., Cochran, J. F., Muir, W. B., Arrot,[a] S., Kirschner, J., Phys Rev. Lett., 65, 1156, 1990.

[20] Néel, M. L., J. PhysRad, 15, 225, 1954.

[21] Jarlborg, T., Freeman, A. J. Phys Ver, B, 47, 810, 1980.

[22] Jarlborg, T., Freeman, A. J. Appl. Phys., 53, 8041, 1982.

[23] Taler, T. J., Ketterson, J. B., Hiliard, J. E., Phys Rev. Lett., 41, 336, 1978.

[24] Blügel, S., Drittler, B., Zeller, R., Dederichs, P. H.,. Appl. Phys., 49, 547, 1989.

[25] Bennett, L. H., Watson, R. E., (Eds.) Magnetic Multílayers, Singapore: World Scientifc, 1994.

[26] Hedin, L., Lundquist, B. L., J. Phys. C, 4, 3065, 1965.

[27] Barth, V. von, Hedin, L., J. Phys. C, 5, 1629, 1972.

[28] Kittel, C., Introduction to Solid State Physics. New York, John Wiley, 1971.

[29] Richter, R., Gay, J. G., Smith, J. R., Phys Rev Lett, 54, 2704, 1990.

[30] Piecuch, M., Nevot, L., Matential Science Fórum, 59/60 1989.

[31] Gumbsch, P., Daw, M. S., Phys Rev 44, 3934, 1991.

[32] Dffiny, B., Speriosu, V. S., Metin, S., Parkin, S. P., Gurney, B., Baumgart, * P., Wilhoit, D. R, J. Appl. Phys. 69, 8, 1991.

[33] Yang, C. L. Abd-Elmequid, M. M., Micklitz, H., Otto, J. W., Jong, Y., Xue, D. S., LI, F. S., J. Magn. Magn. Mat. 151, 19 1995.

[34] Kuhnen C. A., Silva E. Z., da., PhysRev.B 46, 8915, 1992.

[35] George, J. M., tese Université Paris sub-Laboratori de Physique dês Solides Orsay France 1993.

[36] Kuhnen, C. A. Santos A. V. dos., J. Magn. Magn. Mat. 130, 353, 1994.

[37] Cord1er-Robert, C., Foct, J., Eur J. Solid State Inorg. Chem. 29, 39. 1992.

[38] Bohland, J., Kuhnen, C. A., BrazilianJ. Phys 23, 288, 1993.

[39] Willams, A. R., Kubler, J., Gellat Jr. D., Phys Rev. B, 19, 6094 1979.

[40] Santos, A. V. dos., Kuhnen, C. A., Costa Jr., M. I. da, J. Magn. Magn. Mat. 166,223, 1997.

[41] Santos, A. V., Calculating the electronic structure of the multilayer Ni / Ag, Ni / Au, Ni / Cu, Fe / Cu and Fe / Pd bilayers and Fe / Pd. Porto Alegre: UFRGS, 1997. 107p. tesse, (Program for Graduate in Physics). Universidade Federal do Rio Grande do Sul, Porto Alegre 1997.

[42] 63. Santos, A. V. dos., Kuhnen, C. A., SolidState Commu, 95, 123, 1995.

[43] Coehoom, R., J. Magn. Magn. Mat. 159, 55, 1996.

[44] Li, M., Ma, X. D., Peng, C. B., Zhao J. G., Mei, L. M., Gu, Y. S., Chai W. P., Mai, Z. H., Shen, B. G., Liu, Y. H., Daí, D. S., Phys Rev. B, 50, 10323 1994.

[45] Noort, H. M., Van., Broeder, F. A. den.,., J. Magn. Magn. Mat. 51, 273, 1985.

[46] Fiqueiredo R. S., Kuhnen, C. A., Santos, A. V. dos., (aceito para publicação em J. Magn. Magn. Mat.).

Chapter 14

Mechanical Properties of the Thermal Barrier Coatings Made of Cobalt Alloy MAR-M509

Zenon Aleksander Opiekun

Additional information is available at the end of the chapter

http://dx.doi.org/10.5772/61100

Abstract

This manuscript presents the microstructure, geometrical product specification, and results of scratch tests performed on the interlayer of thermal barrier coating (TBC) with Rockwell's intender. The TBC was provided by depositing two layers, metallic interlayer and external ceramic layer, onto a plate coating made of cobalt alloy MAR-M509 in plasma spraying process.

Based on measurements of microhardness made with Berkovitz's indenter using Nano Scratch-Tester (CSM Instruments), it was stated that elastic (Ee) to total energy (Ec) parameters (M_{IT} = Ee/Ec), γ phase matrix of alloy MAR-M509 ($M_{IT\gamma}$), metallic interlayer (45% Ni–22% Co–17% Cr–16% Al–0.3% Y) (M_{ITM}), and ceramic layer (M_{ITZrO2}), are proportion, that is, 0.29:0.22:0.50. The surface of the casting was sandblasted with Al_2O_3 powder in an air stream before the TBC was introduced. Scratches were made along the cross section from a mould material (MAR-M509) through metallic interlayer and external ceramic layer in the TBC. Friction force, friction factor, and acoustic emission were recorded during the test. It has been proven that metallic interlayer in the TBC of ca. 200 μm thickness forms tough coating without pores with good cohesion values and very good adhesion values to the mould.

Keywords: Cobalt alloy MAR-M509, geometrical product specification (GPS), thermal barrier coating (TBC), microstructure, microhardness, interlayer cohesion, adhesion, nanoscratch test

1. Introduction

Thermal barrier coatings (TBCs) are applied via plasma spraying technology (air plasma system [APS]) and build up usually with two layers: metallic interlayer and external ceramic layer [1–3]. For special purposes, when a high density of coating together with its good bonding with the mould is demanded, the TBC is applied in vacuum plasma spraying process (vacuum plasma spray [VPS]) or low-pressure plasma spraying technique (low-pressure plasma spray [LPPS]) [4]. The TBC has an excellent heat resistance and a small thermal conductivity. They are characterized by an erosion and abrasion resistance and resistance to aggressive chemicals [5]. Therefore, TBCs are becoming more commonly used as coatings of jet-propelled parts of aircraft engines (combustor, blade outer air seal [BOAS], and blades of turbine), valves, and parts in chemical reactors [6–7].

Materials with MeCrAlY group single oxides that belong to ceramic materials are usually applied on metallic interlayers. These are most often used on the external layer in the TBC: Al_2O_3, ZrO_2, ThO_2, BeO, MgO, CeO_2, Cr_2O_3i Y_2O_3 [5,6]. Thermal barrier coatings are the layers that separate the surface of metallic material from the stream of hot gases. Simultaneously, they reduce the temperature of metallic items [8]. In the end, such an influence of the TBC causes an improvement of performance characteristic of jet turbine engine, a reduction of fuel consumption, an increase of inlet gas temperature (ca. 100–170°C) (Fig. 1) [9], and a decrease of toxic substance emission from outlet gases because of better fuel consumption [7].

ZrO_2 oxide stabilized by Y_2O_3 is most often used for the external layer in the TBC [10–15]. Functional properties, adhesion, cohesion, and crack resistance depend on the phase composition of this oxide. Solid solutions prepared on the basis of regular ZrO_2 (C) have low resistant to changeable heat load. A considerable improvement of the crack resistance of these solutions is noticed during the presence of tetragonal phase of ZrO_2 (T) in the TBC. Factors that favor the tetragonal phase in the TBC after cooling down to the room temperature from the temperature of plasma spraying are small ZrO_2–8% Y_2O_3 irregular grain sizes, with the introduction of about 8 wt% of Y_2O_3 to the solid solution of ZrO_2–Y_2O_3(Fig. 2) [16].

The aim of this article was to describing the cohesion and adhesion of the metallic interlayer made of 45% Ni–22% Co–17% Cr–16% Al–0.3% Y alloy and to present the geometrical product specification of the ceramic external layer made of ZrO_2 oxides deposited on the casting of plate made of MAR-M509 cobalt alloy.

The main criterion to accommodate metallic interlayer is to compensate stress between the substrate and the external ceramic layer. TBC with plastic interlayer decreases its own stress and enhances connection between harder substrate and fragile hard external ceramic layer [17–19]. Useful properties of TBC (erosion invulnerability, thermal fatigue, scale off, and mastication) depend on fat of layers and their spraying thermal parameters [20–23].

The mechanical properties of TBC layers defined were by microhardness made with Berkovitz's indenter when Nano Scratch-Tester (CSM Instruments) was used. The coating was obtained by single-spraying metallic powder made of 45% Ni–22% Co–17% Cr–16% Al–0.3%

http://dx.doi.org/10.5772/61100

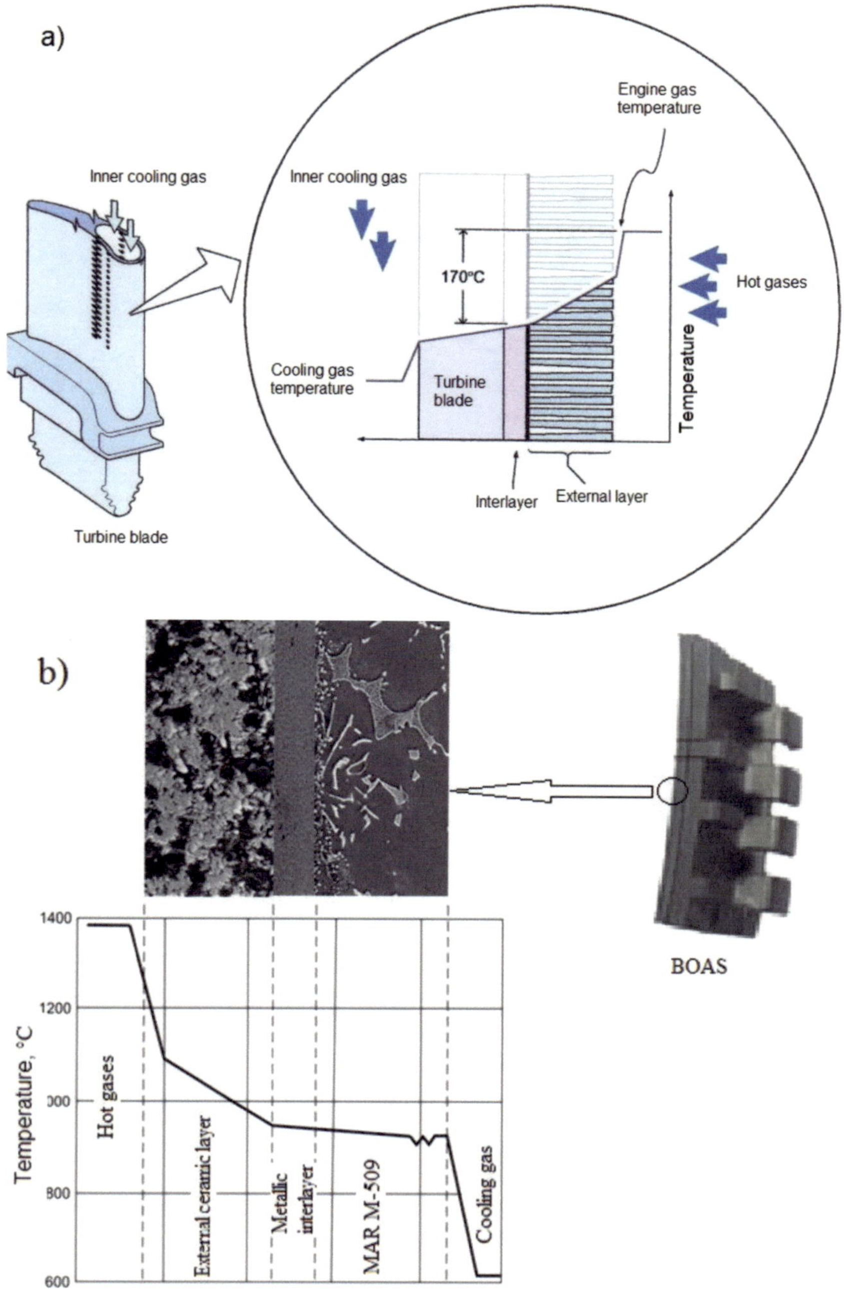

Figure 1. Effect of TBC on reducing temperature of turbine blade [9] (a), and BOAS made of MAR-M509 alloy (b) (schemes).

Y alloy and triple-spraying ceramic powder made of ZrO_2–8% Y_2O_3 oxides onto the surface of the casting with a MultiCoat Plasma Coating System (SulzerMetco).

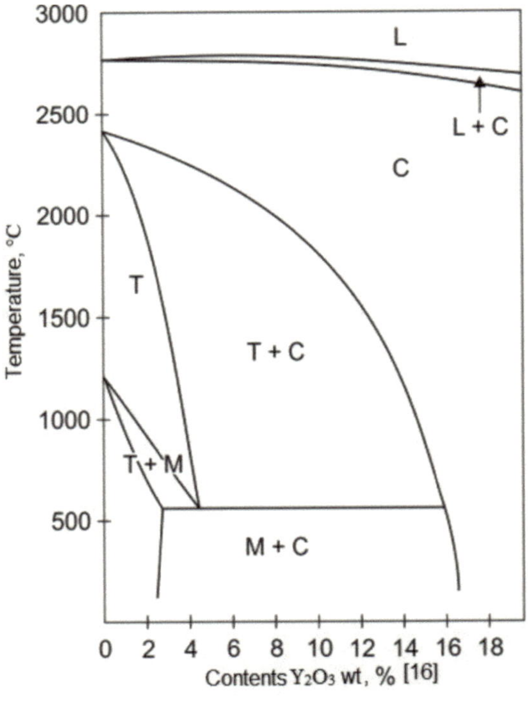

Figure 2. Phase equilibrium system ZrO_2–Y_2O_3.

2. Materials and methods of testing

Plate castings made of cobalt alloy MAR-M509 with a dimension of 4 × 20 × 90 mm were used for tests. The composition of plates was as follows: 0.60% C, 28.83% Cr, 10.0% Ni, 7.12% W, 3.77% Ta, 0.36% Zr, 0.18% Ti, 0.007% B, 0.43% Fe, 0.02% Mn, 0.03% Si, 0.002% S, and the rest is Co. The plates were casted in the multilayer ceramic mould (MCM) at initial temperature ca. 1000°C. The liquid alloy MAR-M509 at temperature ca. 1495 ± 5°C [24,25] was poured over MCM in a medium-frequency induction vacuum furnace. The surface of plate castings was at first sandblasted with Al_2O_3 powder of 200 ÷ 250 μm grain diameter in an air stream of pressure of ca. 0.35 MPa for a few seconds. After a single layer of 45% Ni–22% Co–17% Cr–16% Al–0.3% Y powder alloy was covered, triple layers of ZrO_2–8% Y_2O_3 powder oxides in an argon-hydrogen plasma beam were applied. The plates were heated at a temperature of about 120°C before the layers were covered.

Two spray powders were used in the experiment: one 45% Ni–22% Co–17% Cr–16% Al–0.3% Y powder alloy, which shows fine spherical morphology with particle granulation, ca.75 ± 15 μm, and ZrO_2–Y_2O_3 oxide powder, which shows irregular morphology with particle granulation, ca.53 ± 15 μm.

A deposition of all layers in the TBC was done using the following: primary plasma gas, Ar (40 l/ min); secondary plasma gas, H_2 (10 l/min); electric current, about 600 A; powder feet rate

http://dx.doi.org/10.5772/61100

(40 g/min); and spray distance, about 130 mm. The traverse speed of a spraying gun was constant at 2 mm/min, and the speed of rotating holder was also constant (60 rot/min) (Fig. 3).

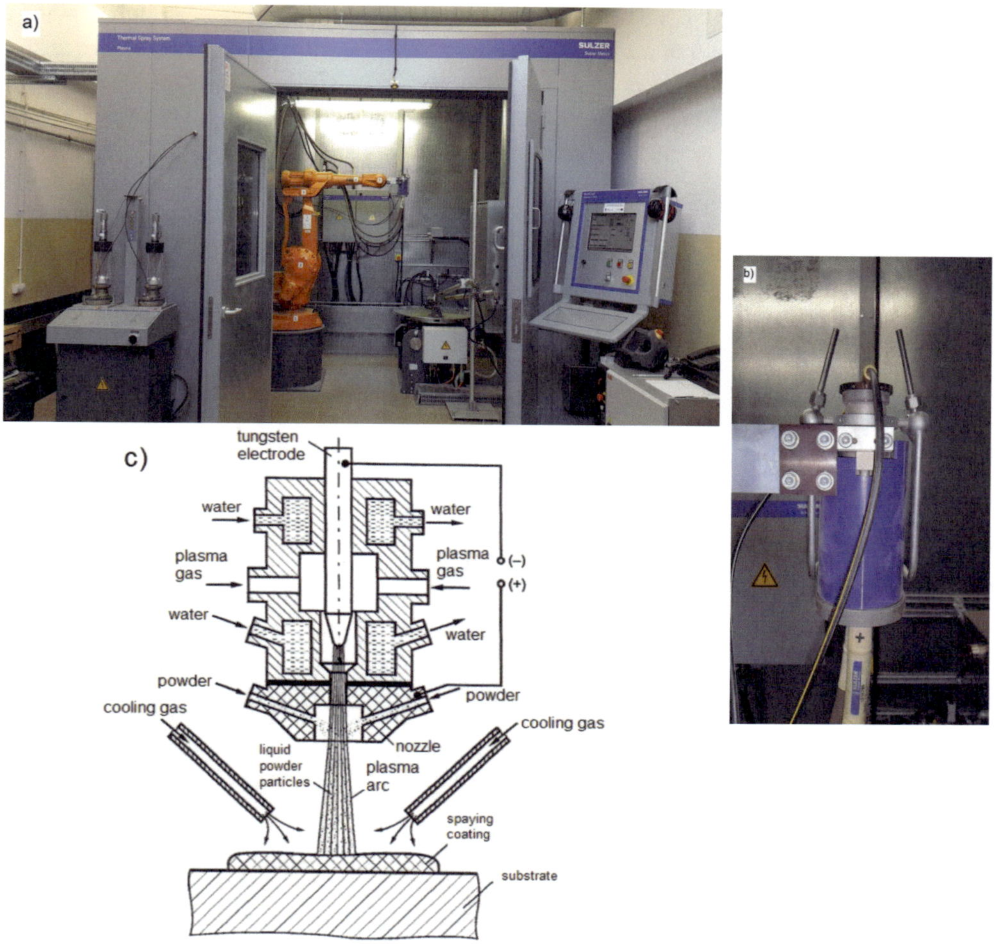

Figure 3. Multicoat plasma coating system. View of system (a), spraying gun F4 MB-HBS type (b), scheme of spraying (c).

The state of surface of plate castings was assessed with the Talyscan 150 (Taylor-Hobson) tool working with the Mountains Map Universal program before and after the ceramic layers was deposited. Six amplitude parameters of geometrical product specification (GPS) were analyzed: S_a (arithmetic mean), S_z (mean of the 5 highest peaks and the 5 lowest points), S_t (total height), S_q (quadratic mean), S_p (highest peak over the mean), and S_v (lowest valley under the mean).

X-ray examination surface of the TBC was performed with the Siemens Kristalloflex D500 X-ray diffractometer.

The testing of microstructure layers in the TBC was conducted with a Berkovitz's indenter and Nano Scratch-Tester (CSM Instruments).

The examination of cohesion and adhesion of the TBC was carried out with the Revetest Scratch Tester CSM Instruments tool. The surface of sample with the TBC was scratched with the diamond Rockwell's intender of a 200-μm top radius. During the test, six different forces were applied: 2, 4, 8, 16, 32, and 40 N onto the surface of a cross section (MAR-M509 mould and TBC). It has been assumed that each scratch is 0.5 mm long. In the interlayer of the TBC and in the MAR-M509 mould, scratches were made at a 5-N intender load along 0.5 mm. All scratches were made at a rate of 1 mm/min—the speed of the intender. The Revetest Scratch Tester tool was calibrated according to reference sample covered with TiN coating. Friction force, friction factor, and acoustic emission were recorded during the test; 100% of acoustic emission scale equals 65-dB value.

3. Results

3.1. Geometrical structure of a surface

The geometrical product specification (GPS) and the surface load–capacity curve of external ceramic layer in the TBC are presented in Figure 4.

3.2. Microstructure and X-ray diffraction pattern

The microstructure of the plate casting made of alloy MAR-M509 with the TBC at the cross section is presented in Figures 5 and 6. The X-ray diffraction patterns of the external ceramic layer in the TBC surface are shown in Figure 7.

3.3. Measurements of microhardness

Nano Scratch Tester (CSM Instruments) equipped with a Berkovitz's indenter (a pyramid with equilateral triangle as the base area) is to measure microhardness and to determine other mechanical properties of single grains, thin layer, and phase boundaries when loading spreads from 0 to 500 mN [26].

A ratio of elastic strain energy (Ee) to total energy (Ec) when forcing the indenter of the hardness tester is a very important feature of the tested materials.

Figure 8 shows example courses of changes of energy plastic deformation (Ep) and energy of elastic deformation (Ee), which describe the forcing of Berkovitz's indenter into a γ phase, into the metallic interlayer (45% Ni–22% Co–17% Cr–16% Al–0.3% Y), and into the external ceramic layer (ZrO_2) [27]. Table 1 shows values of energy of plastic deformation (Ep), energy of elastic deformation (Ee), total energy (Ec), ratio Ee/Ec, longitudinal modulus of elasticity (Young's modulus E), and microhardness HV0.05 for γ phase, metallic interlayer, and external ceramic layer (the average value from five measurements).

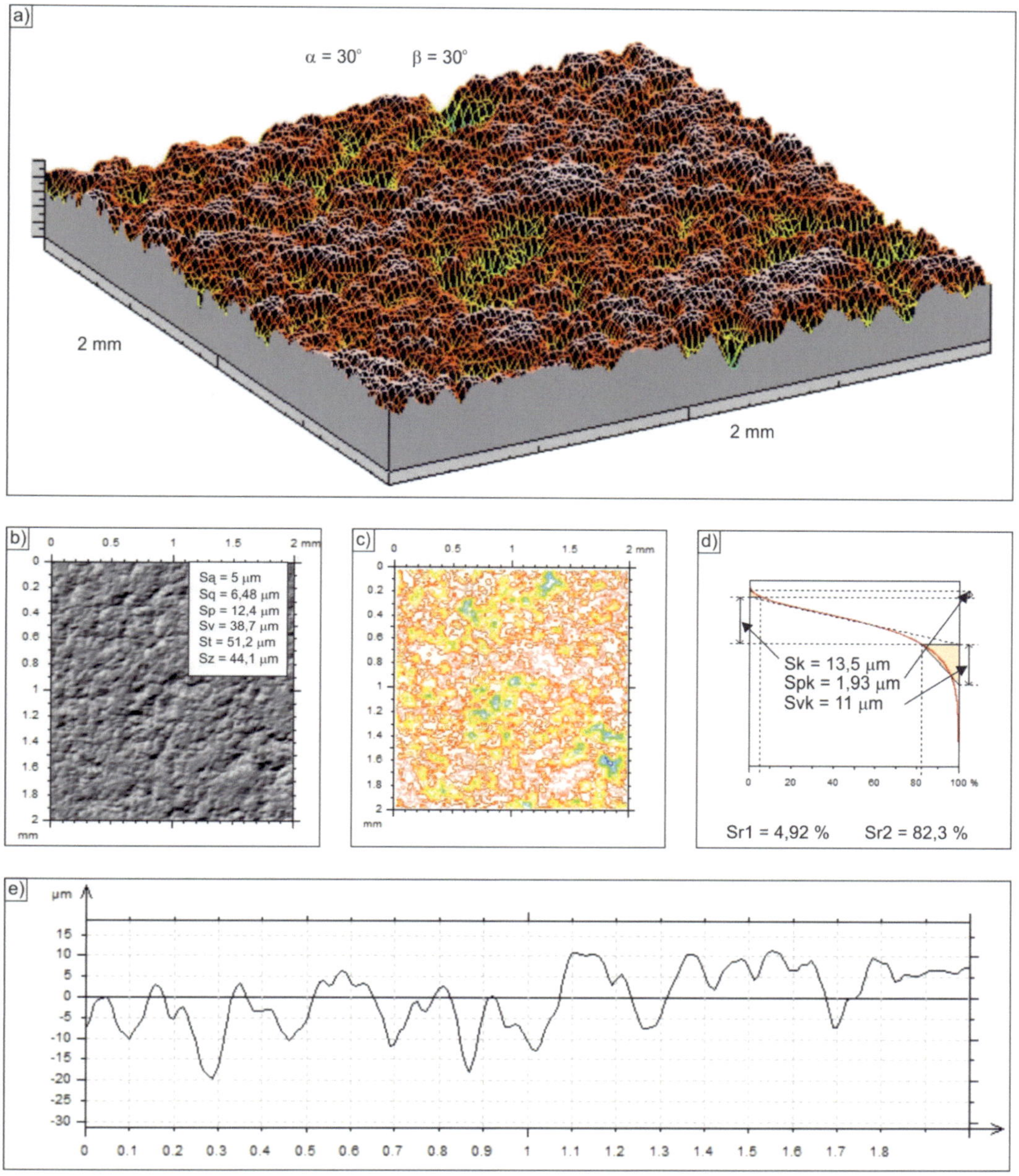

Figure 4. Surface topography of the external ceramic layer in the TBC. 3D view (a), plate surface with amplitude parameters (b), coating map (c), load-capacity curve of GPS and its parameters (d), surface profile (e), and geometrical parameters S_k, S_{pk}, S_{vk}, S_{r1}, S_{r2}.

3.4. Cohesion and adhesion

The control was scratched with the diamond Rockwell's indenter on the surface of a cross-sectional MAR-M509 mould and TBC (Fig. 9).

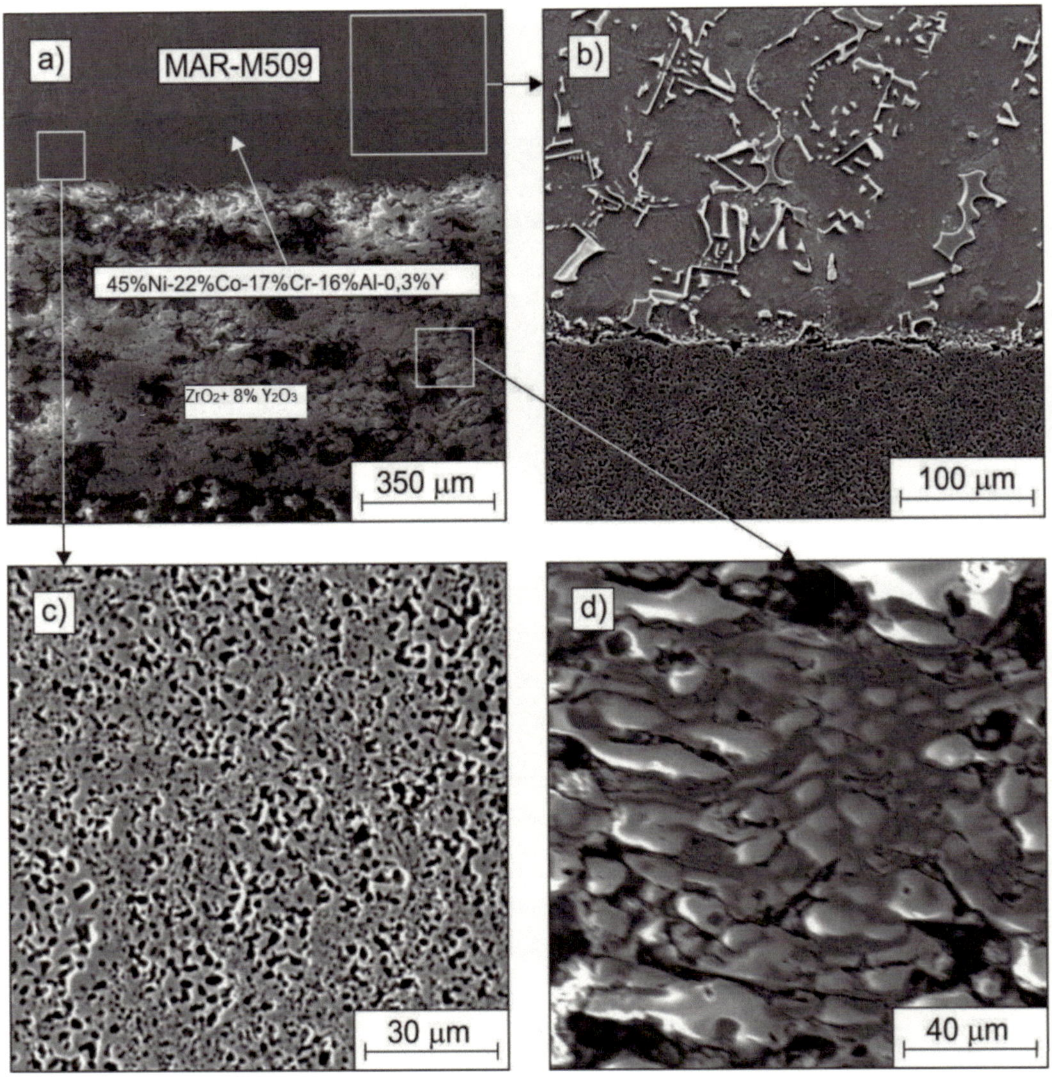

Figure 5. Microstructure at the cross section: mould (MAR-M509) coating (TBC) (a, b), external ceramic layer (ZrO_2) (d), metallic interlayer (45% Ni–22% Co–17% Cr–16% Al–0.3% Y) (c), and SEM micrograph.

Figure 10 shows the surface scratches at the cross section: mould (MAR-M509)–metallic interlayer 45% Ni–22% Co–17% Cr–16% Al–0.3% Y in the TBC together with diagrams of friction force, friction factor, and acoustic emission changes.

The scratches of samples in the direction from the mould (MAR-M509 cobalt alloy) through the metallic interlayer in the TBC at the cross section, which is made at constant intender loads, reveal the presence of typical plastic strain cones (Fig. 10b), crack cones in the coating (Fig. 10a), and "pushes-P" of mould material into this interlayer. The surfaces of longitudinal sections of A_{ca} cones, which characterize the cohesion of the interlayer, are described by a

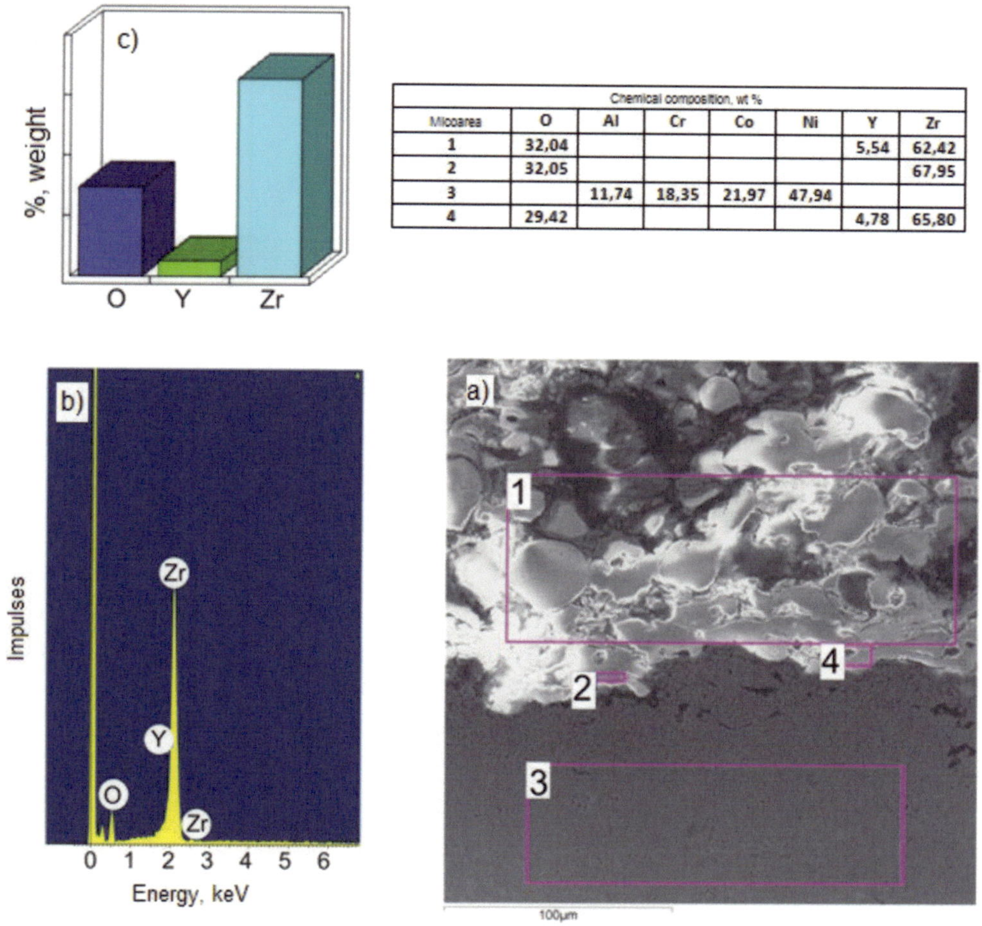

Chemical composition, wt %							
Microarea	O	Al	Cr	Co	Ni	Y	Zr
1	32,04					5,54	62,42
2	32,05						67,95
3		11,74	18,35	21,97	47,94		
4	29,42					4,78	65,80

Figure 6. Microstructure of TBC (a), graph of roentgen dispersion radiation from 1 area (b), share of elements in 1 area (c), chemical composition in microareas (1–4) of TBC (table).

product l_x l_y (Fig. 10e) [28] and are plotted as a function of loading force (Fig. 11). The "pushes" of mould material into a metallic interlayer and the lack of cracks at a boundary–mould interlayer (Fig. 11) indicate a very good interlayer adhesion to the mould.

4. Summary

A TBC with ca. 920 μm thickness was obtained as a result of spraying melted powder with particle granulation ca. 75 ± 15 μm made of 45% Ni–22% Co–17% Cr–16% Al–0.3% Y alloy and ceramic powder made of ZrO_2–8% Y_2O_3 oxides with particle irregular granulation ca. 53 ± 15 μm. The surface of a plate casting was made of the MAR-M509 cobalt alloy in a plasma beam (Figs. 5 and 6). Such a coating consists of two layers of fine grain metallic interlayer 45% Ni–

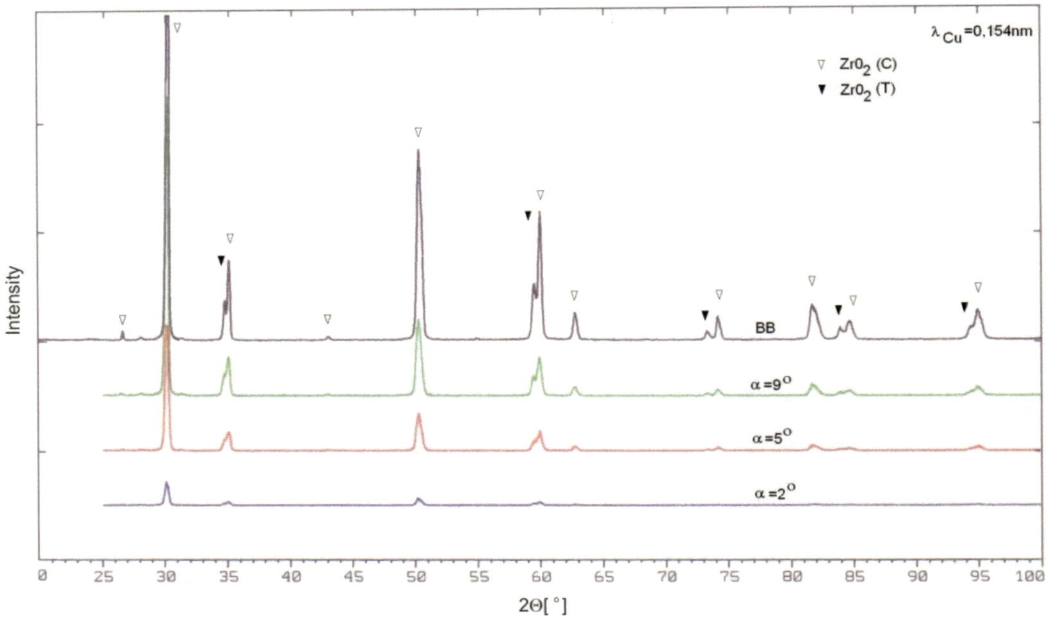

Figure 7. Diffraction patterns of the external ceramic layer in the TBC surface obtained for a grazing incident geometry $\alpha = 2°$, 5°, and 9° and in the Bragg–Brentano geometry (BB). ZrO_2 (C)—regular, ZrO_2 (T)—tetragonal.

Phase	Ep, pJ	Ee, pJ	Ec, pJ	M_{IT} = Ee/Ec	E, GPa	HV0.05
γ	226739 ± 3%	94733 ± 3%	321458 ± 3%	0.29 ± 0.1	122 ± 5	425 ± 5
Metallic interlayer (45% Ni–22% Co–17% Cr–16% Al–0.3% Y)	257307 ± 5%	72309 ± 3%	333381 ± 5%	0.22 ± 0.1	103 ± 7	540 ± 8
Ceramic external layer (ZrO_2)	16187 ± 5%	16124 ± 5%	32311 ± 5%	0.5 ± 0.1	125 ± 16	1220 ± 45

Table 1. The values of energy of plastic deformation (Ep), energy of elastic deformation (Ee), total energy (Ec), ratio Ee/Ec, Young's modulus E, and microhardness HV0.05. The average value from five measurements.

22% Co–17% Cr–16% Al–0.3% Y (ca. 200 μm thickness) and external ceramic layer as a mixture of two phases, regular ZrO_2 (C) and tetragonal ZrO_2 (T) (Fig. 7). The microstructure of the metallic interlayer in the TBC is tough, homogeneous, and deprived of porosity (total porosity less than 2%) (Fig. 5c). The thin layer of the γ phase (about 0.015 μm thick), which adheres to the metallic interlayer, is characterized by similar values of microhardness as the metallic interlayer, ca. 520 HV0.05. This is an effect of hardening of the γ phase with secondary carbides of $M_{23}C_6$ [27].

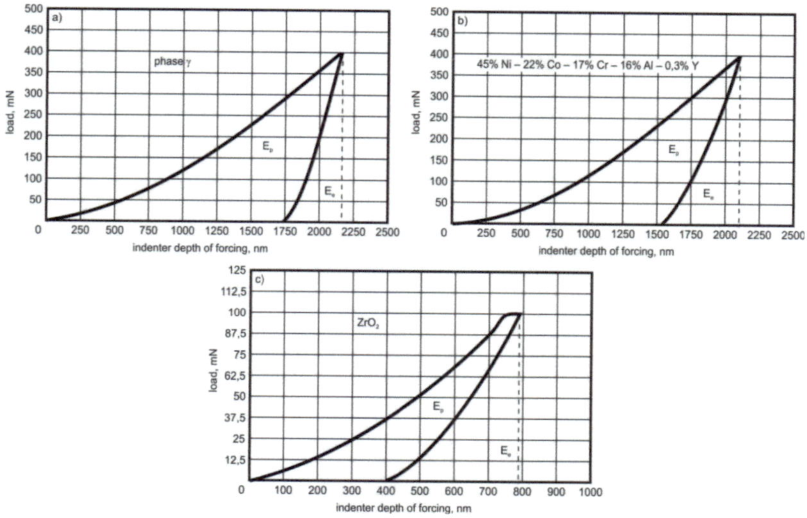

Figure 8. The influence of loading on depth of forcing in the Berkovitz's indenter and energy change for γ phase (a), metallic interlayer (45% Ni–22% Co–17% Cr–16% Al–0.3% Y) (b), and external ceramic layer (ZrO_2) (c).

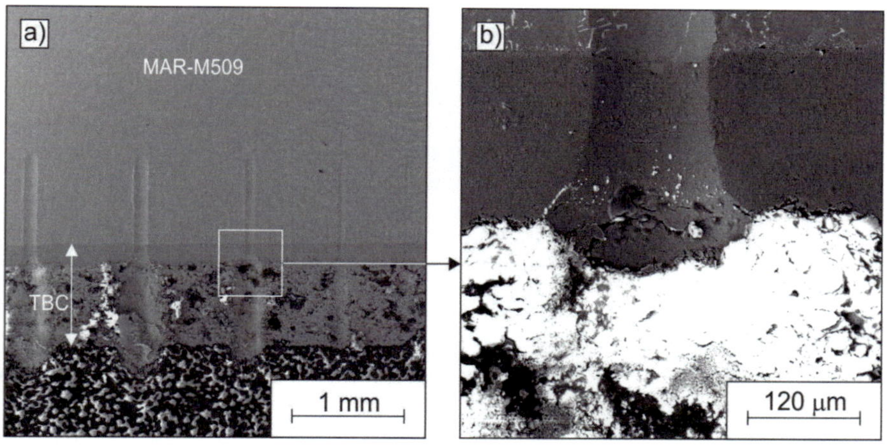

Figure 9. Macrostructure with the picture of surface scratches (a), example of the scratch Rockwell's indenter for the a loading force 8 N of a cross section; MAR-M509 mould, metallic interlayer (45% Ni–22% Co–17% Cr–16% Al–0.3% Y) and external ceramic layer (ZrO_2–8% Y_2O_3) (b), SEM micrograph.

A gradient of elastic strain energy to total energy (M_{IT} = Ee/Ec) when forcing the Berkovitz's indenter is a very good parameter that describes elastic–plastic properties of materials (Fig. 8, Table 1). The M_{IT} parameter equals 0.29 for the γ phase ($M_{IT\gamma}$), 0.22 for the metallic interlayer (M_{ITM}) 45% Ni–22% Co–17% Cr–16% Al–0.3% Y, and 0.50 for the external ceramic layer (MIT_{ZrO2}) (Fig. 12). The metallic interlayer, which has better plastic than the substratum γ phase, was deposited on the plate surface, which was activated by sandblasting with Al_2O_3 particles and by heating at about 120°C. The geometrical product specification (GPS) of the sandblasted plate is isotropic. Its amplitude parameters S_a, S_q, S_p, S_v, S_t, and S_z are 3.1 μm, 3.9 μm, 16.2 μm,

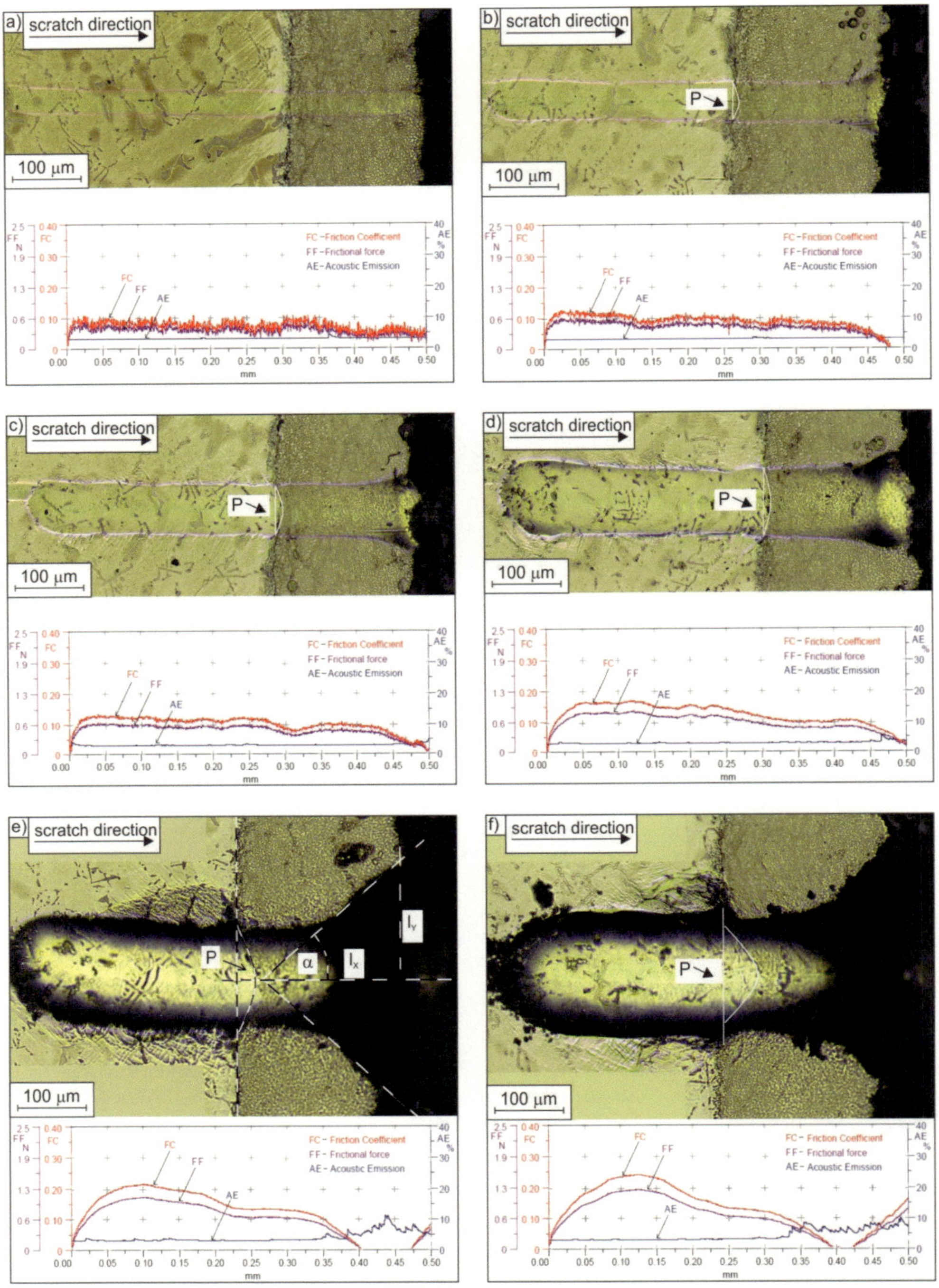

Figure 10. Examples of scratches along the cross section: mould (MAR-M509) interlayer (45% Ni–22% Co–17% Cr–16% Al–0.3% Y) in the TBC, together with diagrams of friction force, friction coefficient, and acoustic emission changes for a loading force2 N (a), 4 N (b), 8 N (c), 16 N (d), 32 N (e), and 40 N (f).

18.3 μm, 34.5 μm, and 31.4 μm, respectively, and functional parameters S_k, S_{pk}, S_{vk}, S_{r1}, and S_{r2} are 10 μm, 4.0 μm, 3.5 μm, 9.6%, and 91.4%, respectively. The GPS of the external ceramic layer in the TBC is also isotropic, but it is rougher than the mould. The total height S_t of the layer has increased by ca. 50%, and the arithmetic mean S_a increased by ca. 60% in comparison to S_t and S_a for the sandblasted surface of the mould. The roughness of the core (S_k) has increased by ca. 35%. The only functional parameter of GPS of ceramic layer, which is lower by ca. 100% in comparison to the mould, is the roughness of peaks (S_{pk}) (Fig. 4).

The scratches of samples along the cross section from the mould (MAR-M509 cobalt alloy) through the metallic interlayer and external ceramic layer (Fig. 9a,b) in the TBC, which were made with Rockwell's cone at different intender loads (Fig. 10a,f), reveal the presence of typical strain and crack cones in the interlayer and "pushes-P" of mould material (γ phase) into the interlayer. A linear surface increase of longitudinal sections of these cones indicates good cohesion of the metallic interlayer (Fig. 11).

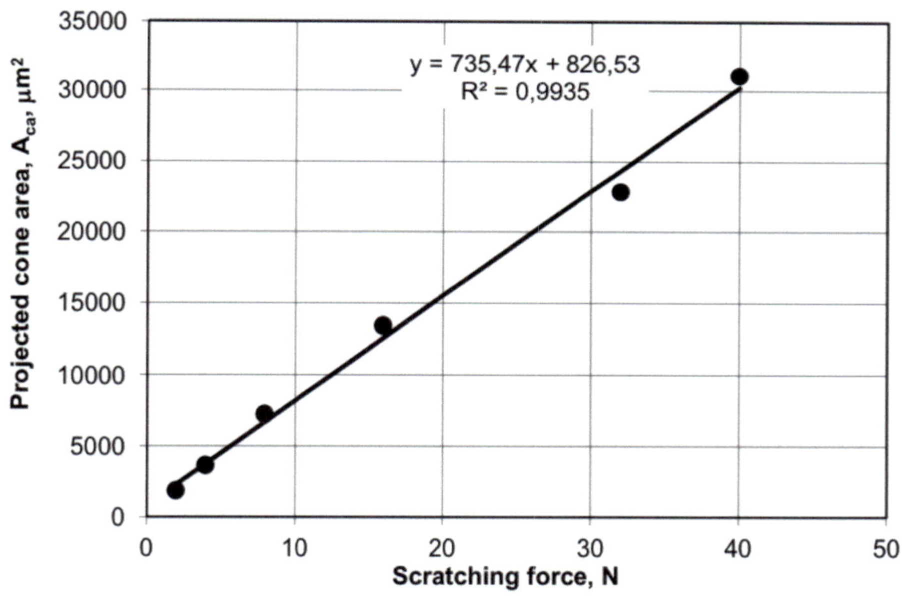

Figure 11. Surfaces of longitudinal sections of A_{ca} cones plotted as a function of loading force in the metallic interlayer 45% Ni–22% Co–17% Cr–16% Al–0.3Y.

A load increase of the Rockwell's intender from 2 N to 40 N (Fig. 10a–f) causes an increase of friction force from ca. 0.3 N to 10 N, together with a minor increase of friction coefficient from ca. 0.1 to 0.2. For all scratches, no increase of acoustic emission effect is observed at boundary mould coating. The constant value of acoustic emission at 3% level (ca. 2 dB) shows that strains at boundary MAR-M509 cobalt alloy–metallic interlayer in the TBC are plastic and do not appear in this crack area (Fig. 10 a–c). The shape of "pushes" of the γ phase is similar to the cones, and its depth increases linearly together with the increase of intender loading force from ca. 3 μm for 2 N force to ca. 53 μm for 40 N force (Fig. 11). Such a course of changes happening at boundary mould (MAR-M509 cobalt alloy), that is, the metallic interlayer in the TBC during an attempt at scratching, proves a very good adhesion for this interlayer made of 45% Ni–22% Co–17% Cr–16% Al–0.3% Y alloy [29].

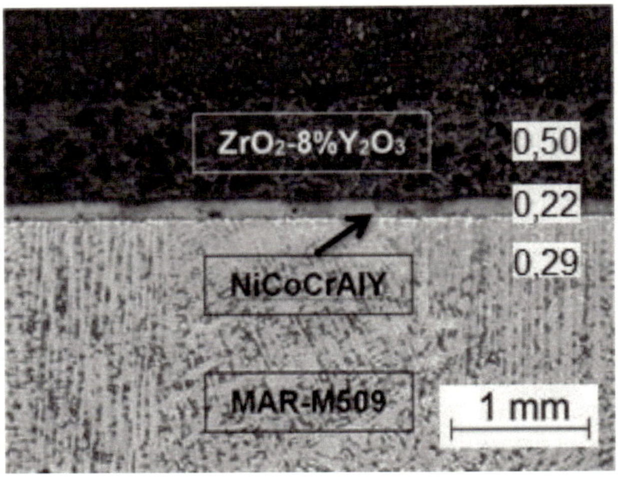

Figure 12. Macrostructure at the cross section: mould (MAR-M509) coating (TBC) with M_{IT} parameters.

Cracks in "cohesion cones" of the metallic interlayer in the TBC are noticed when the loading force of the intender exceeds 16 N (Fig. 10d). Huge differences (even 6.5 dB) between values of acoustic emission are the effect of the interlayer cracking [29] (Fig. 10e,f).

Author details

Zenon Aleksander Opiekun

Address all correspondence to: zopiekun@gmail.com

Department of Casting and Welding, Rzeszów University of Technology, Rzeszów, Poland

References

[1] Pint B.A., Wright J.G., Bridlley W.J.: Evaluation of TBC systems on novel substrates. Journal of Thermal Spray Technology, 9 (6) (2000), 198–203.

[2] Parks W.P. Jr, Hoffman E.E., Lee W.Y. Wright J.G.: Thermal barrier coatings issue in advanced land-based turbines. Journal of Thermal Spray Technology, 6 (2) (1997), 187–192.

[3] Pint B.A., Wright I.G., Lee W.Y., Zhang Y., Prussner K., Aleksander K.B.: Substrate and bond coat compositions. Factory affecting alumina scale adhesion. Materials Science Engineering, A245 (1998), 201–211.

[4] Davis J.R.: Handbook of thermal spray technology ASM. International Materials Park, 2004.

[5] Gugel E., Woetting G.G.: Materials selection for ceramic components in automobiles. Industrial Ceramics, 19 (1999), 196–199.

[6] Masayuki Arai: Damage assessment of tbc sprayed combustor of land-based gas turbine. Journal of the Society of Materials Science, Japan, 57, (3) (2008), 285–291.

[7] Jordan E.H., Gell M.: Solution precursor plasma spray of thermal barrier coatings. Advanced coatings for high temperatures. Processing Turbine Forum, Nicea, 2008.

[8] Stover D., Funke C.: Direction of the development of thermal barrier coatings in energy applications. Journal of Material Processing Technology, (92–93) (1999), 152–202.

[9] Stecura S., Two-layer thermal barrier coatings for high temperature components. Ceramic Bulletin, 56 (1998), 1082–1085.

[10] Coderet Y.: Development of high temperature coatings and TBC bond coats for gas turbine engines. Advanced coatings for high temperatures. Processing Turbine Forum, Nicea, 2008.

[11] Ballard D.J., Schadler L.S., Levis C., Devenport J.: Phase transformation in thermal barrier coatings. Rensselaer Polytechnic Institute Research Experience for Undergraduates, 2000.

[12] Bach W.F., Babiak Z., Duda T., Copitzky T., Josefiak L.A.: Thermal barrier coatings with enhanced properties—a chance for thermal spraying. United Thermal Spray Conference, Dusseldorf, 1999.

[13] Miller A.R.: Thermal barrier coatings for aircraft engines. History and directions. Journal of Thermal Spray Technology (6) (1997), 3.

[14] Haberko K., Pampuch R.: Influence of yttrium content on phase composition and mechanical properties of Y-TZP. Ceramics International (9) (1983), 8–12.

[15] Viancenzini P.: Zirconia thermal barrier coatings for engine application. Industrial Ceramics (10) (1990) 3, 113–126.

[16] Singhal S.C., Bratton R.J.: Stability of a ZrO_2 (Y_2O_3) thermal barrier coating in turbine fuel with contaminants. Transactions of the ASME, 102 (1980), 770–775.

[17] Teixeira Vasco: Numerical analysis of the influence of coating porosity and substrate elastic properties on the residual stresses in high temperature graded coatings. Surface and Coatings Technology, 146–147 (2001), 79–84.

[18] Khor K.A., Gu Y.W.: Effects of residual stress on the performance of plasma sprayed functionally graded ZrO_2/NiCoCrAlY coatings. Materials Science and Engineering, A277 (2000), 64–76.

[19] Guo H.B., Vaßen R., Stöver D.: Thermophysical properties and thermal cycling behavior of plasma sprayed thick thermal barrier coatings. Surface and Coatings Technology, 192 (2005), 48–56.

[20] Vaßen R., Kerkhoff G., Stöver D.: Development of a micromechanical life prediction model for plasma sprayed thermal barrier coatings. Materials Science and Engineering, A303 (2001), 100–109.

[21] Ahmet P., Ozkan S., Erdal C.: Effects of porosity on thermal loadings of functionally graded Y_2O_3-Zr_2O/NiCoCrAlY coatings. Materials and Design, 23 (2002), 641–644.

[22] Traeger F., Ahrens M., Vaßen R., Stöver D.: A life time model for ceramic thermal barrier coatings. Materials Science and Engineering, A358 (2003), 255–265.

[23] Mesratia N., Saif Q., Treheuxb D., Moughilc A., Fantozzic G., Vincent A.: Characterization of thermal fatigue damage of thermal barrier produced by atmospheric plasma spraying. Surface and Coatings Technology, 187 (2004), 185–193.

[24] Opiekun Z.: Analysis of process-related factors that determine structure and properties of cobalt casting alloy intended of aircraft part. Monograph, Poznań, 2007.

[25] Opiekun Z.: Temperature influence of ceramic form on the structure of cobalt alloy MAR-M509 castings. Acta Metallurgica Sinica (English Letters), 24 (1) (2011), 23–33.

[26] ISO 14577 Metallic Materials—Instrumented Indentation Test for Hardness and Material Parameters.

[27] Opiekun Z., Pająk D.: Microstructure and hardness of thermal barrier coating deposited on a high resistant cobalt casting alloy. Rudy iMetaleNieżelazne, 6 (2012), 374–377.

[28] CSM Instruments, Application Bulletin: Characterization of Thermal Spray Instrumented Indentation and Scratch Testing. Part I, No. 28, 2009.

[29] Opiekun Z.A., Trytek A.: Cohesion and adhesion interlayer of the thermal barrier coatings made of cobalt alloy MAR-M509. Archives of Foundry Engineering, 14 (1/2014), 159–164.